Biopsy Pathology of the Liver

BIOPSY PATHOLOGY SERIES

General Editors

Professor Leonard S. Gottlieb, MD, MPH
Mallory Institute of Pathology,
Boston, USA.

Professor A. Munro Neville, MD, PhD, MRC Path.
Ludwig Institute for Cancer Research,
Zurich, Switzerland.

Professor F. Walker, MD, PhD, FRC Path.
Department of Pathology,
University of Aberdeen, UK.

Other titles in the series

Brain Biopsy
J. H. Adams, D. I. Graham and D. Doyle

Biopsy Pathology and Cytology of the Cervix
D. V. Coleman and D. M. D. Evans

Biopsy Pathology of the Oesophagus, Stomach and Duodenum
D. W. Day

Biopsy Pathology of the Small Intestine
F. D. Lee and P. G. Toner

Biopsy Pathology of Bone and Bone Marrow
B. Frisch, S. M. Lewis, R. Burkhardt and R. Bartl

Biopsy Pathology of the Bronchi
E. M. McDowell and T. F. Beals

Biopsy Pathology in Colorectal Disease
I. C. Talbot and A. B. Price

Biopsy Pathology of the Breast
J. Sloane

Biopsy Pathology of Muscle
M. Swash and M. S. Schwartz

Biopsy Pathology of the Lymphoreticular System
D. H. Wright and P. G. Isaacson

Biopsy Pathology of the Liver

R. S. PATRICK
MD, FRCPath
Emeritus Professor of Pathology,
University of Glasgow
and Honorary Consultant Pathologist,
Royal Infirmary,
Glasgow

and

J. O'D. McGEE
MD, PhD, FRCPath, MA (Oxon)
Professor and Head,
Nuffield Department of Pathology and Bacteriology,
Fellow of Linacre College,
University of Oxford;
Honorary Consultant Pathologist,
John Radcliffe Hospital,
Oxford

SECOND EDITION

LONDON
CHAPMAN AND HALL

First published in 1980 by
Chapman and Hall Ltd
Second edition published in 1988 by
Chapman and Hall Ltd
11 New Fetter Lane, London EC4P 4EE

Printed in Great Britain at the
University Press, Cambridge

ISBN 0 412 29240 8

British Library Cataloguing in Publication Data

Patrick, R. S. (Robert S.)
Biopsy pathology of the liver.,
1. Man. Liver. Diagnosis. Needle aspiration biopsy
I. Title II. McGee, J. O'D. (James O'D.)
616.3'6207583

ISBN 0-412-29240-8

Contents

Preface to the second edition

During the past eight years since the publication of the first edition, there have been many additions to our knowledge of liver disease as reflected by an ever increasing number of scientific papers as well as several new textbooks and journals devoted to hepatology. A lot of this new knowledge is concerned with matters that do not fall within the scope of this book, but there is now much to add to our descriptions of hepatic histopathology in order that the book may continue to provide an up-to-date account of the subject of value to hospital histopathologists. The more important changes deal with adequacy and procurement of biopsy specimens, intrahepatic biliary disease from various causes, viral hepatitis, alcoholic liver disease, chronic hepatitis and drug-induced liver injury. There is one new chapter concerned with liver damage in malnutrition and in various gastrointestinal diseases. There are a number of additional photomicrographs and the opportunity has been taken to replace some other illustrations with better examples. This has led to a modest increase in the size of the book but we have attempted to maintain its purpose, providing a well illustrated but otherwise concise account of liver histopathology necessary for biopsy interpretation.

In addition to the colleagues who helped with material for illustrations to the first edition, we would like to thank Miss Lesley Watts for typing the manuscript and the late Dr Crucioli, Department of Histopathology, John Radcliffe Hospital, Oxford, who provided fine needle aspirates of liver tumours.

Preface to the first edition

With the growing awareness of the value of histopathology in the diagnosis of human hepatic disease, liver biopsy interpretation has become an important component of the work of most general hospital pathology departments. It follows that many histopathologists must now become proficient in the assessment of these specimens and this book has been written to assist them in this task. It is by no means a textbook of liver pathology, its purpose being the provision of illustrated descriptions of most human liver diseases which may be encountered in hospital practice. For those with limited experience of the subject it includes an early chapter on definitions and a terminal chapter which draws together some clinical and pathological correlations.

The great majority of photographs have been produced from our own collections of biopsies supplemented with a few autopsy specimens. We are indebted to our colleagues for their kindness in providing microscopic slides for the following illustrations: Figs 2.7, 12.47, 12.48 (Dr A. A. M. Gibson, Royal Hospital for Sick Children, Glasgow); Fig. 12.49 (Dr Jean Keeling, John Radcliffe Hospital, Oxford); Figs 4.20, 5.10, 5.13, 5.14, 6.18 (Dr F. D. Lee, Royal Infirmary, Glasgow); Figs 12.20, 12.21, 12.22 (Dr W. G. S. Spilg, Victoria Infirmary, Glasgow). We must record our thanks also to our secretaries, Mrs M. Thomson, Glasgow, and Mrs V. Macintosh, Oxford, for their work on the typescript. The photomicrographs were prepared by Mr T. Parker, University of Glasgow, Department of Pathology, Royal Infirmary, Glasgow, and by Dr T. Parry and Mr R. Holton, University of Oxford, Department of Pathology, John Radcliffe Hospital, Oxford.

1 Introduction

Liver biopsy plays a most important role in the investigation of human hepatic disease and is indeed necessary for the correct assessment of the majority of cases. Apart from its diagnostic value in individual patients, it has contributed considerably to our knowledge of liver pathology in general, and many recent advances in this subject have followed the introduction of liver biopsy as a routine investigative procedure; there can be little doubt that future developments will continue to depend, to a large extent, on the study of fresh hepatic tissue obtained in this way.

Liver biopsy by itself may be sufficient for a final diagnosis, although in many cases it is essential not to treat the result in isolation but to take into account the clinical findings and reports of laboratory tests of liver function. Thus, the detailed history of drug administration is often necessary in assessing cholestatic hepatitis, and the results of non-specific serum antibody tests can be helpful in differentiating some cases of primary biliary cirrhosis from chronic active hepatitis. Repetition of liver biopsy after a few weeks or months is sometimes justified, for example in the differential diagnosis between persistent viral hepatitis and chronic active hepatitis, and in assessing a particular form of therapy.

With the introduction of needles of the 'Trucut' variety aspiration biopsy specimens are usually very adequate for assessing any diffuse hepatic condition. Ideally one should obtain a core of tissue about 2.5 cm long, but a reliable diagnosis of diffuse conditions such as viral hepatitis and leukaemic infiltration can be made with small amounts of tissue 0.5 cm long (Holund *et al.*, 1980). By contrast, focal lesions such as secondary cancer can be missed even in three aspiration specimens when the result is compared with autopsy findings (Abdi *et al.*, 1979). In such cases it may be possible to direct the needle to particular targets revealed by radiological and isotope screening or located by peritoneoscopy. At laparotomy, a wedge biopsy is usually taken and this can, with advantage, be supplemented at the same time by single or multiple needle biopsies, as the subcapsular tissue in the former type of specimen may not be typical of the liver as a whole.

Accuracy of diagnosis will depend, of course, on the skill of the observer as well as the adequacy of the specimen. There have been very few investigations on this subject, but in one in which one of us (R.S.P.) was a participant (Theodossi *et al.*, 1980) it was apparent that there was considerable variation in the interpretation of histological appearances among pathologists, although a high level of consistency in reaching a correct diagnosis was achieved by those members of the group who had considerable experience and a special interest in liver pathology.

1.1 Complications of aspiration biopsy

Aspiration needle biopsy is a safe procedure when undertaken by an experienced operator. This is borne out by Conn (1975) in his personal knowledge of several thousand biopsies, and most hepatologists would agree that the very small risk of serious complication is far outweighed by the value of obtaining a correct diagnosis by this method. Severe internal haemorrhage is the main complication and the operation should of course be avoided when there is any risk of serious internal bleeding, as in patients with a coagulation defect or Weil's disease. Rarely there is haemobilia and a large intrahepatic haematoma may form. The production of a hepatic arteriovenous fistula is another unusual event. Dilatation of intrahepatic bile ducts caused by obstruction to a large biliary channel is regarded by some as a contraindication to needle biopsy because of the risk of biliary peritonitis which could follow escape of bile from damage to such a duct. This matter is discussed in Chapter 3. Septicaemia has been reported in similar cases in which biliary infection was also present.

1.2 Special investigations and types of biopsy

Light microscopy of paraffin sections prepared in conventional ways continues to be the main method for examination, while increasing use is being made of immunohistochemistry, electron microscopy, microchemical analysis and other specialized techniques. These will be referred to only where they provide important additional information or are essential for diagnosis. Accordingly, there is little reference in this book to electron microscopy except where ultrastructural changes have diagnostic value.

Special biopsy procedures are employed occasionally:

(a) Transvenous (transjugular) liver biopsy

This involves the passage of a catheter from the right jugular vein to a branch of the right hepatic vein and recovery of hepatic tissue obtained by a fine needle passed along the catheter. The method is described by Gilmore *et al.* (1978). In experienced hands an adequate amount of tissue is obtained in a high proportion of cases (Bull *et al.*, 1983). It is useful in patients with a tendency to bleeding and when there is severe ascites. However, it is unlikely to become a routine procedure as the amount of tissue obtained is generally less compared with conventional biopsies.

(b) Fine needle aspiration biopsy

This has been used in the investigation of nodular lesions, usually secondary cancer within the liver. A short wide-bore needle is used to penetrate the abdominal wall and through this is introduced the longer and finer aspiration needle. This can be visualized by scanning devices and directed towards the lesion which has also been visualized. Smears are made of the aspirated material and examined by a cytopathologist. For the detection of liver tumours (Fig. 1.1) the results are good in experienced hands, but the method has very limited value in the diagnosis of non-neoplastic diseases (Frable, 1983).

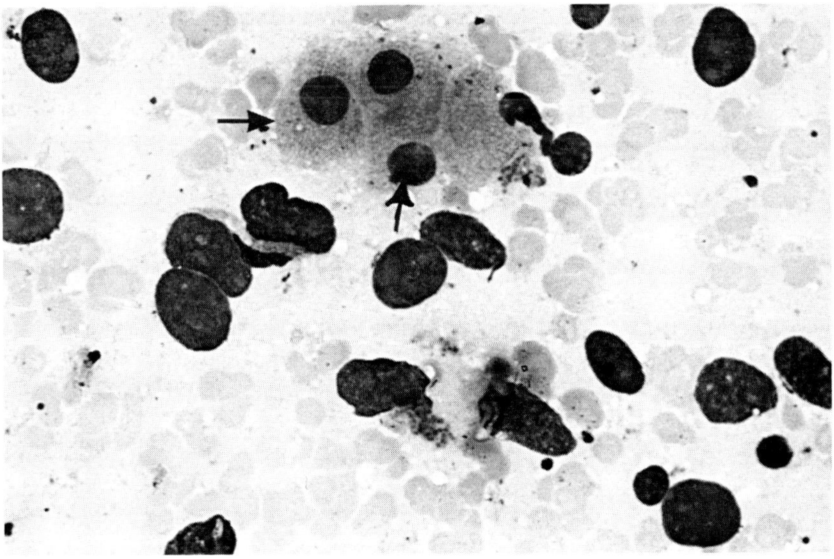

Fig. 1.1 Cells aspirated from a secondary cancer in liver. Malignant cells from a primary bronchial cancer are intermixed with hepatocytes (arrows) (× 980).

(c) Fetal liver biopsy

Fetal hepatic tissue has been obtained successfully and without serious complications by needle aspiration via the maternal abdominal wall and has value in the early detection of liver damage associated with an inherited metabolic disorder (Rodeck *et al.*, 1982).

1.3 Laboratory handling of biopsies

1.3.1 *Record of gross appearance*

This is rarely of much value and in most cases hardly necessary when the entire specimen will be processed for histological examination. Cirrhosis or tumour may be observed, especially in open wedge biopsy material. Some preliminary information can be obtained from the colour: pallor in severe fatty change (in extreme cases the tissue may float in fixative); green in cholestasis (especially after fixation); brown in haemochromatosis; slate grey in Dubin–Johnson syndrome.

The gross appearance of partial hepatectomy specimens is, by contrast, of considerable importance. Usually there is a tumour, when adequacy of resection, vascular invasion, encapsulation and changes in surrounding liver should all be noted, and blocks of tissue taken for the further investigations of these points. The gross appearance of the tumour itself may point to the correct diagnosis, such as bile staining in some hepatocellular adenomas or cancers and a central stellate scar in focal nodular hyperplasia.

1.3.2 *Fixation for light microscopy*

Kinks in needle biopsies are prevented by arranging the specimens on filter paper before fixation. If unusually long they may be cut into a few segments with a sharp razor and the parts laid out parallel to each other. Slices about 2 mm thick are prepared from wedge biopsies, cut at right angles to the capsule. All are fixed in 10% neutral formalin for at least 6 hours and preferably overnight, when they can be passed to paraffin wax in the usual way.

1.3.3 *Staining*

Sections (6 μm) should be stained routinely by haematoxylin and eosin, and for reticulin fibres and haemosiderin. A section stained by Masson's trichrome or Van Gieson's method is often useful in assessing the relationship between developing fibrous tissue and adjacent parenchyma. Periodic acid-Schiff-diastase sections may give prominence to macrophage activity which can be the only remaining evidence of recent liver injury, and this is a good method for revealing alpha-1-antitrypsin inclusions. It is useful to retain a few unstained sections for special investigations, e.g. orcein or immunoperoxidase staining of hepatitis B surface antigen, rubeanic acid and rhodanine demonstration of copper, autoradiography for the detection of thorotrast, etc.

1.3.4 *Serial sections*

These are routinely indicated if the specimen is unusually small or if it is important to uncover some suspected small focal lesion, such as a granuloma in sarcoidosis or in primary biliary cirrhosis, or diffuse metastatic disease.

1.3.5 *Frozen sections*

Hepatocyte cytological detail which is so often important in diagnosis is frequently distorted or lost completely by freezing. Accordingly, it is wise to resist requests for rapid frozen sections of liver biopsies unless to confirm the presence of tumour or other gross lesion or unless, as with wedge biopsies, sufficient material is also available for the preparation of paraffin sections.

1.3.6 *Fixation for electron microscopy*

For the liver especially, meaningful electron microscopy depends on good fixation which can be achieved only by the pathologist or the technician being present at the operation, so that appropriate blocks are placed in chilled 2.5% buffered glutaraldehyde as soon as possible, preferably within 1 minute of removal from the patient. The specimen should be laid on dental wax and a sharp scalpel or razor blade used to obtain small cubes not more than 1 mm thick, and preferably thinner than this, for rapid transfer to fixative. A very thin slice of tissue can be placed immediately in a drop of fixative and then diced into tiny cubes. Further embedding, cutting and staining are by conventional methods, but if this is delayed it is preferable to store the tissue in cacodylate or phosphate buffer after 1–3 hours' fixation. It follows that the pathologist must be notified in advance of cases which might require electron microscopy to facilitate the diagnosis, as in certain metabolic disorders.

1.4 Normal liver morphology

1.4.1 Microanatomy

Judged by histochemical studies of normal hepatic parenchyma and the distribution of ischaemic and toxic liver damage, the acinus is the functional unit of the liver, each consisting of plates of hepatocytes radiating from portal tracts which contain terminal branches of the portal vein, hepatic artery and bile ducts (Fig. 1.2). As seen in three-dimensional reconstructions, the peripheral part of the acinus interdigitates with its neighbours and is in contact with hepatic venous radicles. For the assessment of two-dimensional sections it is more convenient to retain the long-established lobular concept of the liver by which each unit

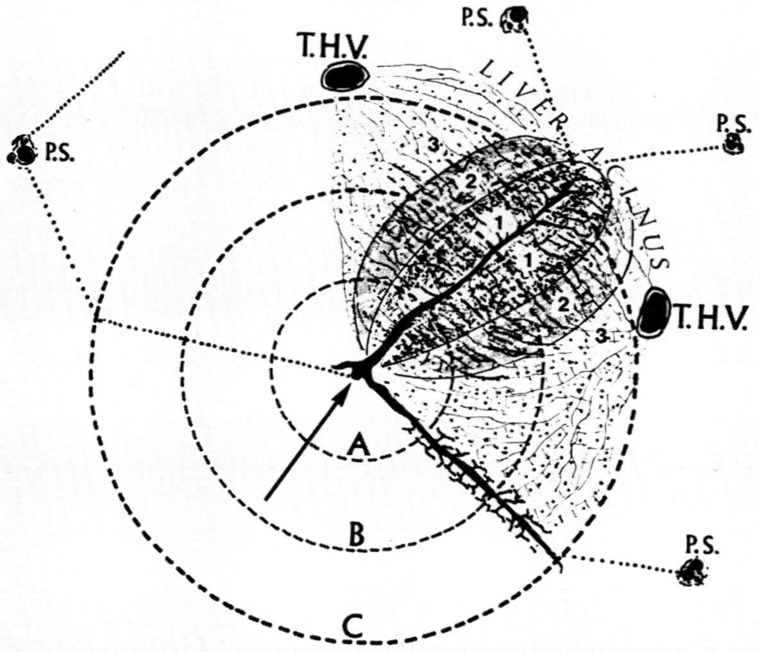

Fig. 1.2 Diagrammatic representation of the liver acinus. The acinus occupies adjacent sectors of two hexagonal liver lobules. Its backbone is formed by terminal afferent vessels in the portal spaces (arrow and PS). Zones 1, 2 and 3 are centred around these vessels and each extend into the 'periportal' area (A). The terminal hepatic vein (THV) – central vein – drains adjacent acini. The circles B and C represent peripheral circulatory areas, as traditionally conceived, around the periportal zone (A).

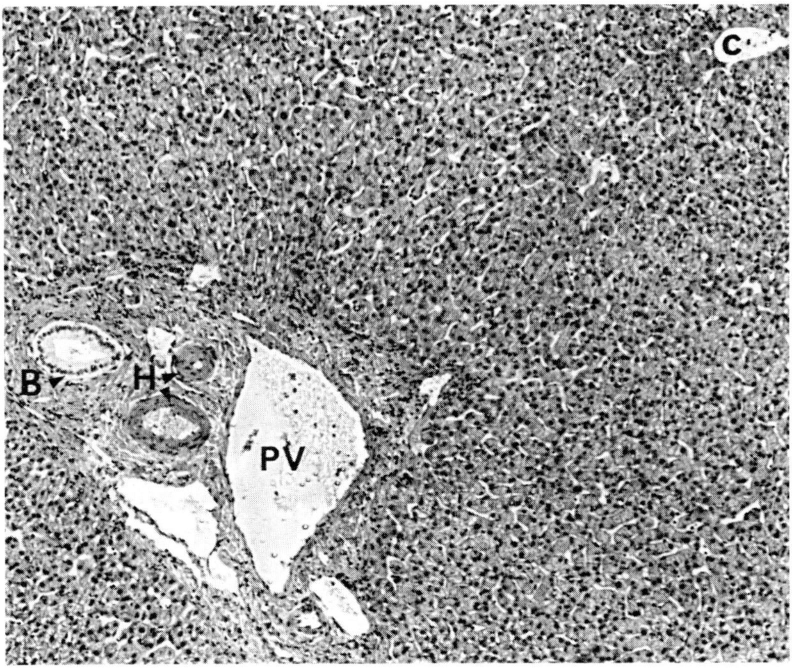

Fig. 1.3 Normal liver showing centrilobular vein (C) and large portal tract with interlobular bile duct (B), hepatic artery branches (H) and portal vein (PV). Smaller branches of these entities are also present (HE, × 90).

consists of cords of cells radiating from the central vein to peripheral portal tracts; it will be retained in this book as a basis for our histopathological descriptions (Fig. 1.3).

Each radiating plate consists of hepatocytes, one cell thick, and merges with similar plates which border upon portal tracts and hepatic veins. Separating these plates are the liver sinusoids along which blood flows from portal tracts to central veins. Sinusoids are lined by cells which have indistinct cytoplasm and small elongated and darkly staining nuclei. Between this lining and the hepatocytes there is a space or potential space (the space of Disse) within which tissue fluid drains peripherally to lymphatics in the portal tracts. Disse's space also contains reticulin fibres, which constitute the normal framework of the liver, and scanty perisinusoidal cells (Ito cells; fat-storage cells) which are probably modified fibroblasts. The reticulin framework can be demonstrated readily by standard silver impregnation methods (Fig. 1.4) but a basement membrane is not present in the sinusoidal walls of normal human liver.

Between the adjacent surfaces of hepatocytes are tiny capillary channels or canaliculi into which bile is excreted and along which it flows

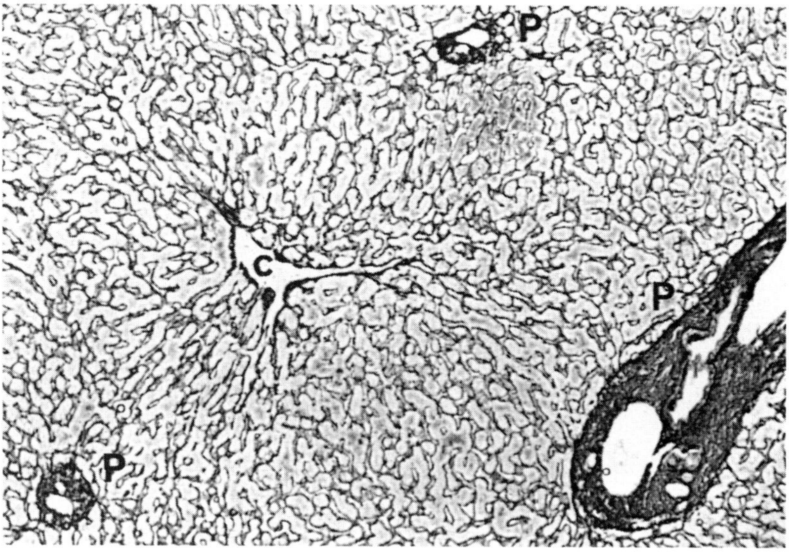

Fig. 1.4 Reticulin framework of normal liver. There is a central vein (C) together with one large and two small portal tracts (P) situated peripherally (reticulin stain, × 90).

peripherally towards the portal tracts. These channels are not readily recognized in paraffin sections unless distended in intrahepatic cholestasis, but are seen readily in 1 μm plastic sections. They communicate with small acinar structures or ductules lined either partly by hepatocytes and partly by cuboidal (ductular) cells, or entirely by the latter type. These canals of Hering, which have no accompanying blood vessels, are very scanty in the liver lobules but may be seen adjacent to some portal tracts. They in turn drain into other more conscpicuous interlobular ducts lying within the portal tracts and lined by cuboidal epithelium (Fig. 1.5). This epithelium rests on a distinct basement membrane. Septal bile ducts are even larger channels lined by columnar epithelium with basal nuclei which is often thrown into folds, and in the walls of which there is a distinct fibroelastic coat (Fig. 1.6); smooth muscle fibres may be included, especially towards the hilum of the liver.

The vascular components of the portal tracts require no special description. Small autonomic nerves may be seen, particularly in the septa and larger portal tracts close to the porta hepatis. They permeate into the smallest tracts and can be demonstrated also in the walls of hepatic venous radicles.

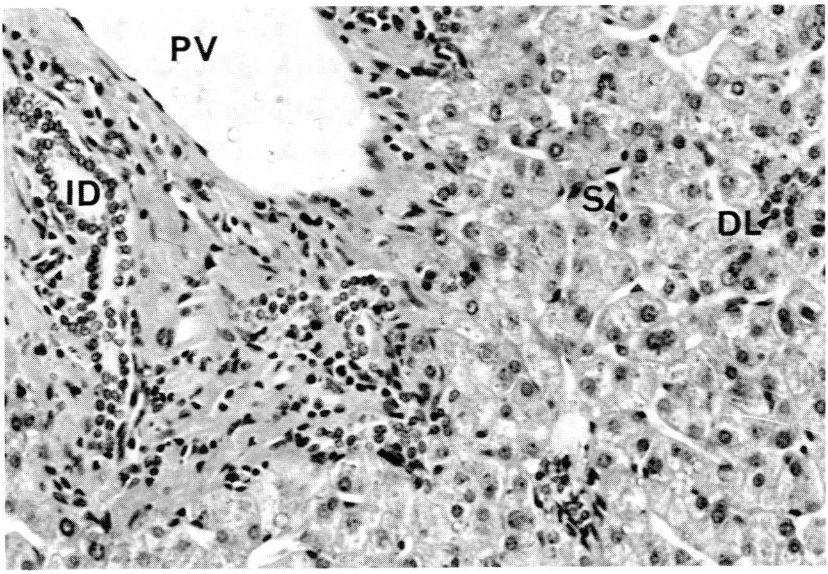

Fig. 1.5 Part of portal tract (top left) with portal vein (PV), interlobular bile duct (ID) and a number of small bile ductules at the periphery. Note the small round nuclei of ductular cells including those within the liver lobule (DL) and compare with the more elongated nuclei of sinusoidal lining cells (S) (HE, × 280).

1.4.2 Cytology

Hepatocytes are polygonal cells, each with a clearly defined round nucleus containing one or more prominent nucleoli. Binucleate forms sometimes occur but mitotic activity is rare (Fig. 1.7). Some nuclei, particularly in the elderly, are larger than others indicating polyploidy, the normal cell being diploid. Hepatocyte cytoplasm is also distinct and stains readily with eosin. Indistinct basophilic granules may be seen, an appearance that is accentuated in sections treated with toluidine blue, thionin or Romanowsky stains. These granules represent aggregates of rough endoplasmic reticulum and disappear quickly in liver damage associated with dislocation of ribosomes from membranes. Scanty droplets of lipid especially in a few periportal hepatocytes are usually unimportant. Cytoplasmic glycogen is seen readily in appropriately stained sections of fresh tissue, particularly in the centrilobular zones. A fine dusting of brown pigment in centrilobular hepatocyte cytoplasm indicates lipofuscin in lysosomes but has little significance.

Compared with hepatocytes, bile ductular epithelium consists of much smaller cuboidal cells with scanty cytoplasm and distinct round nuclei.

Reference has already been made to sinusoidal and perisinusoidal cells

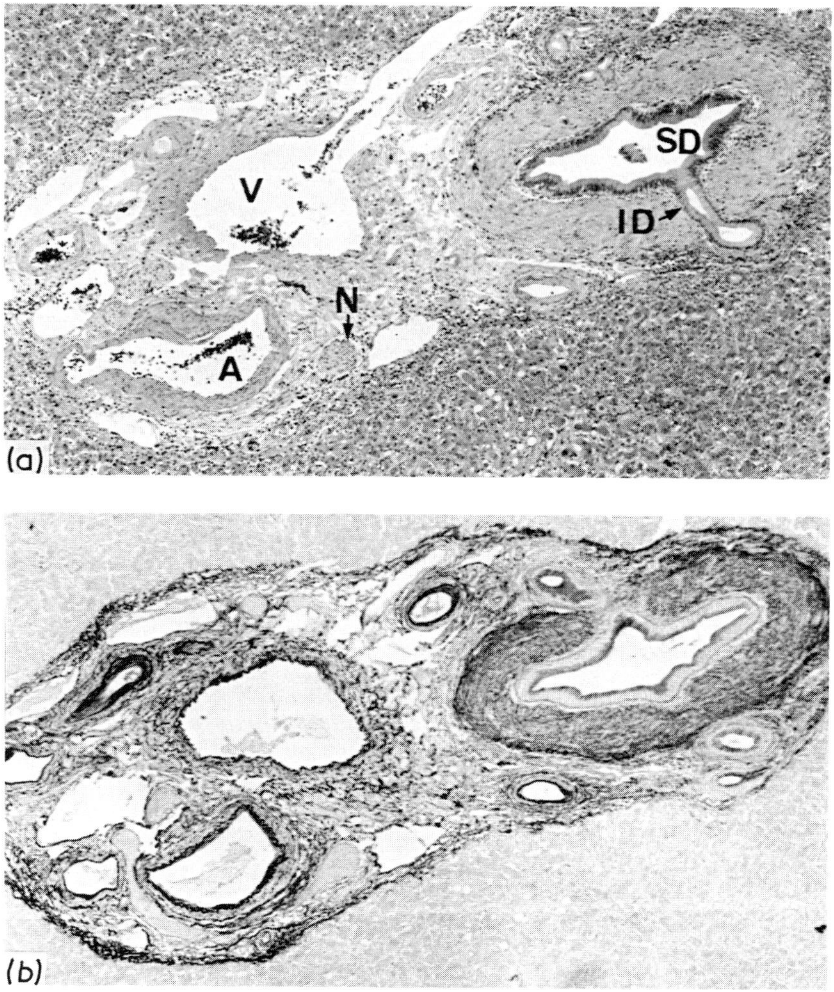

Fig. 1.6 (a) Hepatic septum containing nerve (N) and several blood vessels including a prominent hepatic artery (A) and portal vein (V). On the right an interlobular bile duct (ID) enters a much larger septal duct (SD). (b) Fibroelastic tissue is present in the wall of the septal duct but not in smaller interlobular ducts. ((a) HE, × 56; (b) orcein, × 56.)

which are readily differentiated only in electron-microscopic preparations. However, the perisinusoidal variety can be detected with the light microscope when there is accentuation of their lipid storage capacity (Fig. 1.8), as in hypervitaminosis A. Larger Kupffer cells are evident as scanty histiocytes on sinusoidal surfaces. Small numbers of similar macrophages may be seen in the stroma of otherwise normal portal tracts together with

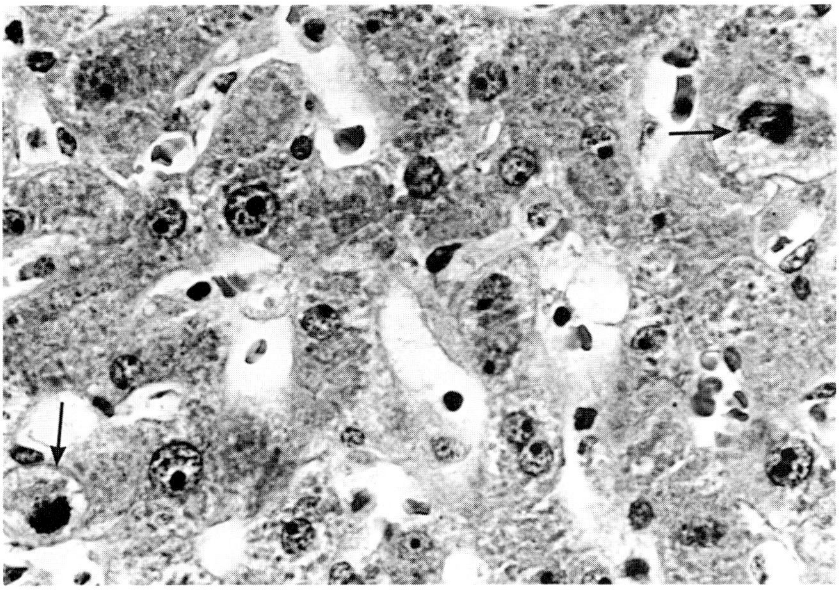

Fig. 1.7 Mitotic activity in regenerating liver. Two mitotic figures are present (arrows) (HE, × 720).

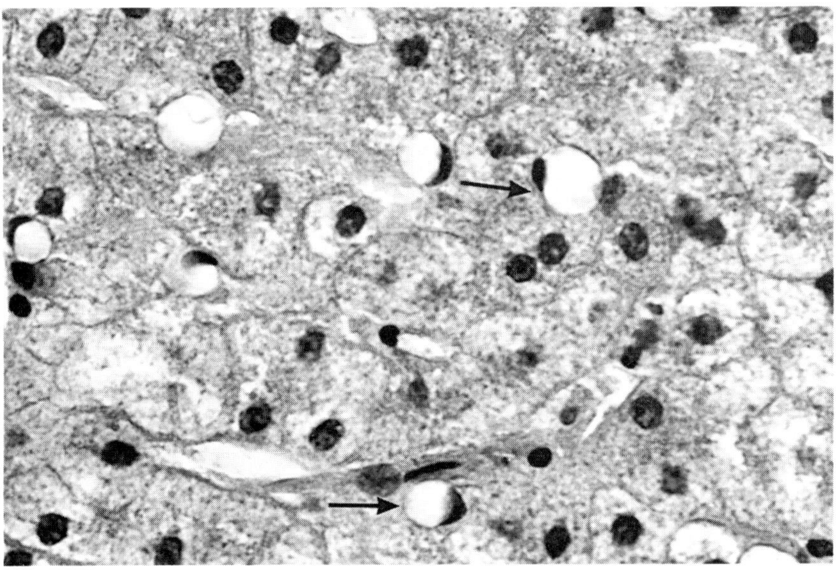

Fig. 1.8 Unusual prominence of perisinusoidal cells which contain small droplets of fat (arrows) (HE, × 720.)

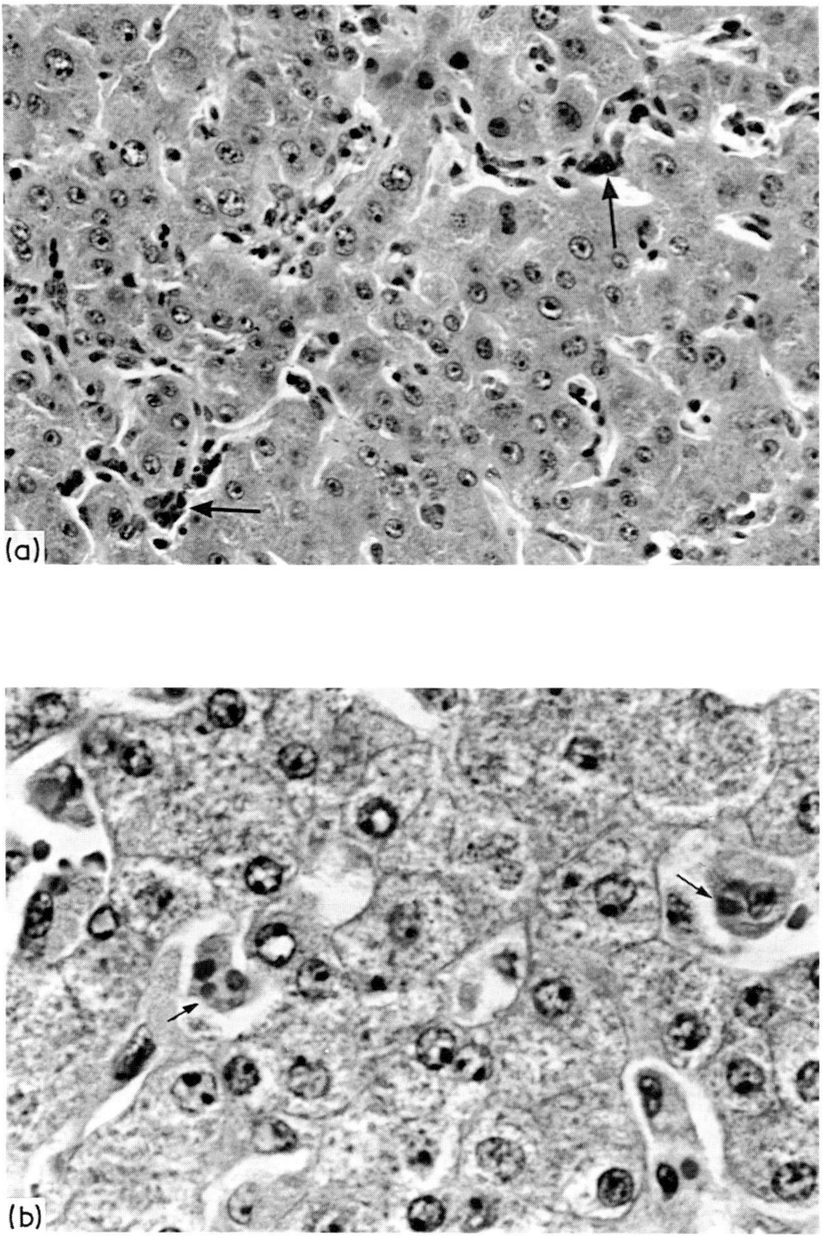

Fig. 1.9 (a) Unusual prominence of Kupffer cells (arrows). From a case of coliform bacillary septicaemia. (b) Erythrophagocytic activity by prominent Kupffer cells is evident (arrows). From a case of haemolytic anaemia. ((a) HE, × 450; (b) HE, × 900.)

a few lymphocytes. Prominence of Kupffer cells is a common feature of hepatic injury and may occur also in generalized infections and other conditions associated with histiocytic activity unrelated to primary liver disease (Fig. 1.9).

1.4.3 *Normal histological variations*

Rarely the pathologist may be asked to examine embryonic or fetal liver for the investigation of certain inherited metabolic disorders. Even at a few weeks of embryonic life it is possible to recognize hepatocyte plates separated by capillaries. By the ninth week or earlier bile ductular channels appear, possibly arising from hepatocyte limiting plates at the margins of portal tracts which, at this stage, consist of loose mesenchymal tissue and capillary vessels. A prominent feature is the presence of large numbers of cells with small round deeply staining nuclei; these are haemopoietic cells, mainly erythropoietic, which are extravascular and closely adjacent to hepatocytes. They are most numerous during the third to fifth month of embryonic life. Thin reticulin fibres can be demonstrated between the liver cell plates (Fig. 1.10).

In normal infant liver and as a regenerative phenomenon after partial hepatectomy or destruction by disease, the parenchymal plates may be several cells thick. In biopsies from elderly patients hepatocyte pleomorphism with hyperchromatic nuclei due to increased polyploidy may be conspicuous and should not be misinterpreted as regenerative activity (Fig. 1.11). Lipofuscin pigment may also be excessive, especially in elderly subjects (see Fig. 2.19); this again is of little significance provided that the Dubin–Johnson syndrome is kept in mind.

Golden-brown granules of haemosiderin pigment are abundant in the hepatocyte cytoplasm of infants by the first week of life (Fig. 1.12). This is gradually reduced in amount and should be absent in normal liver after 6–9 months. In small amounts it is most conspicuous in periportal cells. It may reappear in association with high iron intake or excessive blood destruction. Foci of extramedullary haemopoiesis are a normal feature of the newborn infant liver and persist for a few weeks; in older patients they may also reappear and are associated with various blood disorders (Fig. 1.13). In the neonatal liver granulocyte differentiation occurs preferentially in portal tracts.

Strands of fibrous tissue may be seen in wedge biopsies linking Glisson's capsule with superficial portal tracts (Fig. 1.14). These must not be regarded as characteristic of the organ as a whole, and should not be interpreted as evidence of cirrhosis.

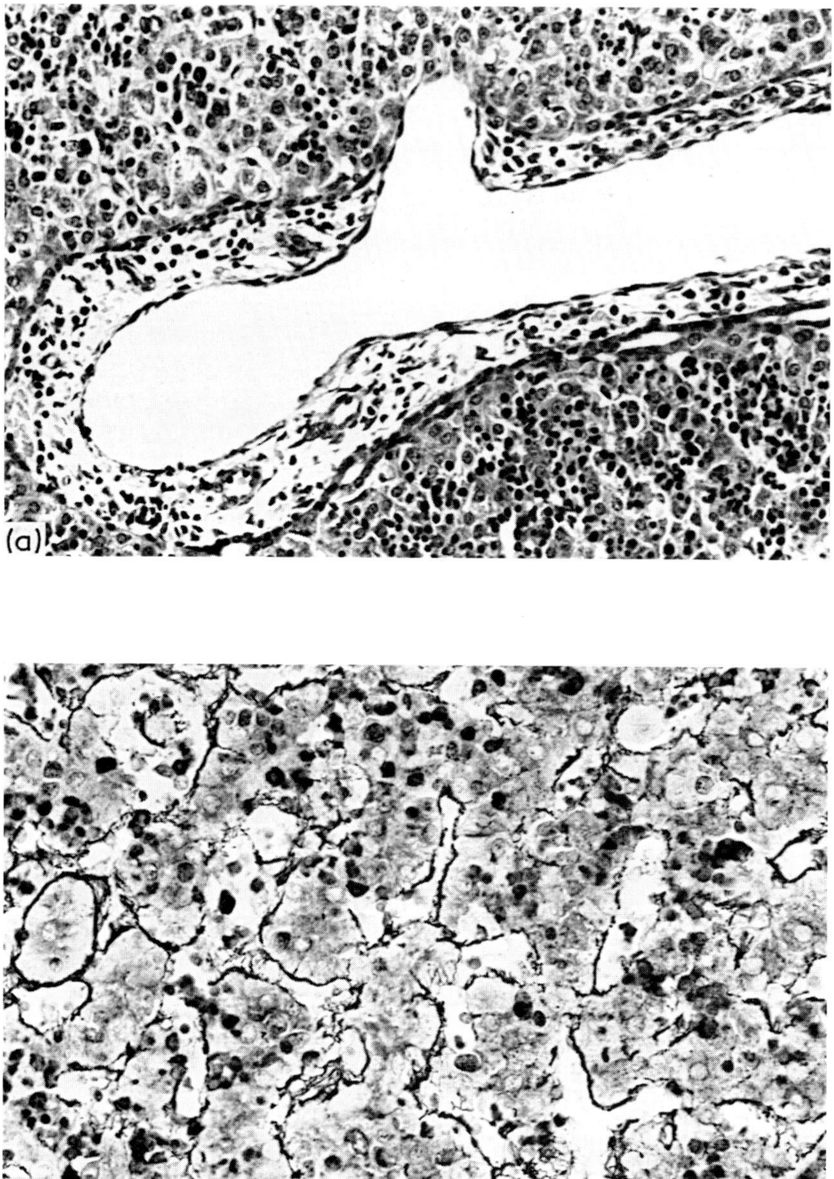

Fig. 1.10 (a) Twelve-week old embryonic liver. There are numerous deeply stained haemopoietic cells scattered diffusely between larger hepatocytes. At the margin of the elongated portal tract are a number of ductular structures which are developing biliary channels. (b) The same specimen stained to show reticulin fibres. ((a) HE, × 247; (b) reticulin stain, × 395.)

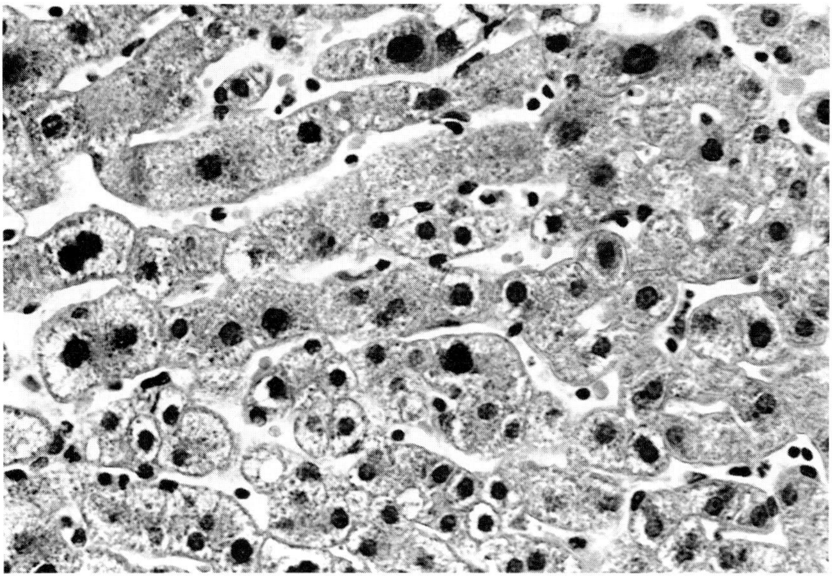

Fig. 1.11 Pleomorphic, hyperchromatic and binucleate hepatocytes in an otherwise normal liver from an elderly patient (HE, × 450).

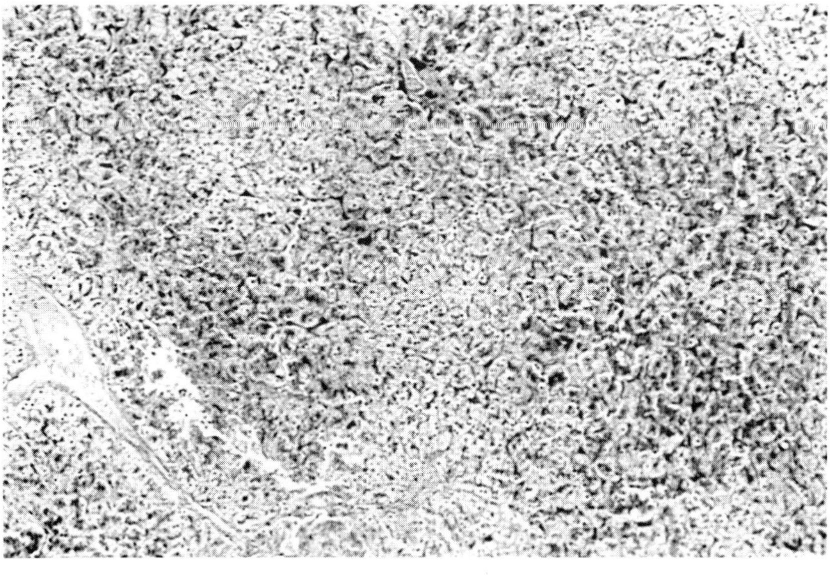

Fig. 1.12 Liver biopsy from an infant aged 5 weeks. Dark staining represents haemosiderin pigment especially in periportal and midzonal hepatocytes and in some Kupffer cells (Perls' stain, × 90).

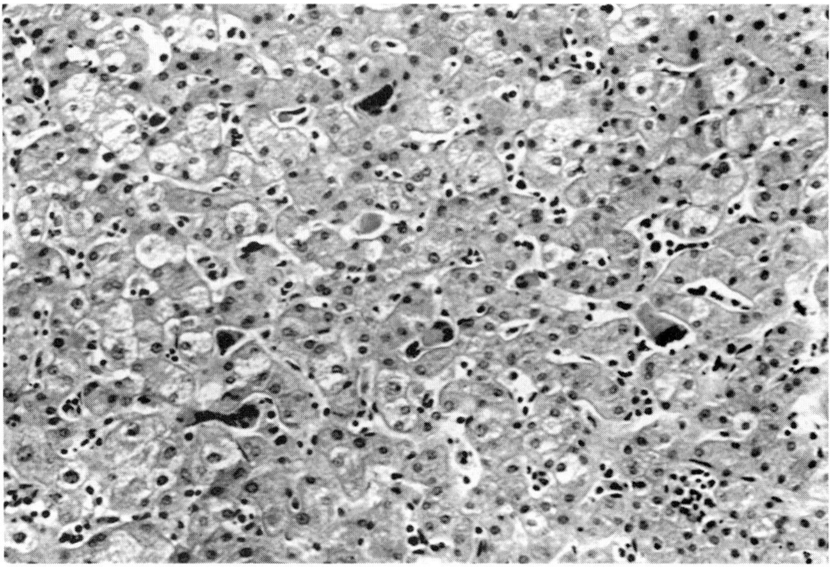

Fig. 1.13 Haemopoiesis seen in liver biopsy. Small groups of myeloid cells are present within sinusoids and include a few prominent megakaryocytes. From a case of myelofibrosis (HE, × 225).

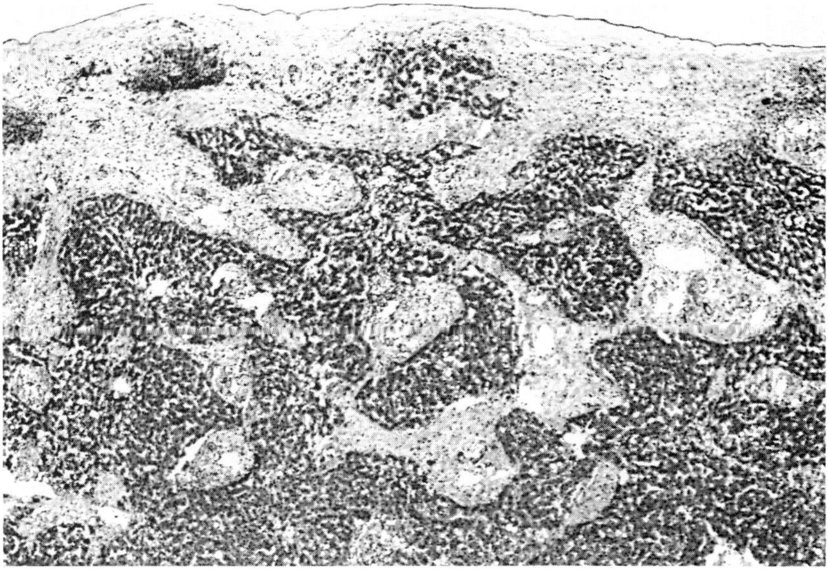

Fig. 1.14 Wedge biopsy of otherwise normal liver showing portal fibrosis confined to the subcapsular region (HE, × 45).

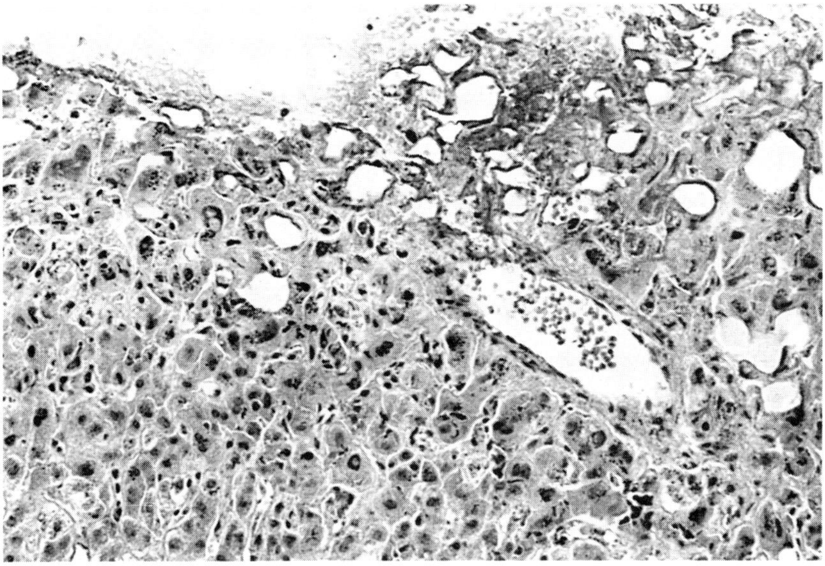

Fig. 1.15 Dilated sinusoids near the surface of this biopsy are lined by shrunken desiccated liver cells (HE, × 225).

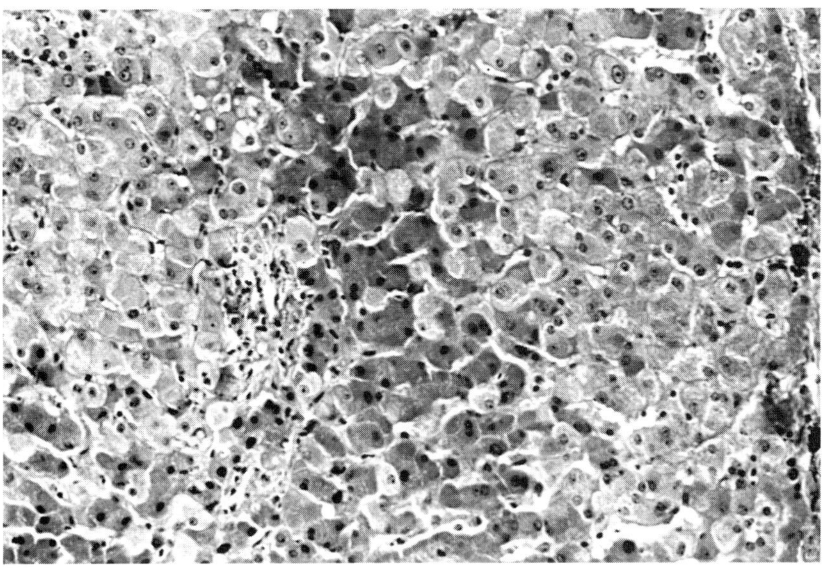

Fig. 1.16 Variable intensity of hepatocyte staining due to faulty fixation (HE, × 225).

1.4.4 *Artefacts in histological preparations*

The liver is rich in proteolytic enzymes which remain active and cause a loss of cellular detail especially in degenerate hepatocytes if there is serious delay in fixation. This may explain conspicuous zonal necrosis in autopsies compared with the biopsy appearances of certain diseases. Delay may cause surface desiccation (Fig. 1.15), while irregular staining of hepatocytes throughout the specimen (Fig. 1.16) can also be attributed to faulty fixation. Poor histological technique may give rise to false-positive results in Perls' Prussian blue method for the demonstration of haemosiderin; it is suspected by an irregular distribution of staining or by staining which is accentuated at the edge of the section. In wedge biopsies, neutrophil polymorphonuclear leucocytes may be conspicuous in sinusoids, even to the extent of resembling microabscesses (see Fig. 8.18); this is attributed to minor trauma from handling the liver at operation.

References

Abdi, W., Millan, J. C. and Mezey, E. (1979), Sampling variability on percutaneous liver biopsy. *Arch. Intern. Med.*, **139,** 667–9.

Bull, H. J. M., Gilmore, I. T., Bradley, R. D. *et al.* (1983), Experience with transjugular liver biopsy. *Gut*, **24,** 1057–60.

Conn, H. O. (1975), Liver biopsy in extrahepatic biliary obstruction and in other 'contraindicated' disorders. *Gastroenterology*, **68,** 817–21.

Frable, W. J. (1983), Fine needle aspiration biopsy: a review. *Hum. Pathol.*, **14,** 9–28.

Gilmore, I. T., Bradley, R. D. and Thompson, R. P. H. (1978), Improved method of transvenous liver biopsy. *Br. Med. J.*,ii, 249.

Hølund, B., Poulsen, H. and Schlichting, P. (1980), Reproducibility of liver biopsy diagnosis in relation to the size of the specimen. *Scand. J. Gastroenterol*, **15,** 329–35.

Rodeck, C. H., Patrick, A. D., Pembrey, M. E. *et al.* (1982), Fetal liver biopsy for prenatal diagnosis of ornithine carbamyl transferase deficiency. *Lancet*, **ii,** 297–300.

Theodossi, A., Skene, A. M., Portmann, B., Patrick, R. S. *et al.* (1980), Observer variation in assessment of liver biopsies including analysis by kappa statistics. *Gastroenterology*, **79,** 232–41.

2 Terminology in liver histopathology

This chapter presents a brief review of the more important histological features likely to be found in the vast majority of biopsies. Some will be readily apparent at low magnification, e.g. the increased cellularity of hepatitis, the loss of normal lobular architecture in cirrhosis and the presence of a gross lesion such as tumour or abscess. Many others require for their detection more careful scrutiny at higher magnifications, sometimes supplemented by the examination of slides treated by special staining techniques.

2.1 Hepatic necrosis

Apart from destruction by inflammatory and neoplastic lesions distinct necrosis of hepatic parenchyma may be recognized and various patterns are described.

2.1.1 Zonal necrosis

The same zone of each liver lobule is affected. In most cases this is centrilobular (Fig. 2.1), but very rarely necrosis is confined to periportal or midzonal cells depending on the aetiological agent. The extent of lobular damage may vary in different parts of the liver, being greatest in the relatively ischaemic subcapsular region.

2.1.2 Bridging necrosis

This refers to bands of necrotic tissue passing between adjacent portal tracts or centrilobular zones or transversing individual lobules from centre to periphery (Fig. 2.2). Such lesions may appear as extensions of a stellate pattern of centrilobular zonal necrosis and are seen especially in viral hepatitis and chronic active hepatitis.

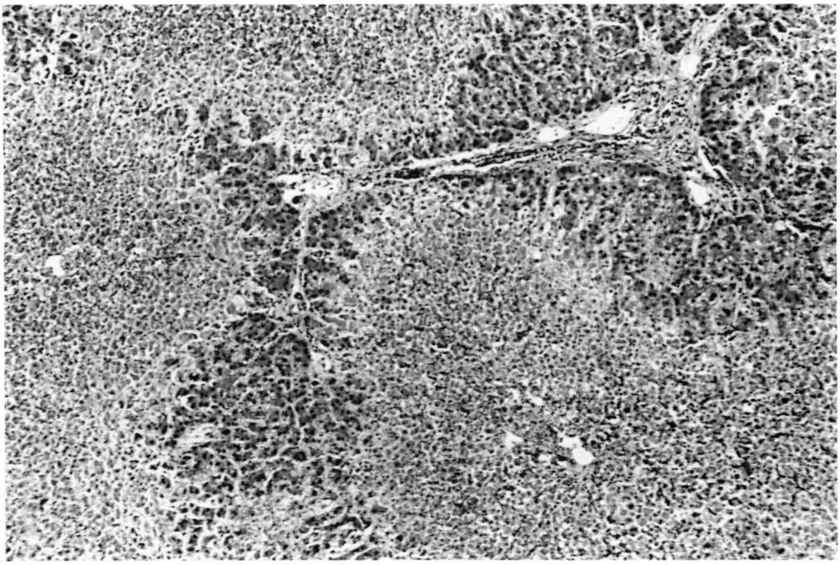

Fig. 2.1 Zonal necrosis of liver induced by carbon tetrachloride poisoning. Surviving liver parenchyma stains darkly and is situated at the periphery of liver lobules adjacent to portal tracts. This is basically a centrilobular zonal lesion but in this severe case the pale necrotic process extends to the midzonal parts of each lobule (HE, × 72).

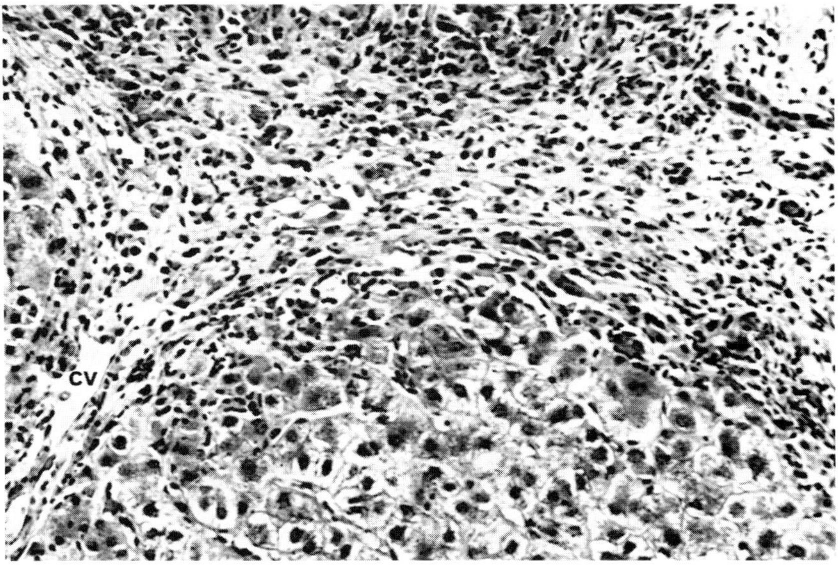

Fig. 2.2 Bridging necrosis. A band of pale necrotic tissue infiltrated with inflammatory cells extends from the portal area on the right, where bile ductular proliferation is evident, to the small centrilobular vein (CV). From a case of viral hepatitis (HE, × 225).

2.1.3 Massive and submassive necrosis (multilobular necrosis)

These terms imply necrosis of large areas of liver or at least several adjacent lobules in their entirety (Fig. 2.3). Massive necrosis may be an extreme form of zonal necrosis. More localized irregularly shaped lesions are usually infarcts.

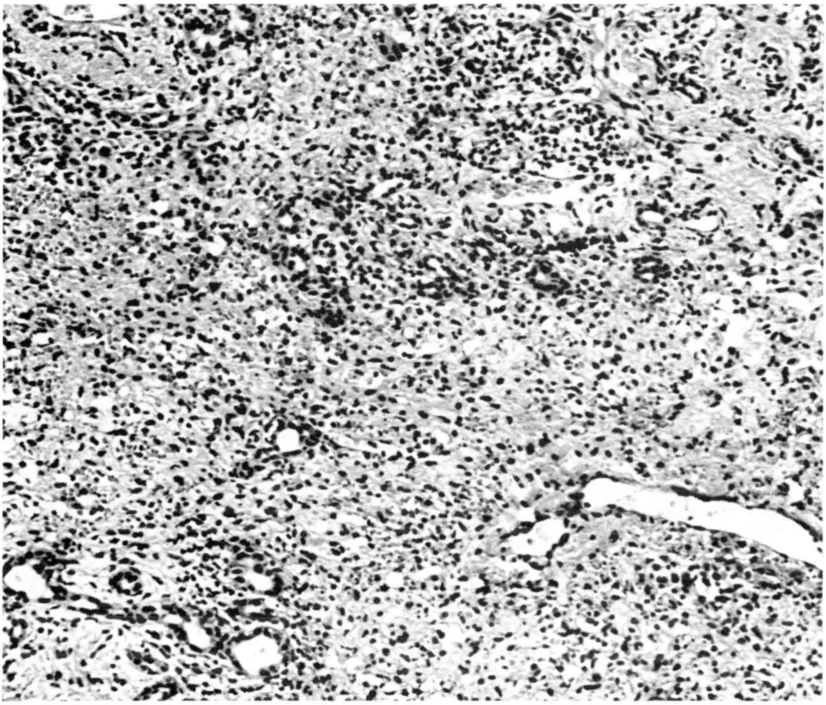

Fig. 2.3 Submassive necrosis in viral hepatitis. There is an elongated centrilobular vein (lower right) and a few surviving hepatocytes (left). The remaining nucleated cells are viable mesenchyme, bile duct epithelium and inflammatory cell infiltrate. The necrotic process extended through several adjacent lobules in a wedge biopsy, the remainder of the specimen showing bridging necrosis (HE, × 180).

2.1.4 Focal necrosis

Single necrotic hepatocytes ('pin-point' necrosis) are made conspicuous by the intense eosinophilic staining of cytoplasm while the nucleus is pyknotic or absent (Fig. 2.11). Larger lesions involving groups of hepatocytes may occur within individual lobules and have an irregular distribution throughout the liver (Fig. 2.4). In cases of biliary obstruction they tend to be situated periportally and show bile staining ('bile infarcts').

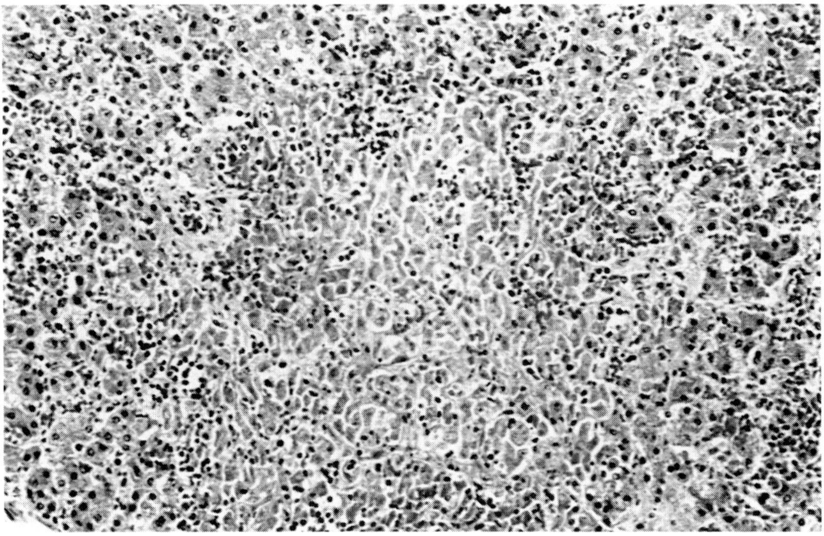

Fig. 2.4 Focal necrosis seen post-mortem in an infant dying with herpes hepatitis. A focus of pale necrotic liver cells occupies the centre of this field. This was one of several lesions of similar size having a haphazard distribution in the liver (HE, × 180).

Small aggregates of inflammatory cells may persist within liver lobules at sites of 'cell dropout' where necrotic hepatocytes have been removed by macrophage activity.

2.1.5 *Piecemeal necrosis*

There is loss of cells in limiting plates and adjacent parenchyma in the periportal zones (Fig. 2.5). This is a slow continuous process associated with chronic hepatitis, degenerate hepatocytes being surrounded or replaced by inflammatory cells, fibroblasts and young connective tissue. It might be regarded as a form of periportal zonal necrosis but this term is hardly appropriate to a chronic condition in which frank necrosis is often inconspicuous despite extensive loss of hepatocytes.

2.2 Hepatocyte degeneration

2.2.1 *Fatty change*

This is the commonest form of hepatocyte degeneration, particularly the type with large cytoplasmic droplets of fat and displacement of cell nuclei.

The small discrete droplet type with central nuclei is less common but

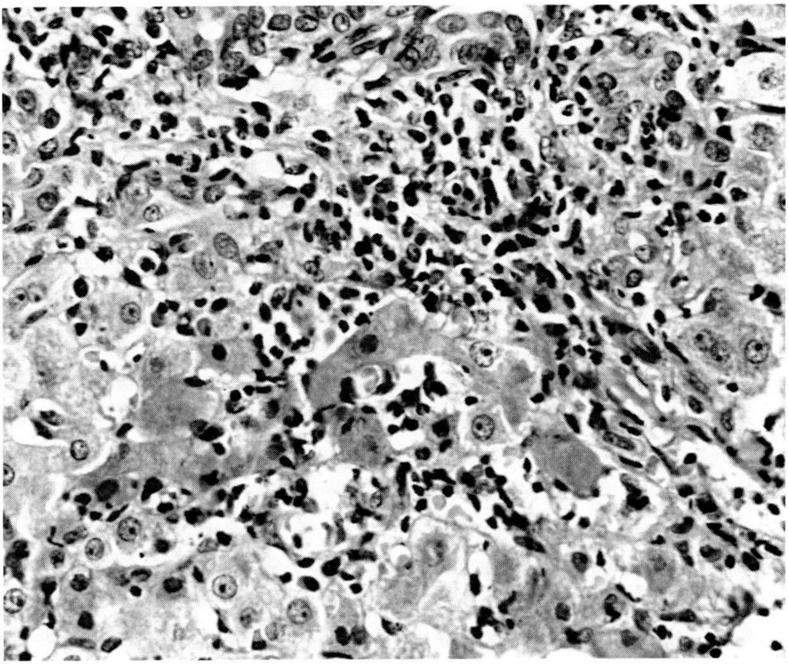

Fig. 2.5 Piecemeal necrosis in chronic active hepatitis. Inflammatory cells spread into the liver parenchyma from the portal tract (top). There is loss of the limiting plate where inflammatory cells surround degenerate and necrotic hepatocytes (HE, × 450).

has greater clinical significance. Both large and small droplet fatty change are seen occasionally in the same specimen (Fig. 2.6). Large droplets of fat in adjacent hepatocytes may coalesce to form 'fat cysts' (see Fig. 6.2). This fat is liberated by further disintegration of hepatocyte cytoplasm and induces focal inflammation. Fat-laden macrophages are seen in these 'fat granulomas' (see Fig. 6.3) and in adjacent portal tracts. It is also possible that ruptured fat cysts are a source of fat emboli. Focal scarring may follow the resolution of fat granulomas but it is very doubtful if fatty change is ever a cause of cirrhosis in man, although this seems to be true in certain experimental conditions.

2.2.2 *Glycogen infiltration*

Cytoplasmic glycogen in excess of that found in normal fresh hepatic tissue causes an hydropic appearance which affects the liver cells diffusely (Fig. 2.7). This may occur in diabetes mellitus and certain uncommon disorders of carbohydrate metabolism.

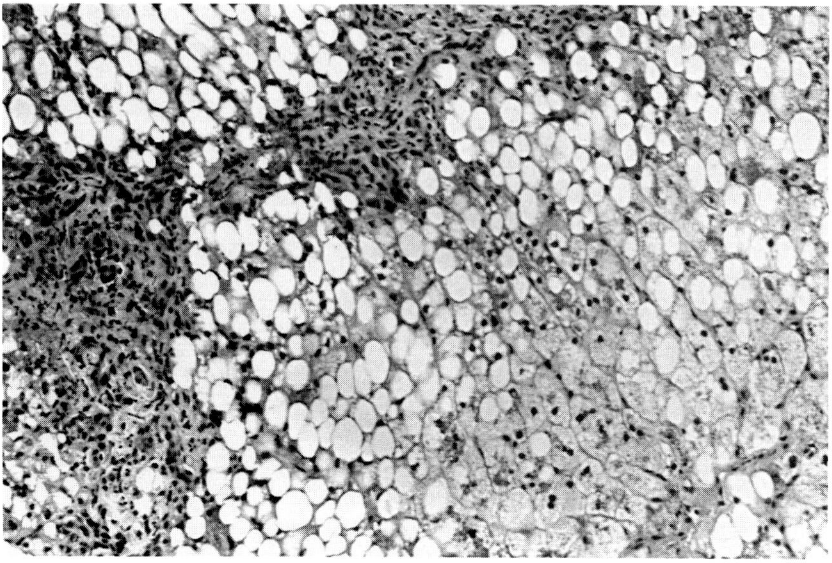

Fig. 2.6 Fatty change. Large droplets of fat with nuclear displacement involve most hepatocytes adjacent to the portal tracts which are fibrotic. Small droplet fatty change without nuclear displacement involves centrilobular parenchyma (bottom right) (HE, × 180).

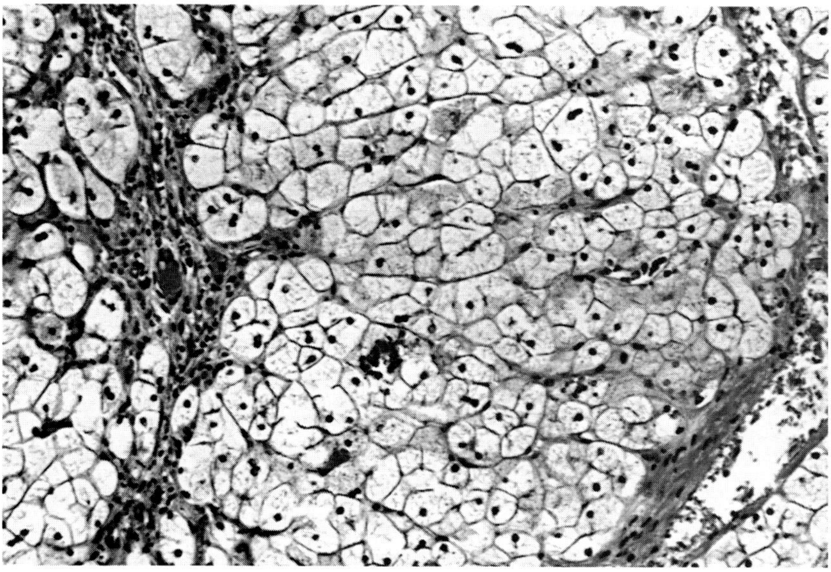

Fig. 2.7 Glycogen storage disease involving infant liver. All hepatocytes are swollen with excessive cytoplasmic glycogen, while the nuclei are small and centrally placed (HE, × 180).

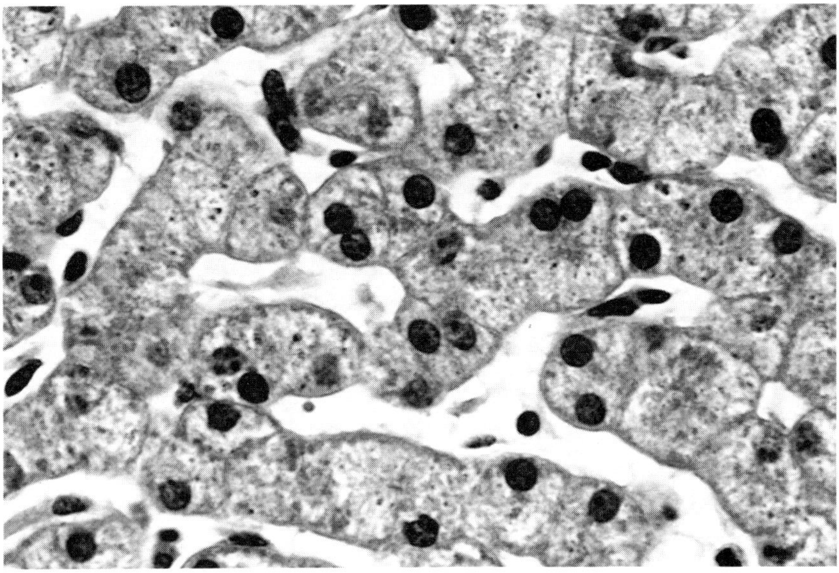

Fig. 2.8 Normal liver parenchyma for comparison with Figs 2.9–2.13 at the same magnification (HE, × 720).

2.2.3 *Hydropic degeneration*

This term is generally applied to single hepatocytes or groups of these cells with granular but pale-staining cytoplasm. Compared with normal (Fig. 2.8) they are enlarged, sometimes to a marked degree when the term 'ballooning degeneration' is used (Fig. 2.9). The condition is associated with liver dysfunction and is probably reversible.

2.2.4 *Feathery degeneration*

The cells have faintly staining cytoplasm which has a reticulated appearance and is usually bile stained. Their outer membranes are clearly defined and they need not be swollen (Fig. 2.10). They usually occur in groups near portal tracts in patients with obstructive jaundice. They may undergo necrosis and form bile infarcts.

2.2.5 *Acidophilic degeneration*

These hepatocytes are shrunken but conspicuous because of the intense eosinophilic staining of their cytoplasm (Fig. 2.11). Nuclear changes indicative of necrosis supervene, the condition becoming that of focal pinpoint necrosis already referred to.

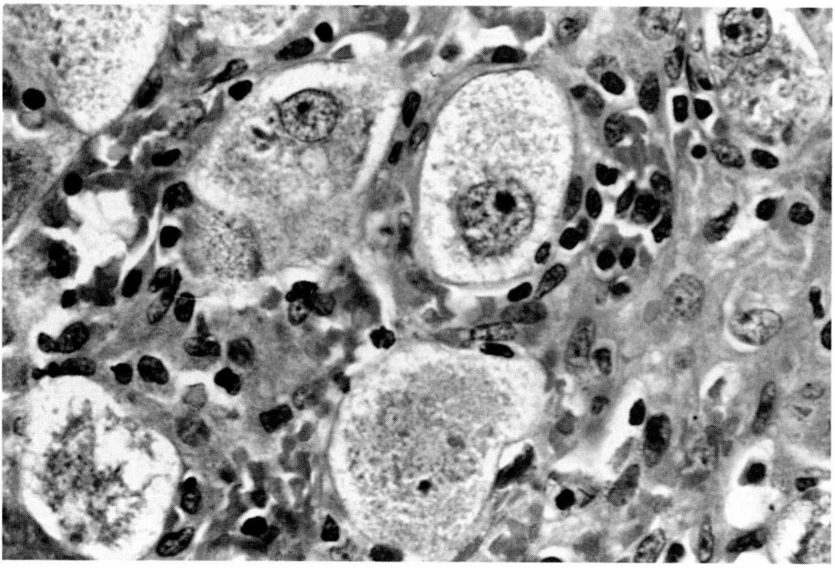

Fig. 2.9 Hydropic degeneration of hepatocytes in viral hepatitis. Several ballooned cells are conspicuous. There is also lymphocytic infiltration (HE, × 720).

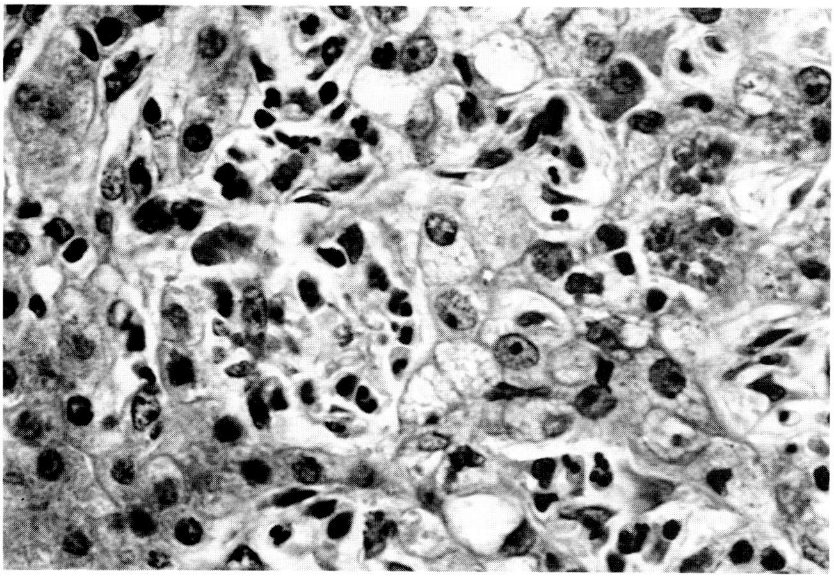

Fig. 2.10 Feathery degeneration. Compared with normal cells (bottom left) the abnormal hepatocytes are slightly swollen and have pale reticulated cytoplasm. They are also well demarcated. Some have undergone necrosis where there are aggregates of inflammatory cells at sites of cell dropout. Bile staining was present but is not evident in this photograph (HE, × 720).

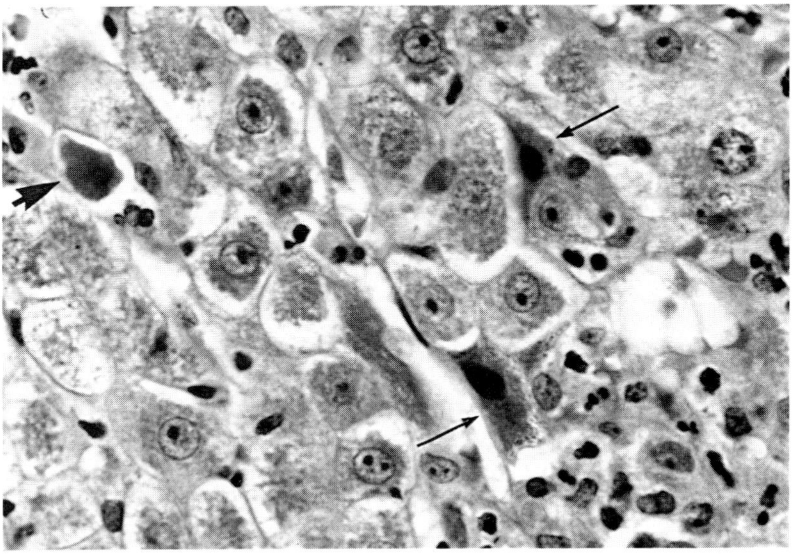

Fig. 2.11 Acidophilic degeneration and pin-point necrosis of hepatocytes. Two shrunken cells with this form of degeneration are seen (thin arrows). A third cell has undergone necrosis with loss of a recognizable nucleus (bold arrow). From a case of viral hepatitis (HE, × 720).

2.2.6 *Hyaline degeneration*

The cytoplasm of affected hepatocytes contains perinuclear hyaline bodies (Mallory bodies) of rounded or irregular shape which may not be clearly defined (Fig. 2.12). Sometimes there is an acute inflammatory reaction around affected cells which become necrotic while retaining visible hyalin. There is a common association with alcoholic hepatitis, when the degenerative change is usually best seen in centrilobular hepatocytes. A peripheral distribution is suggestive of some other condition such as primary biliary cirrhosis or Wilson's disease.

2.2.7 *Ground-glass hepatocytes*

This distinctive cytoplasmic appearance affects liver cells singly or in groups and indicates the presence of hepatitis B surface antigen in large amounts (Fig. 2.13). This may be confirmed by appropriate immunohistochemistry or with non-specific stains such as orcein and toluidine blue.

A somewhat similar but less striking change which affects hepatocytes diffusely throughout the liver may be seen in asymptomatic hepatomegaly induced by certain drugs (see Fig. 8.7).

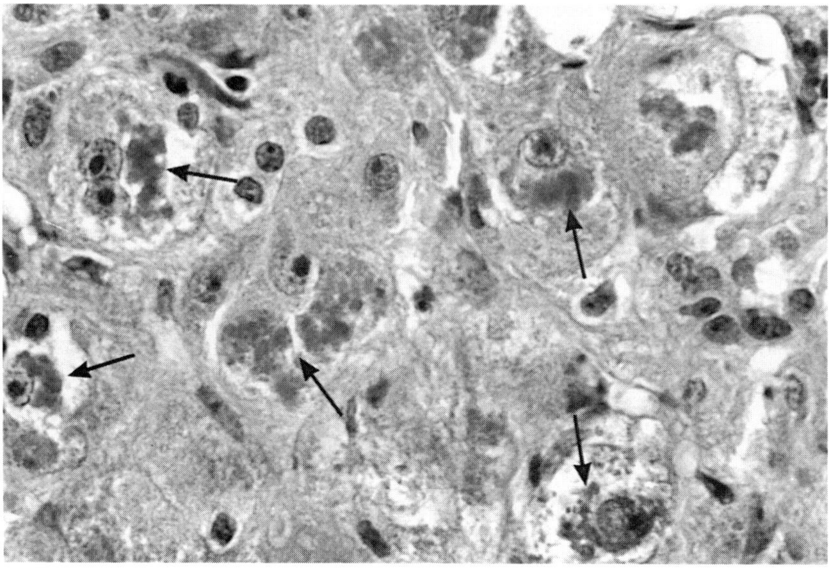

Fig. 2.12 Hyaline degeneration. Several hepatocytes contain cytoplasmic dense hyaline bodies of irregular shape (arrows). From a case of alcoholic hepatitis (HE, × 720).

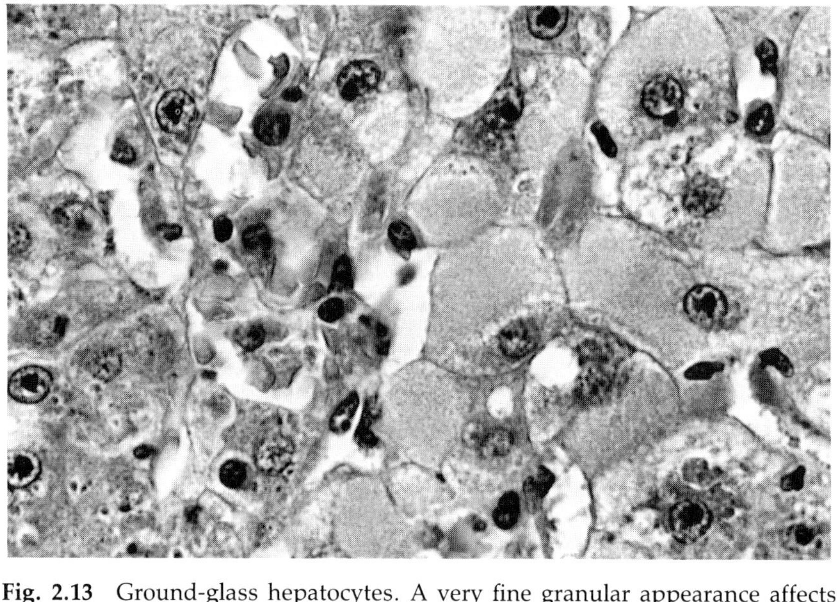

Fig. 2.13 Ground-glass hepatocytes. A very fine granular appearance affects part or all of the cytoplasm of most hepatocytes seen in the right half of this figure. Several of these cells also show hydropic degeneration (HE, × 720).

2.2.8 *Oncocytic hepatocytes*

The presence of mitochondria in large numbers causes swelling of hepatocytes which have a finely granular eosinophilic cytoplasm. Groups of such cells are seen very rarely in chronic hepatitis (Fig. 2.14) or as components of certain primary liver tumours.

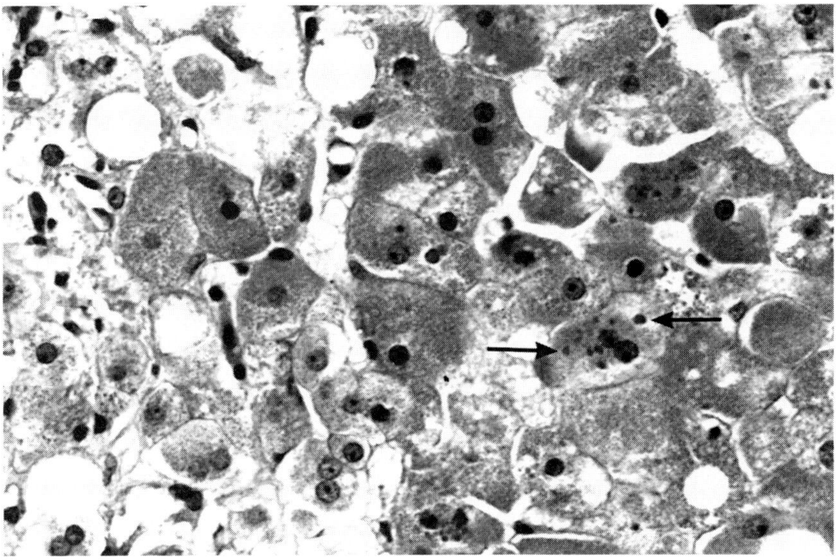

Fig. 2.14 The swollen hepatocytes with deeply staining and finely granular cytoplasm (centre and right) are oncocytes. A few giant mitochondria are visible (arrows). From a case of alcoholic hepatitis (HE, × 450).

2.2.9 *Hepatocyte inclusion bodies*

These may be aggregates of virus as in cytomegalic inclusion disease. Cytoplasmic inclusions may represent abnormal storage of a liver cell product which can occur in certain inherited metabolic disorders such as the PAS-positive bodies found in alpha-1-antitrypsin deficiency (Fig. 2.15). Giant mitochondria (see Fig. 6.8) can be seen with the light microscope in biopsies of regenerating liver, alcoholic hepatitis and scleroderma and should not be confused with cytoplasmic inclusion bodies or Mallory bodies. Nuclear inclusions are usually due to glycogen and are prominent in diabetes mellitus. This glycogen is labile and may be demonstrable only in frozen sections, but spaces are formed which are recognized readily in routine preparations (Fig. 2.16).

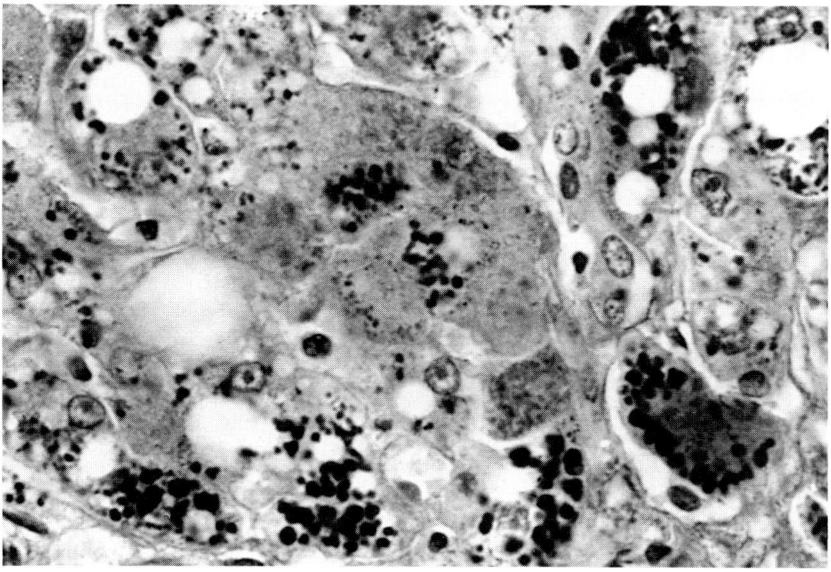

Fig. 2.15 Liver in alpha-1-antitrypsin deficiency. Darkly staining inclusion droplets of variable size are conspicuous in hepatocyte cytoplasm (PAS-diastase, × 720).

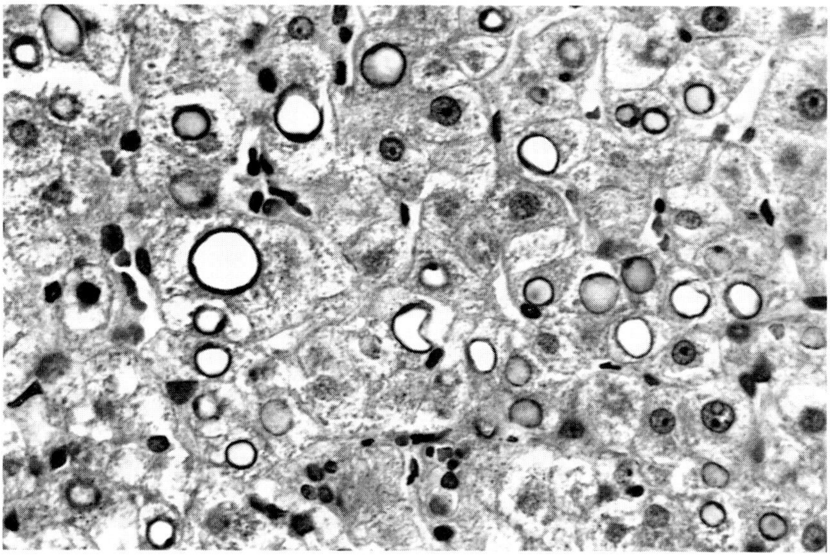

Fig. 2.16 Glycogen inclusions of hepatocyte nuclei in diabetes. The affected nuclei are swollen and clearly defined (HE, × 520).

2.2.10 Giant cell transformation

These large multinucleated liver cells may occur in biopsies from infants (Fig. 2.17). However 'giant cell hepatitis' is not a specific disease being found, for example, in various viral infections, rhesus incompatibility and biliary atresia. We have seen it very rarely in adult biopsies.

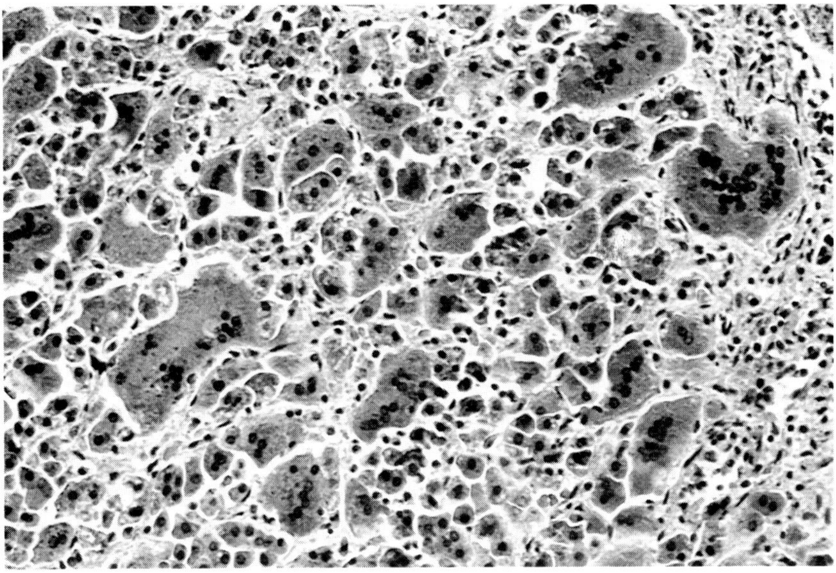

Fig. 2.17 Giant cell transformation of infant liver in hepatitis (HE, × 225).

2.3 Hepatic pigmentation

2.3.1 Bile pigment

This forms small granules in hepatocyte cytoplasm or more conspicuous plugs filling bile capillaries (Fig. 2.18) and larger distended ductules and ducts. Bile capillary cholestasis is usually most conspicuous in the centrilobular zones but a peripheral distribution can occur in association with chronic portal inflammation. Prominent pigmentation may involve foci of degenerate or necrotic hepatocytes or there may be small lakes of extracellular bile pigment. The significance of these changes is discussed in Chapter 3.

The colour of bile pigment may change from dark brown to olive green through the action of tissue fixatives. Green coloration is usually prominent with dichromate fixation and in sections stained by Van Gieson's method.

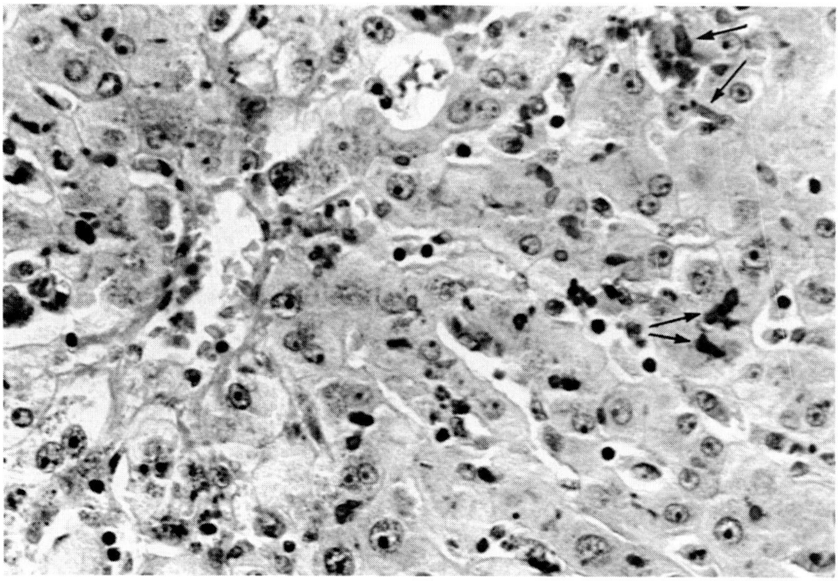

Fig. 2.18 Intrahepatic cholestasis. Bile pigment is present in the centrilobular zone, mainly within bile capillaries (arrows) and macrophages (Van Gieson's stain, × 450).

2.3.2 Lipid pigmentation

The following classification refers to groups of pigmented lipoproteins and not single compounds:

(a) Lipofuscin

Lipofuscins form brown intracytoplasmic granules, especially in centrilobular hepatocytes (Fig. 2.19). There may be confusion with bile and the two pigments can coexist in the same specimen. Unlike bile, lipofuscin remains intracellular and is never green in colour but gives a strongly positive Schmorl's ferric ferricyanide reduction test. This pigmentation is a normal feature of ageing and is exaggerated in wasting diseases and following ingestion of certain drugs.

Morphologically similar pigmentation is seen to a gross degree in Dubin–Johnson syndrome when it imparts a brick red or grey colour to the organ evident on naked-eye examination.

(b) Ceroid

This group of pigments causes brown discoloration to liver macrophages which may become conspicuous within portal tracts (Fig. 2.20). Unlike

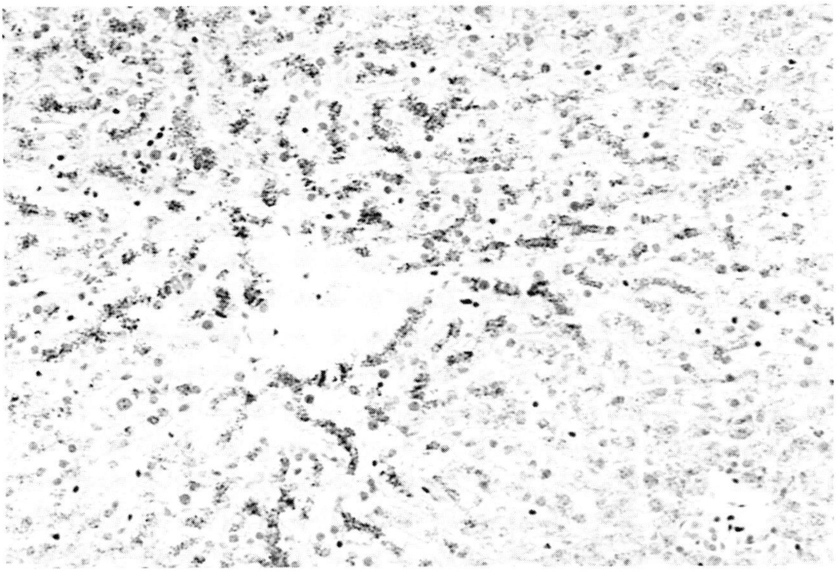

Fig. 2.19 Lipofuscin granules are present in the cytoplasm of many centrilobular hepatocytes (PAS and H, × 225).

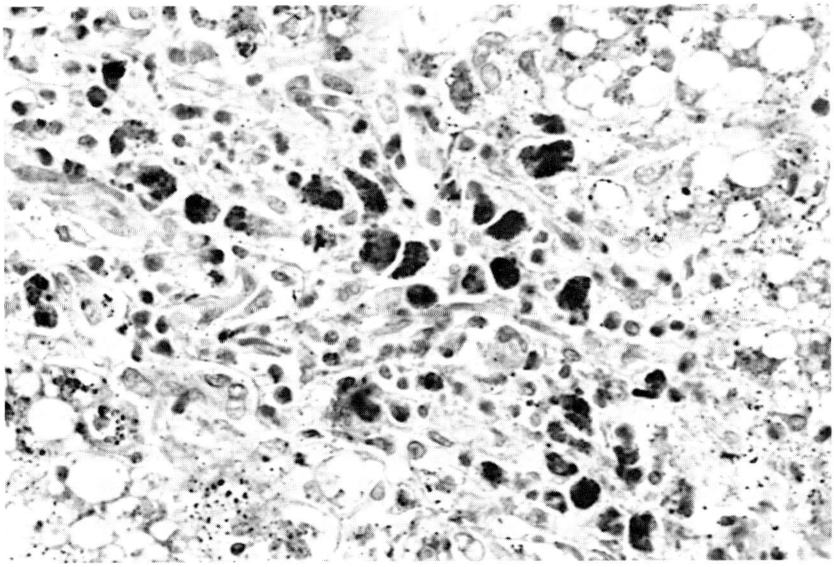

Fig. 2.20 Ceroid pigment is seen in many macrophages. From a case of tetracycline-induced liver injury (PAS and H, × 560).

lipofuscin the Schmorl test is negative but in paraffin sections ceroid is usually acid fast and gives positive staining by the PAS method and with Sudan stains (see Fig. 8.5). It may include some haemosiderin pigment. With ultraviolet light there is brownish-yellow autofluorescence.

Ceroid pigmentation is not a specific condition, being a manifestation of various forms of experimental liver injury. In man it is associated most frequently with liver diseases which include fatty change.

2.3.3 *Haemosiderin*

This forms yellow-brown granules of pigment which give a positive Prussian blue reaction.

(a) In hepatocytes

Cytoplasmic granules are most prominent in periportal cells (Fig. 2.21) and their presence indicates saturation of plasma transferrin. It may follow ingestion of excessive amounts of iron and is seen in normal young infants. It is a feature of some cases of alcoholic liver disease and is prominent in idiopathic haemochromatosis and porphyria cutanea tarda. The pigment may be present also in Kupffer cells and on collagen fibres where cirrhosis accompanies haemochromatosis.

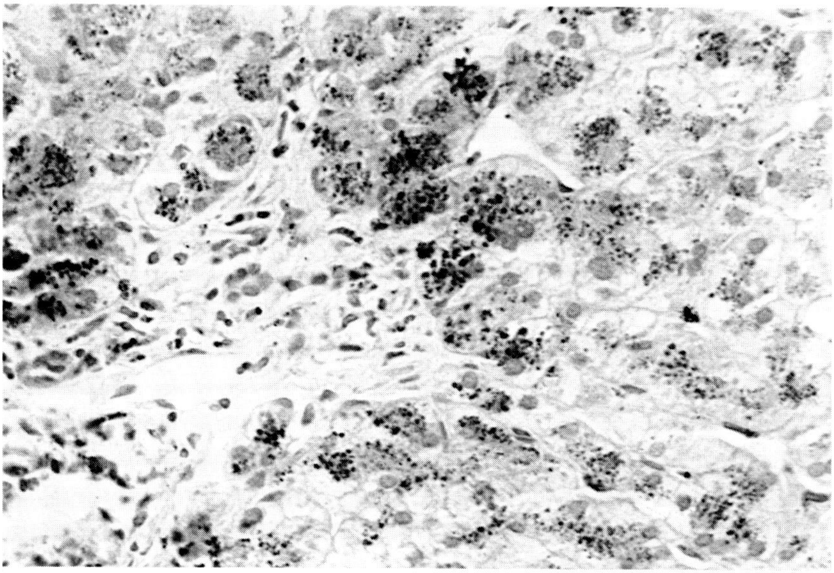

Fig. 2.21 Alcoholic hepatitis with haemosiderin granules in periportal hepatocytes (Perls' stain, × 450).

(b) In macrophages

Haemosiderin may be confined mainly or entirely to these cells in cases of excessive haemolysis (Fig. 2.22). It may indicate recent parenchymal damage and is a common feature of persistent viral hepatitis.

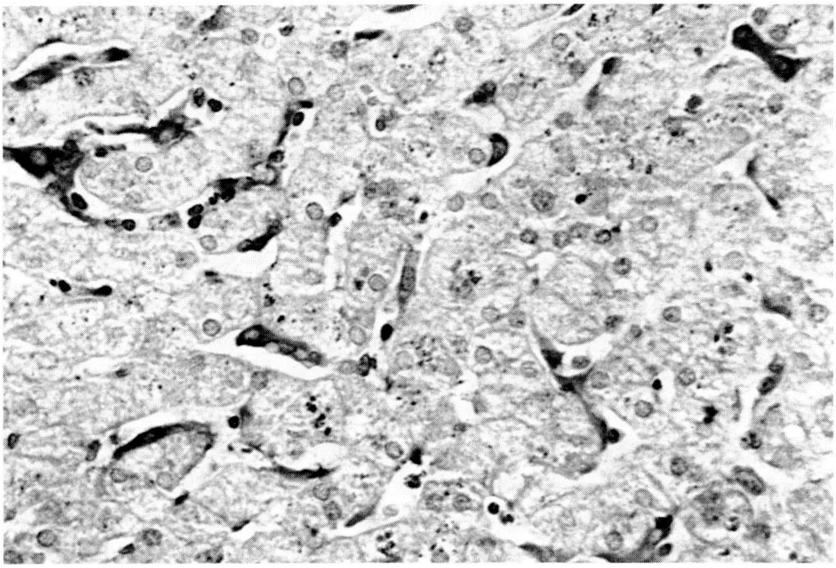

Fig. 2.22 Kupffer cells are conspicuous because of their content of haemosiderin pigment (Perls' stain, × 450).

2.3.4 Copper

Granular deposits of protein-bound copper salts may be present in the cytoplasm of peripheral hepatocytes. Compared with haemosiderin these granules (Fig. 2.23) are brownish-red in colour and may stain with rhodanine or rubeanic acid. Histochemical staining for copper can be unreliable and in children it tends to be distributed diffusely in hepatocyte cytoplasm rather than as granules. Orcein stains copper-binding protein (CBP) (see Fig. 3.7) and so might be regarded as an alternative method for the demonstration of copper in liver, but this does not apply in some cases of Wilson's disease in which there may be deficiency of CBP. Both copper and CBP may be prominent in chronic biliary disease and in Indian childhood cirrhosis. The subject is extensively reviewed by Sumithran and Looi (1985).

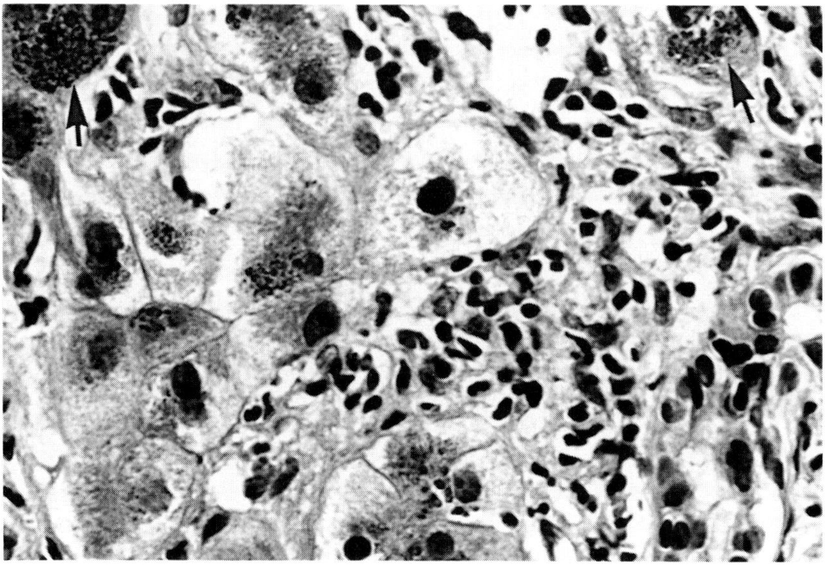

Fig. 2.23 Copper in hepatocytes. Part of a portal region in a case of primary biliary cirrhosis. There is a damaged bile duct (lower right) and infiltration with chronic inflammatory cells. The adjacent hepatocytes are hydropic and some contain numerous small dark granules of copper containing pigment in their cytoplasm (arrows) (HE, × 720).

2.3.5 *Miscellaneous pigmentation*

In addition to fixation artefacts the following may be seen:

(a) Haemozoin

This very dark brown or black granular form of haematin which is iron negative is seen within Kupffer cells in malaria and in bilharziasis.

(b) Calcium salts

Rarely in calcinosis haematoxylin and eosin sections of the liver show bluish-black granules of calcium salts especially in the centrilobular zones (see Chapter 9).

(c) Porphyrins

Conclusive evidence for the presence of these pigments in liver biopsies can best be obtained by the examination of fresh smears with ultraviolet light. Porphyrins show a red fluorescence which is greatest in hepatocyte nuclei.

(d) Pigmented tumours

Certain tumours may show appropriate pigmentation such as melanin and enterochromaffin granules.

(e) Carbon pigmentation

Coarse black granules of carbon pigment within Kupffer cells and portal tract macrophages are seen rarely in severe anthracosis (Fig. 2.24).

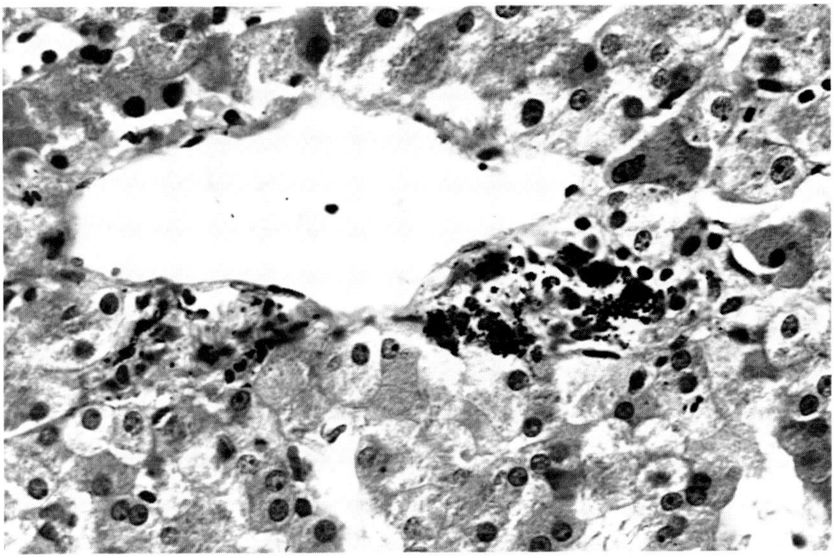

Fig. 2.24 Granules of black carbon pigment are seen in portal tract macrophages. The patient was a coal miner with anthracosis (HE, × 460).

2.4 Liver parenchymal regeneration

In the adult, liver plates of hepatocytes which are more than one cell thick are suggestive of regenerative activity (Fig. 2.25). This feature is best seen in reticulin preparations. Small acinar structures may be formed which are conspicuous in some cases of chronic active hepatitis (Fig. 2.26); in infant liver they are characteristic of galactosaemia. Mitotic activity of hepatocytes and a high proportion of binucleate cells may be seen, particularly in chronic liver injury and in large duct obstruction (see

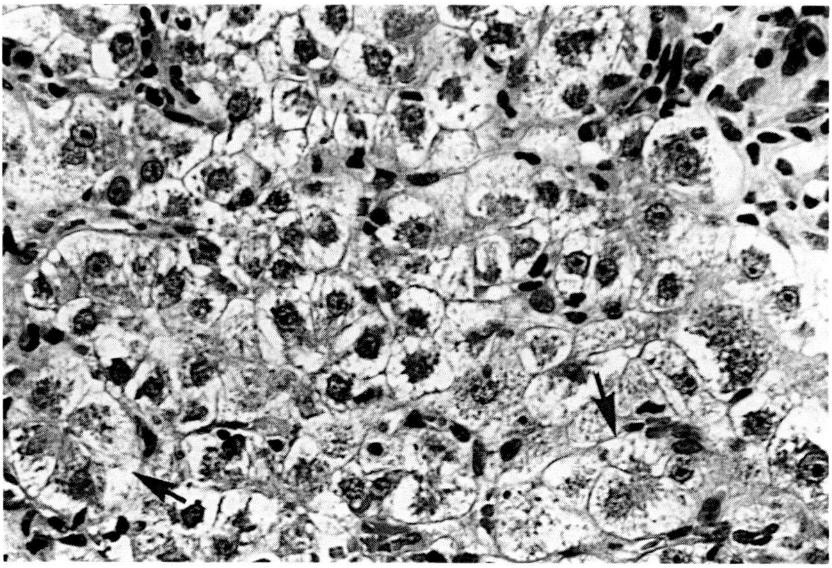

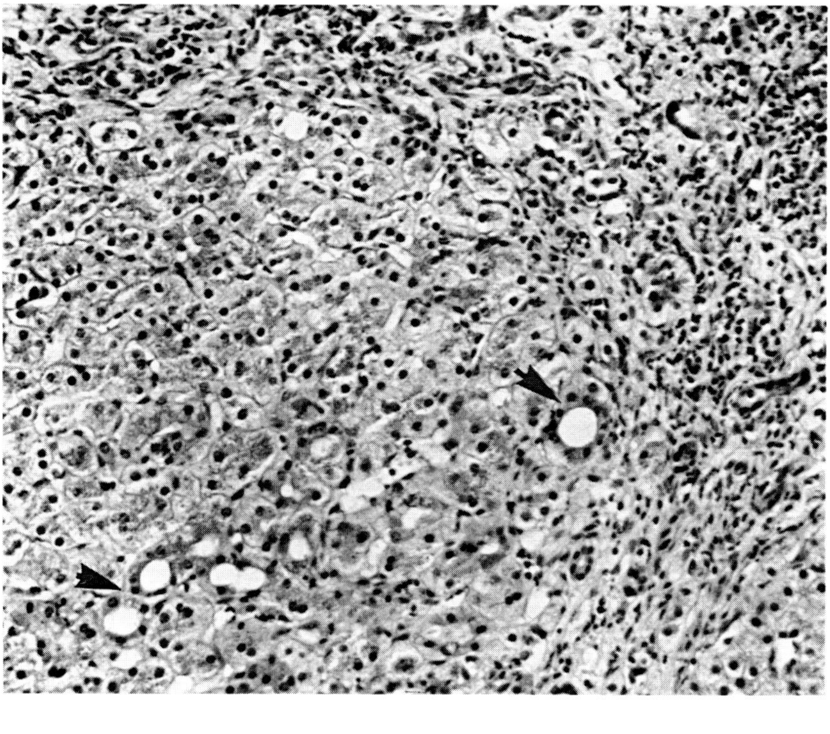

Fig. 8.10). It has to be remembered that increased ploidy of liver cells may be no more than an ageing phenomenon. Dysplasia, sometimes regarded as a premalignant change in hepatocytes, is recognized by nuclear pleomorphism and increase in cytoplasm (see Fig. 13.25).

Proliferation of bile ductules in portal tracts is seen in a variety of chronic liver diseases. It is often prominent in cases of large bile duct obstruction (Fig. 2.33).

2.5 Hepatitis and cholangitis

Infiltration of the liver with acute or chronic inflammatory cells can be the only conspicuous histological abnormality or accompany many of the necrotic and degenerative parenchymal changes already described. Hepatitis is best classified according to its cause but where this is not apparent the following terms are sometimes used.

2.5.1 *Portal and periportal hepatitis*

The portal tracts are the commonest sites for infiltration with inflammatory cells and this is seen in a variety of liver diseases. The infiltrate may be confined within the tracts or spill over into the adjacent parenchyma (Fig. 2.27). When this extension is marked with substantial piecemeal loss of liver cells, as in chronic active hepatitis and cholangiohepatitis, the term 'periportal hepatitis' may be used. Intense lymphocytic infiltration of portal tracts, sometimes with germinal centre formation and readily apparent at low power examination, is characteristic of primary biliary cirrhosis (Fig. 2.28). Leukaemic and lymphomatous infiltration should not be mistaken for hepatitis (see Figs 13.59 and 13.60).

2.5.2 *Lobular hepatitis*

Inflammatory cells are numerous within sinusoids and may form focal collections adjacent to degenerate or necrotic hepatocytes (Fig. 2.29). Portal and periportal inflammation are usually present but may be less conspicuous. The condition is best seen in viral hepatitis.

◀ **Fig. 2.25** Liver parenchymal regeneration in chronic active hepatitis. Hepatocyte plates are more than one cell thick and there is a rosette pattern of these cells in places (arrows) (HE, × 560).

◀ **Fig. 2.26** Liver parenchymal regeneration with pseudoacinar formation in a case of chronic active hepatitis. Hepatocytes lining these acini (arrows) are larger than the cells lining bile ductules which are numerous in this field (top and right) (HE, × 225).

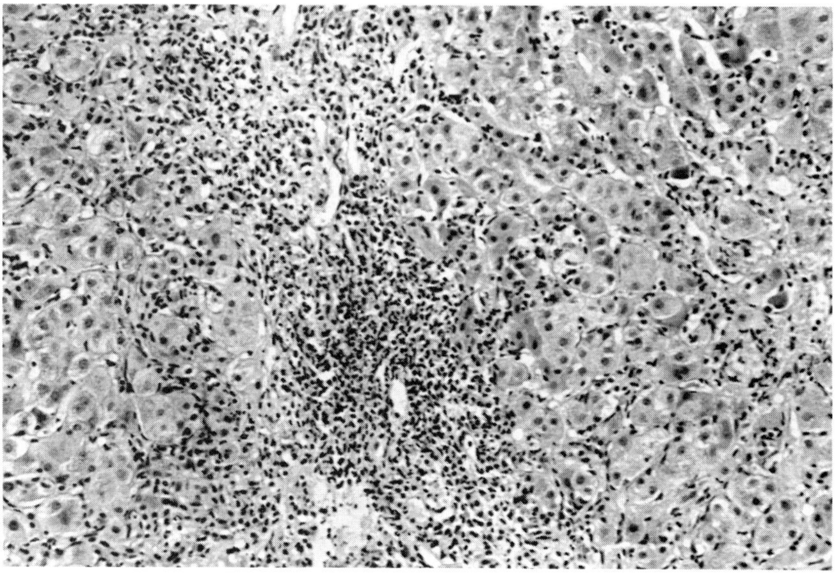

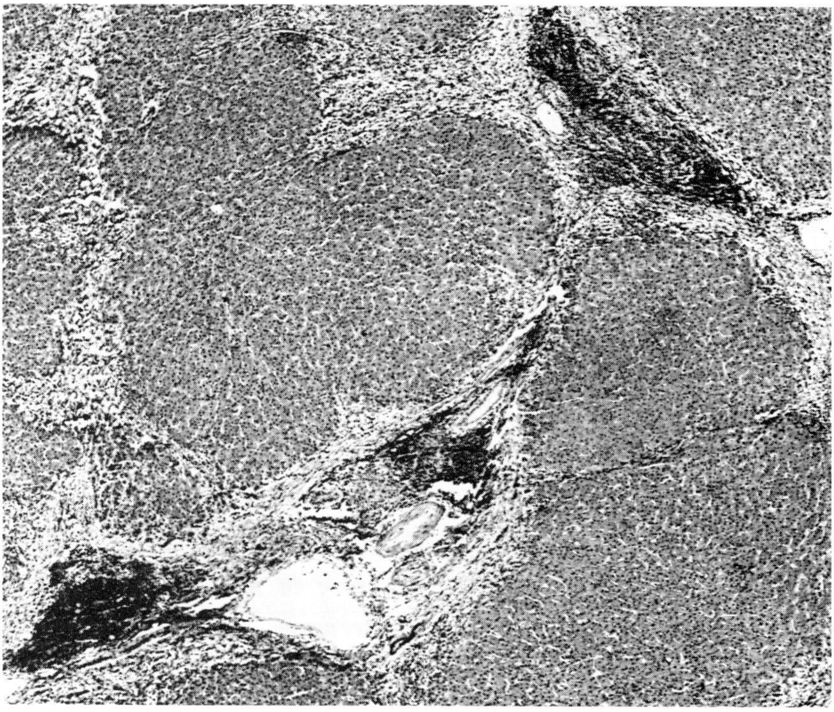

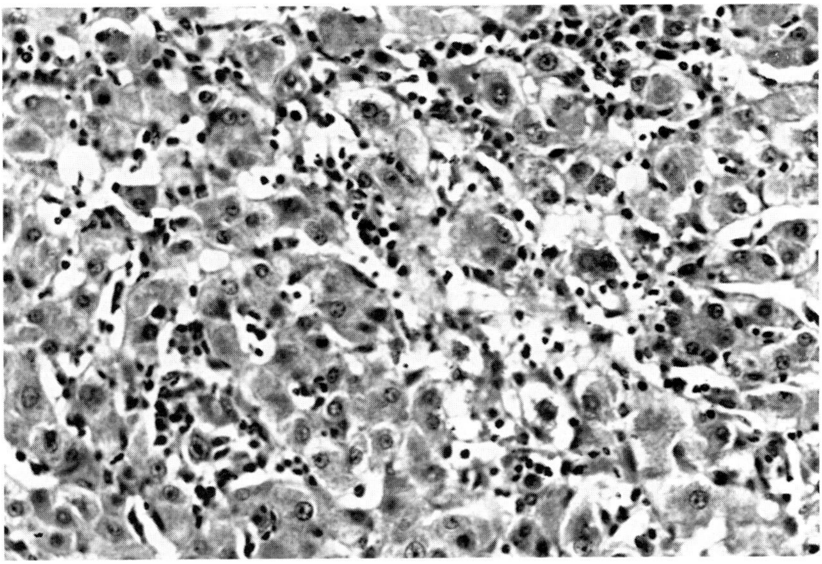

Fig. 2.29 Lobular hepatitis. Many chronic inflammatory cells are present in sinusoids and in close association with damaged hepatocytes (HE, × 285).

2.5.3 *Non-specific reactive hepatitis*

This is generally a mild condition characterized by portal hepatitis and small foci of lobular inflammation (Fig. 2.30). The distribution of these features is irregular, leaving parts of the liver with little or no abnormality. It may be part of a general response to various infective diseases or may be induced by certain drugs.

2.5.4 *Granulomatous hepatitis*

There are many causes for prominent or indistinct granulomas (see Section 5.7) scattered throughout the liver and some cases are idiopathic (Fig. 2.31). When confined within portal tracts the possibility of primary biliary cirrhosis should be considered.

◀ **Fig. 2.27** Portal and periportal inflammation in chronic active hepatitis. The portal tract (centre) is infiltrated by lymphocytes that extend into the adjacent liver parenchyma (HE, × 180).

◀ **Fig. 2.28** Portal hepatitis in primary biliary cirrhosis. There is very dense leucocytic infiltration of portal tracts where fibrosis is also advanced (HE, × 45).

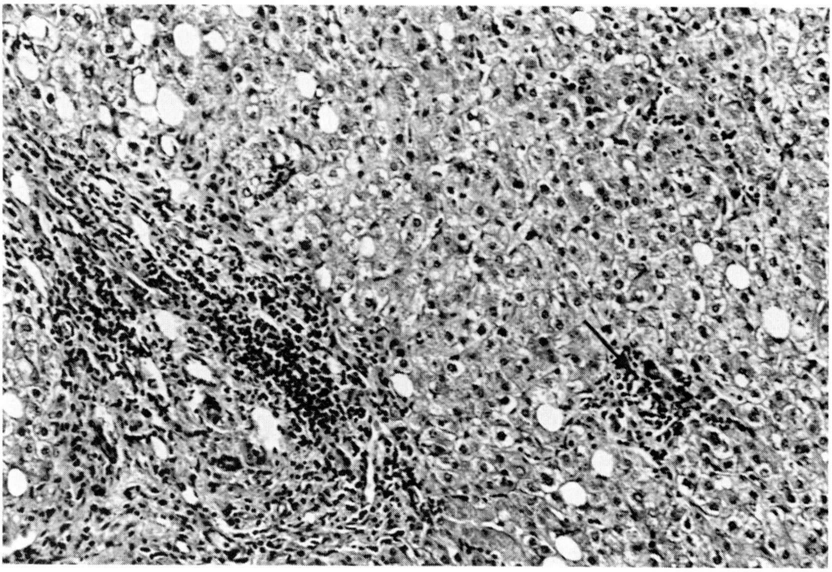

Fig. 2.30 Non-specific reactive hepatitis. Chronic inflammatory cells are present in the portal tract and form a small focus in the liver lobule at a site of cell dropout (arrow). The remainder of this wedge biopsy showed only minimal hepatitis (HE, × 180).

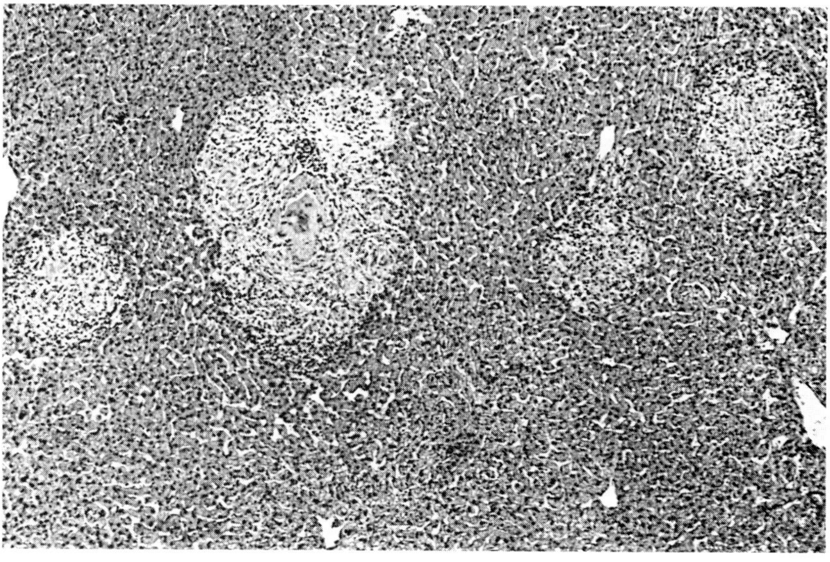

Fig. 2.31 Granulomatous hepatitis due to sarcoidosis (HE, × 72).

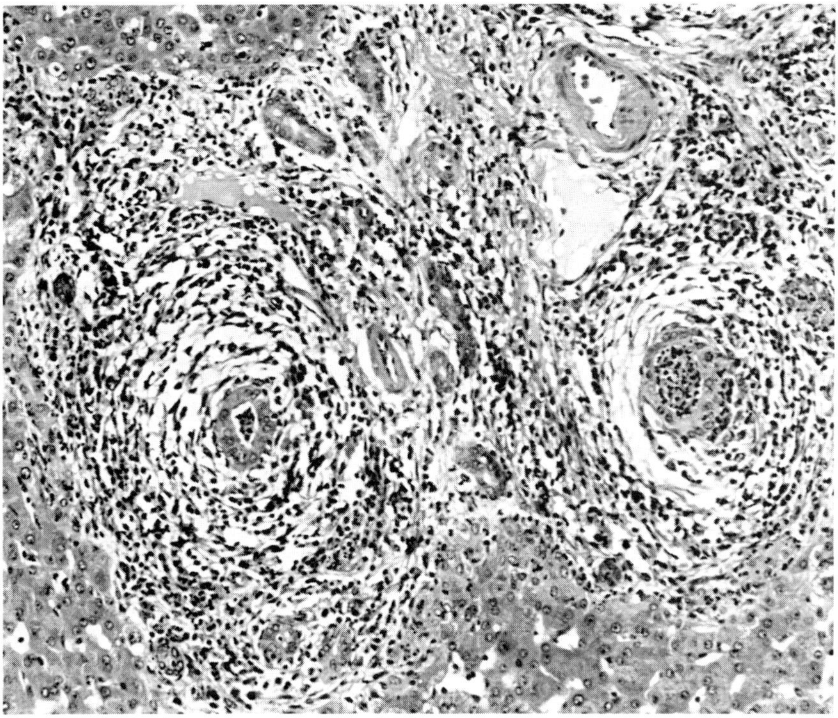

Fig. 2.32 Acute cholangitis. Small bile ducts contain and are surrounded by acute inflammatory cells and by a concentric arrangement of young connective tissue (HE, × 180).

2.5.5 Cholangitis and cholangiohepatitis

The presence of acute inflammatory cells within and around bile ducts is usually due to bacterial infection complicating an obstructive biliary lesion, in most cases extrahepatic (Fig. 2.32). Widening of portal tracts with inflammatory oedema is also characteristic and may be the main feature in less acute infections (Fig. 2.33). The inflammatory infiltrate may include numbers of foamy macrophages. In the majority of cases the inflammation spreads into the adjacent liver parenchyma and the term 'cholangiohepatitis' is used.

The term 'pericholangitis' is sometimes applied to inflammatory changes, usually chronic, which encircle smaller bile ducts within portal tracts. Some of these cases are related to chronic inflammatory bowel disease. A more chronic variety with fibrous obliteration of ducts is referred to as 'sclerosing cholangitis' and may have a similar aetiology (see Chapter 7).

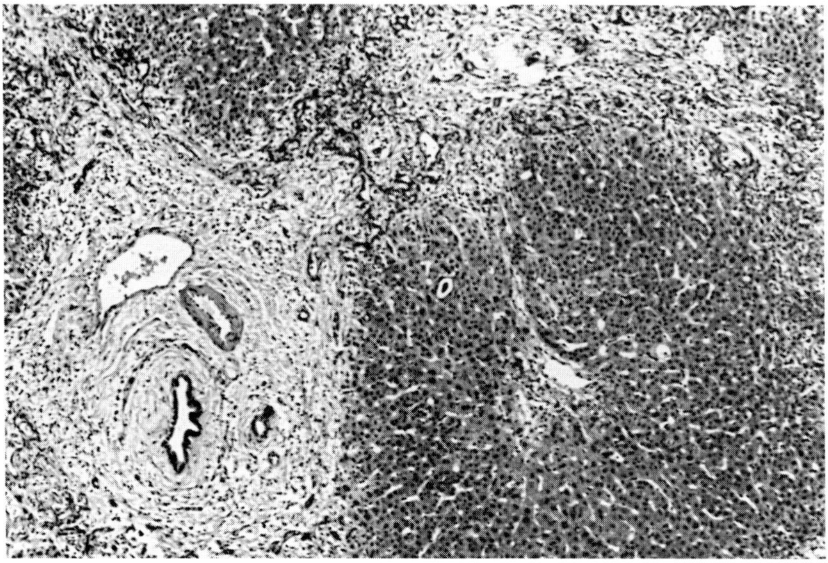

Fig. 2.33 Subacute 'cholangiohepatitis'. There is widening of the portal tract by a chronic inflammatory process which extends in places into the adjacent liver parenchyma. At the junctional zone there is proliferation of small deeply staining bile ductules. A larger interlobular duct shows fibrous thickening of its wall (HE, × 90).

2.5.6 *Liver abscesses*

These may be examples of suppurative cholangiohepatitis. Others arise from blood-borne infection, usually from an intra-abdominal inflammatory process. Microscopic features of specific infections are dealt with in Chapter 5.

2.6 Hepatic fibrosis

This is distinguished from true cirrhosis by persistence of a recognizably normal lobular and vascular pattern. In many cases it is a component of chronic active inflammation, and prominence of fibroblasts will indicate the persistence and progression of a chronic lesion which could terminate in cirrhosis. A trichrome stain is often useful in assessing this fibrogenic activity, while reticulin staining is better for demonstrating the general pattern of fibrosis. The following terms are used to describe special varieties.

2.6.1 *Non-cirrhotic portal fibrosis*

Connective tissue septa extend into the live parenchyma from fibrous portal tracts (Fig. 2.34), where there is often thickening of the walls of portal veins. Chronic inflammation may be present (p. 229).

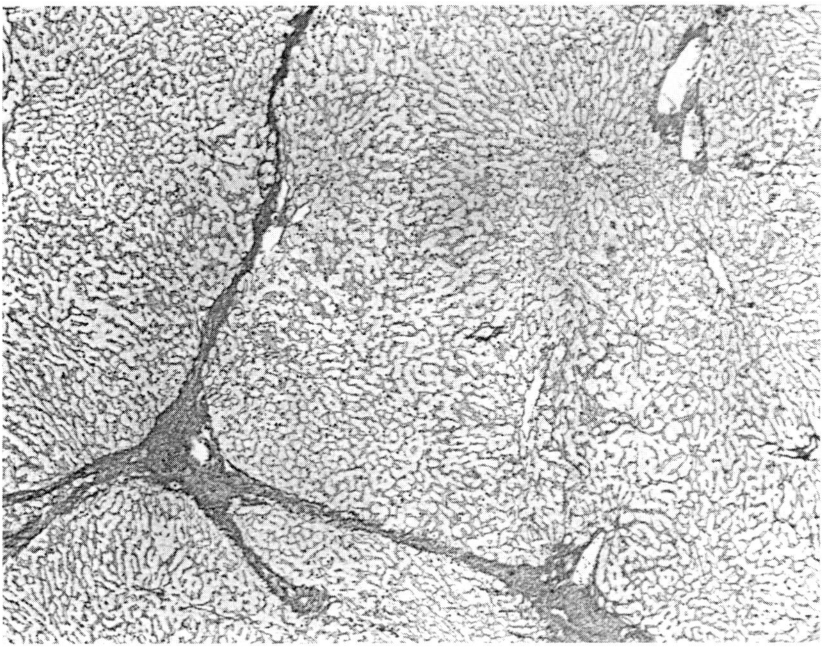

Fig. 2.34 Hepatic fibrosis. Thin fibrous septa form bridging lesions between portal tracts and central veins but in much of the liver the normal lobular architecture is preserved (reticulin stain, × 45).

2.6.2 *Post-necrotic scarring*

Irregular scars are the result of extensive parenchymal damage from various causes. Depending on the extent of the lesion and the presence of parenchymal regeneration this condition may or may not be a form of macronodular cirrhosis as defined below.

2.6.3 *Central venous sclerosis and hyaline sclerosis*

In alcoholic hepatitis, sclerosis is typically rather diffuse with thickening of sinusoidal walls. This is most marked in the centrilobular zones and in severe cases it is associated with prominent Mallory bodies or necrosis of hepatocytes and obliteration of central veins (see Fig. 6.14).

2.6.4 *Congenital hepatic fibrosis*

Bands of fibrous tissue traverse the liver which is otherwise normal (Fig. 12.12). These septa contain numerous small bile ductular structures which are characteristic features of this uncommon familial disease (p. 292).

2.7 Cirrhosis of the liver

This is defined as a condition in which there is loss of normal liver architecture, the parenchyma consisting of hyperplastic nodules separated by fibrous septa. Two main types are recognized: 'micronodular' (Fig. 2.35) in which nodules are mostly less than 3 mm in diameter, and 'macronodular' (Fig. 2.36) in which the majority are much larger. Not infrequently it is impossible to classify cases in this way as nodule size can vary from place to place in the same liver, and the term

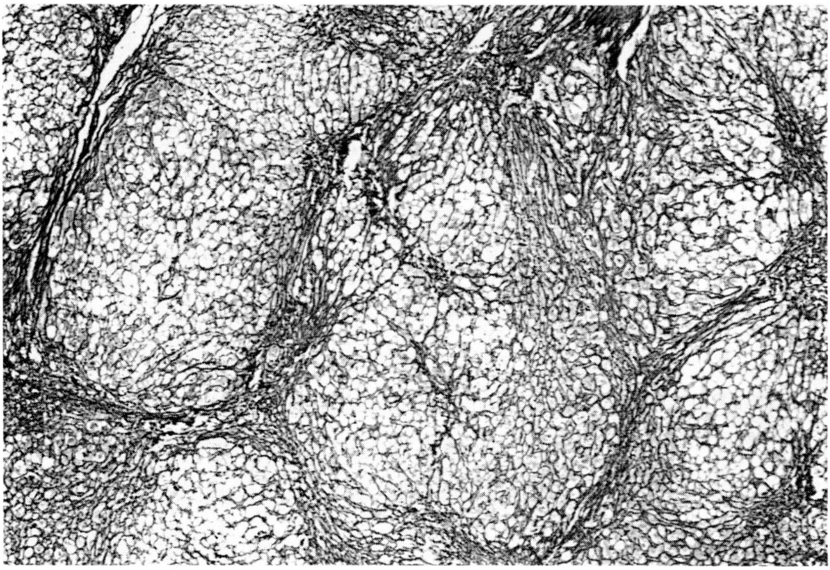

Fig. 2.35 Micronodular cirrhosis. Small nodules of regenerating parenchyma are separated by fibrous tissue. The parenchyma shows fatty change, this being an example of alcoholic cirrhosis (reticulin stain, × 45).

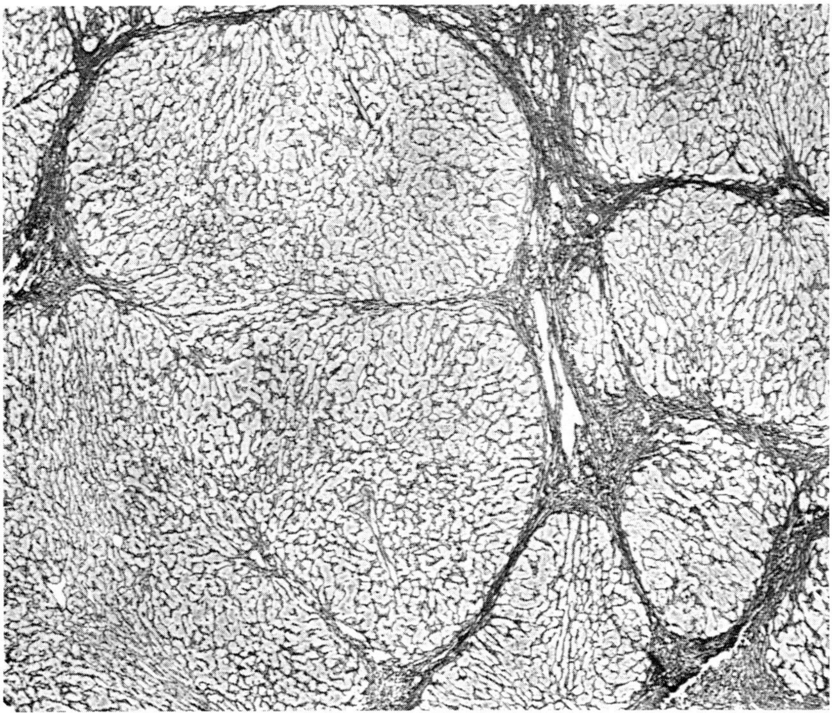

Fig. 2.36 Macronodular cirrhosis. Compared with Fig. 2.35 the regenerating nodules are much larger and contain small vascular channels. There was a history of viral hepatitis in this case (reticulin stain, × 45).

mixed cirrhosis' is then used. Cirrhosis can be classified also as 'active' when there is continuation of hepatitis and hepatocyte damage, especially at the periphery of nodules (Fig. 2.37), and 'inactive' when such changes are minimal or absent and the nodules clearly defined. Cirrhosis is discussed fully in Chapter 12 and has been reviewed by a group sponsored by the World Health Organization (Anthony *et al.*, 1978).

2.8 Miscellaneous histological features

Degenerative changes in bile duct epithelium, vascular lesions, infiltration with tumours, etc. should be sought in all liver biopsies. These are dealt with in Chapters 3, 10 and 13.

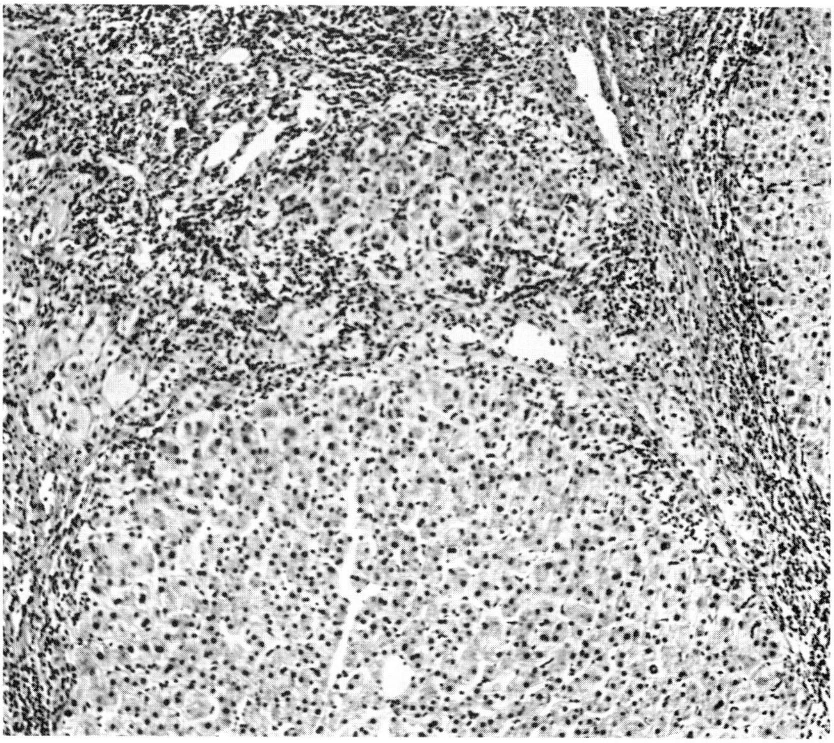

Fig. 2.37 Active cirrhosis. A case of post-hepatic macronodular cirrhosis showing a part of two large parenchymal nodules defined by fibrous septa which are infiltrated by inflammatory cells. Elsewhere (top and left) there appears to be destruction of hepatocytes by an intense inflammatory reaction (HE, × 100).

References

Anthony, P. P., Ishak, K. G., Nayak, N. C. *et al.* (1978), The morphology of cirrhosis: recommendations on definition, nomenclature, and clasification by a working group sponsored by the World Health Organization. *J. Clin. Pathol.*, **31**, 395–414.

Sumithran, E. and Looi, L. M. (1985), Copper binding protein in liver cells. *Hum. Pathol.*, **16**, 677–82.

3 Cholestatic jaundice and biliary obstruction

Liver biopsy has a very valuable role in the investigation of cholestatic jaundice. Cases due to blockage of a large duct, usually extrahepatic, require surgical exploration and management. Removal of an obstruction is essential to avoid the risk of severe or progressive cholangiohepatitis, while bile duct cancer may be detected at the stage when it can be successfully resected as a result of the proper investigation of obstructive jaundice. It must be stressed, however, that the important distinction between these extrahepatic cases and intrahepatic cholestasis due to a variety of primary liver parenchymal disorders cannot always be made by liver biopsy alone.

The importance of histopathology in assessing these cases must be set against the risk of biliary peritonitis from injury to dilated ducts in extrahepatic obstruction by the biopsy needle. The risk is probably quite small (Morris *et al.*, 1975) but, ideally, a case of cholestasis should be investigated in the first instance by endoscopy and cholangiography, ultrasonography and, if necessary fine needle percutaneous transhepatic cholangiography; these procedures can be followed by aspiration biopsy if they reveal no visual or radiological evidence of a large duct lesion. There will be, of course, the opportunity for histological investigation of the extrahepatic cases when wedge or needle biopsies are obtained under proper surgical control during laparotomy.

3.1 Histopathology

3.1.1 Intrahepatic cholestasis

Inspissated bile is present within bile capillaries and may outline the pattern of these structures (Fig. 3.1). This is seen in the centrilobular zone where, in addition, granules of bile pigment or diffuse bile staining are often present in the hepatocyte cytoplasm. Estimates of severity and duration of cholestasis from intensity of bile pigmentation are unreliable,

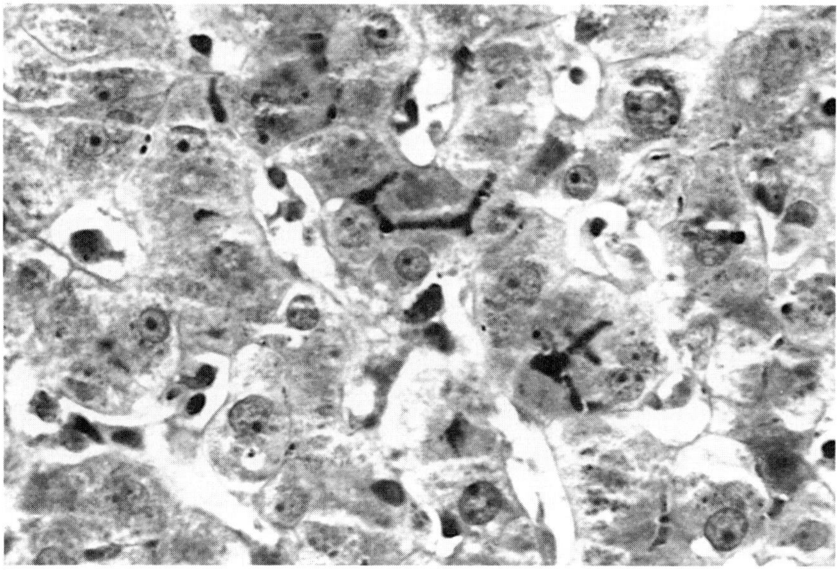

Fig. 3.1 Intrahepatic cholestasis. Some canaliculi are outlined by inspissated bile (HE, × 720).

as pigment can be lost by solution in fixation fluid or by digestion with hepatocyte lysosome enzymes. In longstanding cases there may be prominent bile thrombi within small acini (Fig. 3.2) formed by hepatocytes. The pigment is dull brown or olive green in paraffin sections, the latter colour usually being a feature of those stained by Van Gieson's method. Granules of cytoplasmic lipofuscin which retain a more distinct brown colour may also be conspicuous.

Centrilobular cholestasis cannot be explained simply by mechanical obstruction of bile flow along canaliculi, even when there is blockage of a large duct. Instead, it should be regarded as evidence of disordered hepatocyte function related to disturbance of normal bile salt metabolism or impairment of active sodium transport across canalicular membranes, both of which are necessary for the excretion of bile. In addition, it is postulated that there may be some damage to the microfilamentous system located in the pericanalicular part of the hepatocyte cytoplasm, the intermittent contraction of which could provide the propulsive energy for the flow of bile along these small channels. The process of cholestasis can be self-perpetuating as the retained bile is in itself cytotoxic. The cytoplasm of hepatocytes so affected is bile stained and may be hydropic (Fig. 3.3) or present a fine reticulated appearance known as 'feathery degeneration' (Fig. 3.18). Inflammatory cells, especially

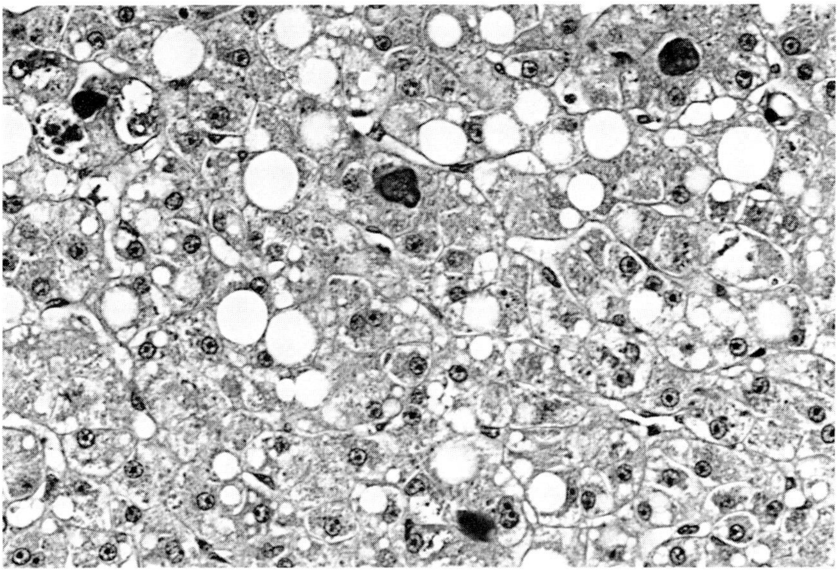

Fig. 3.2 Several dark bile thrombi are seen within acinar structures formed by hepatocytes. Fatty change is also prominent (HE, × 320).

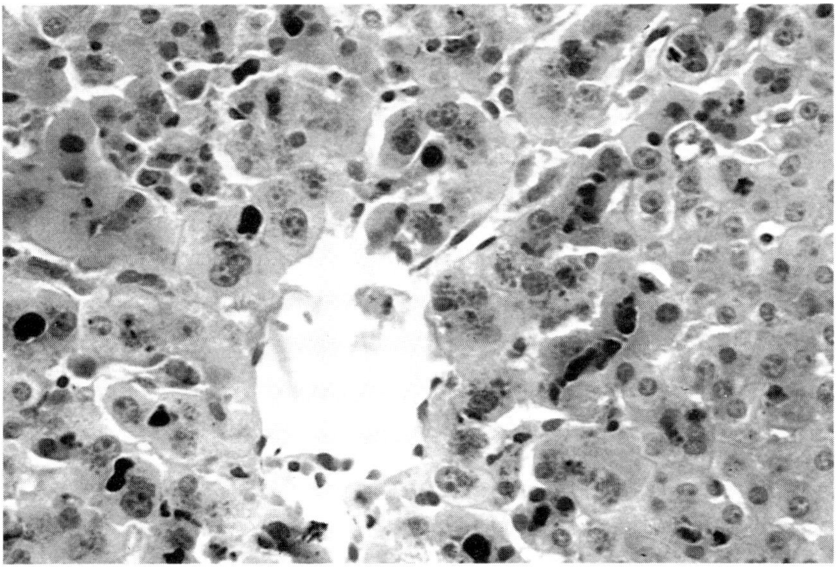

Fig. 3.3 Intrahepatic centrilobular cholestasis. Bile is present in canaliculi and as granular deposits in the cytoplasm of hepatocytes around the centrilobular vein. Some of these cells are swollen or necrotic (HE, × 450).

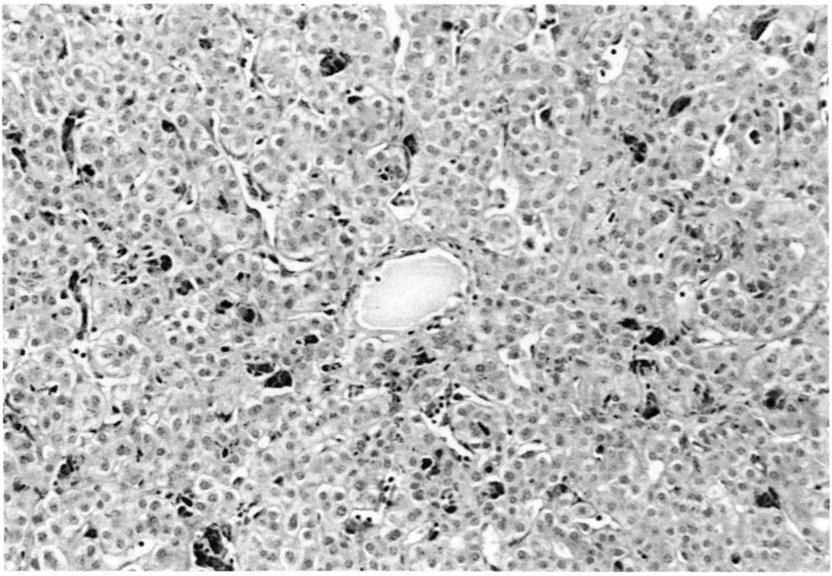

Fig. 3.4 Intrahepatic cholestasis. Many Kupffer cells contain bile-stained and PAS-positive debris (PAS stain, × 180).

macrophages, become prominent in persistent cases because of their ingested bile-stained material which is also PAS positive and diastase resistant; these remain for some weeks after resolution of cholestasis (Fig. 3.4) following removal of the cause.

Abnormal bile salt metabolism can also affect the canals of Hering and the larger bile channels in the portal tracts, causing proliferation or degenerative changes in their epithelial lining with altered permeability that allows conjugated bilirubin to pass through into vascular channels, and induces portal inflammation and periductal fibrosis. Obstruction to the onward flow of bile into these portal ducts may follow, causing cholestasis in the peripheral bile capillaries and pseudoacini (Fig. 3.5) and this becomes a feature of chronic cases. Where there is a primary defect of bile duct epithelium without initial damage to hepatocytes, the essential lesion of primary biliary cirrhosis, portal inflammation and peripheral cholestasis can develop in the absence of centrilobular cholestasis. The same parenchymal degenerative changes which accompany centrilobular cholestasis are found also in relation to the peripheral variety; in addition, some hepatocytes in these periportal zones may show Mallory bodies similar to those seen in alcoholic liver damage (Fig. 3.6). Protein-bound copper may be demonstrated in hepatocyte cytoplasm by orcein staining (Fig. 3.7); it is often associated with primary biliary cirrhosis but may be

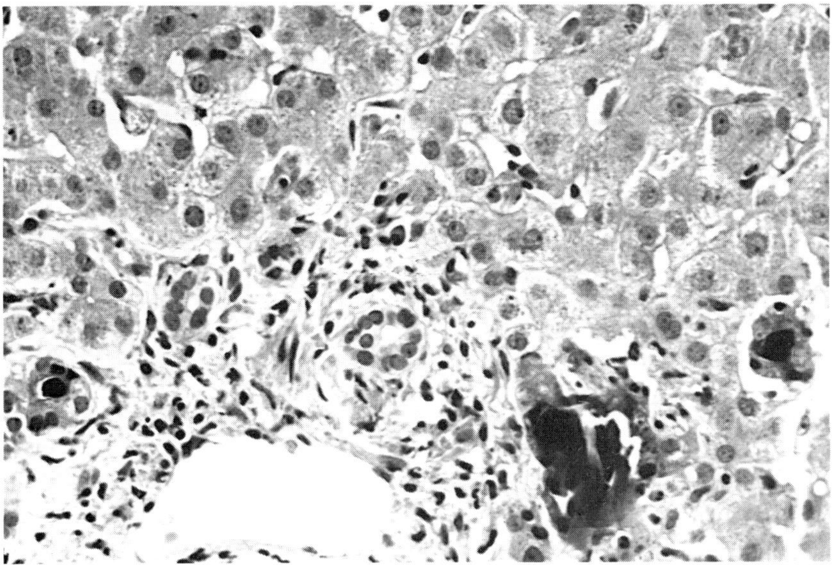

Fig. 3.5 Intrahepatic peripheral cholestasis. Inspissated bile is seen in acinar structures adjacent to the portal tract and in bile canaliculi. There was common bile duct obstruction due to a calculus and cholestasis also involved the centrilobular zones (HE, × 395).

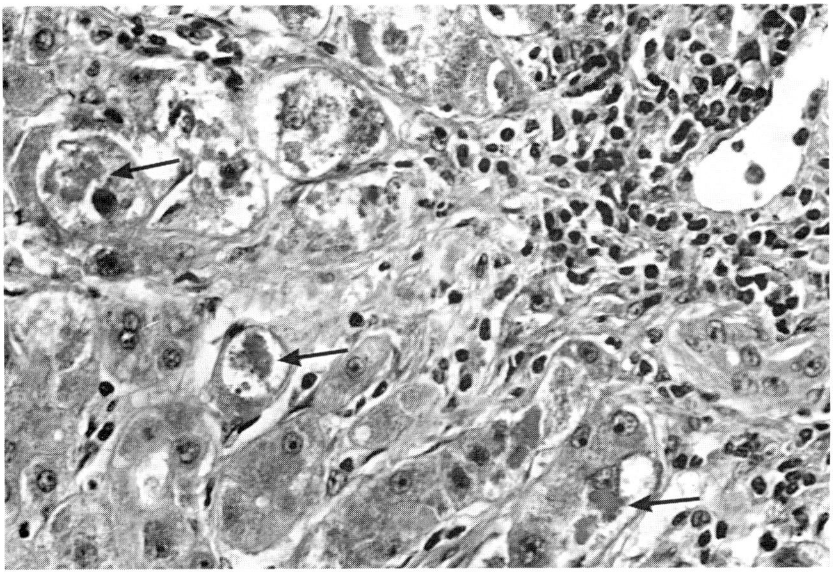

Fig. 3.6 Hydropic and hyaline degeneration of peripheral hepatocytes in extrahepatic biliary obstruction with chronic cholangiohepatitis. Cytoplasmic hyalin (arrows) resembles Mallory bodies but there was no history of alcoholism in this case (HE, × 450).

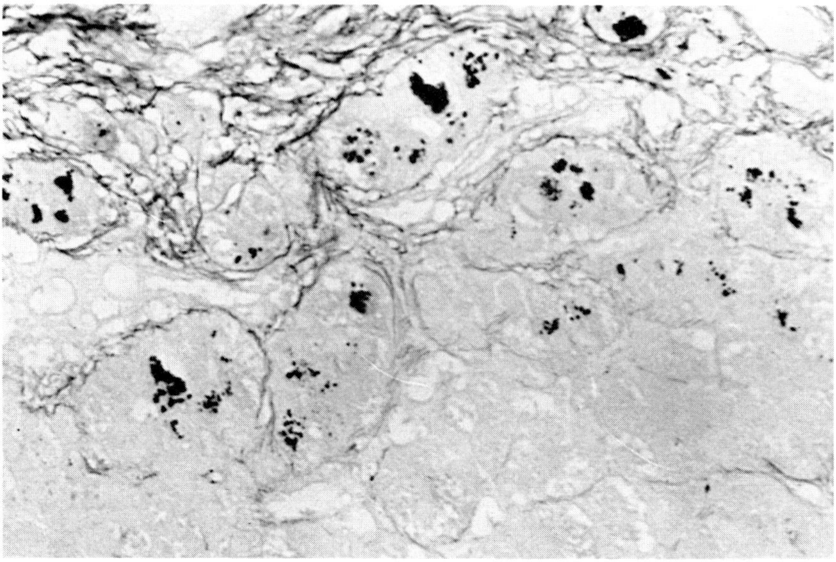

Fig. 3.7 Black granules of copper-binding protein are seen within hepatocyte cytoplasm adjacent to an area of chronic portal inflammation with fibrosis. From a case of secondary biliary cirrhosis (orcein, × 560).

present in any chronic disease with impaired biliary excretion of copper. It could be an additional cause of hepatocyte damage in biliary disease, but this is uncertain. When cholestasis persists for many months or years hyaline change can involve biliary epithelium.

Mechanical obstruction may be a factor in the induction of bile capillary cholestasis in cirrhosis of any type, not necessarily biliary cirrhosis (Fig. 3.8). Inspissated bile may be present also in large tortuous ducts in cirrhosis and in developmental anomalies such as congenital hepatic fibrosis. Duct cholestasis in these cases is irregular in distribution and can be found in the absence of jaundice. Inspissation from fluid loss may explain the more diffuse ductular cholestasis seen in certain patients with severe shock (Fig. 3.25).

The histological appearances associated with bile canalicular cholestasis are non-specific with regard to aetiology and do not help to distinguish extrahepatic from intrahepatic cases. There is a large list of causes of intrahepatic cholestasis and many of these are noted in various parts of the book. Some examples can be recognized by the additional histological features of the primary condition, such as cholestatic viral and alcoholic hepatitis and liver involvement in Hodgkin's disease. The greatest difficulty usually arises with drug-induced cholestatic jaundice with or without hepatitis, especially if the clinical history is inconclusive,

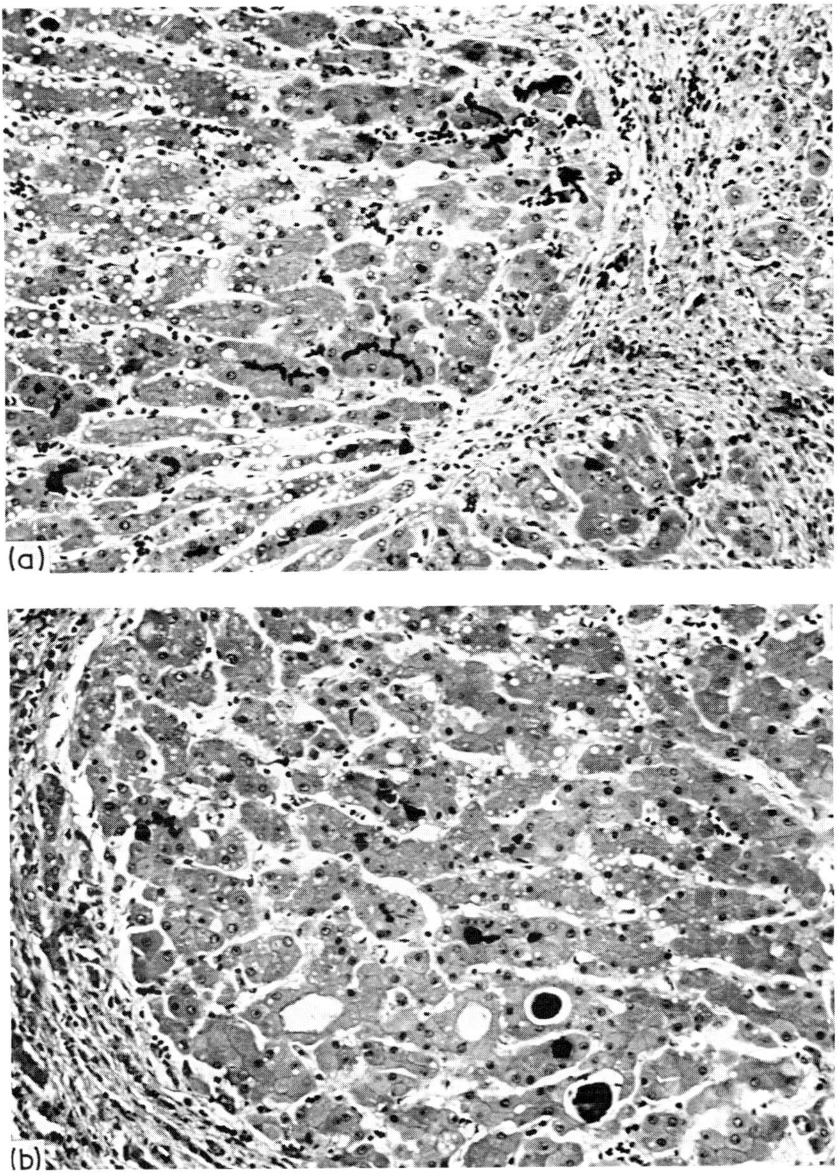

Fig. 3.8 (a) and (b), intrahepatic cholestasis in post-hepatitis cirrhosis. No extrahepatic obstruction was detected at autopsy. ((a) and (b) HE, × 180.)

as it often is. If large duct obstruction cannot be excluded readily it is essential to undertake further investigation of the biliary tract by cholangiography and endoscopy if not already done and, if necessary, exploratory laparotomy.

3.1.2 Extrahepatic bile duct obstruction

Centrilobular cholestasis can be the only abnormality seen in liver biopsy, especially during the first week or two of illness, and Fig. 3.9 illustrates one such example. Fortunately in many cases there are helpful positive diagnostic features, such as prominent bile ductular proliferation and dilatation of interlobular and septal ducts, usually with evidence of cholangitis. While portal inflammation can be the result of abnormal bile salt metabolism (as already noted), there is frequently the additional factor of secondary infection, often of low virulence. Both are causes of progressive widening of the portal tracts with prominent fibroblasts and variable numbers of acute inflammatory cells set in loose oedamatous connective tissue (Fig. 3.10) which is typically arranged in a concentric fashion around bile ducts. With spread of inflammation into adjacent parenchyma there is loss of the limiting plate of hepatocytes, but this is

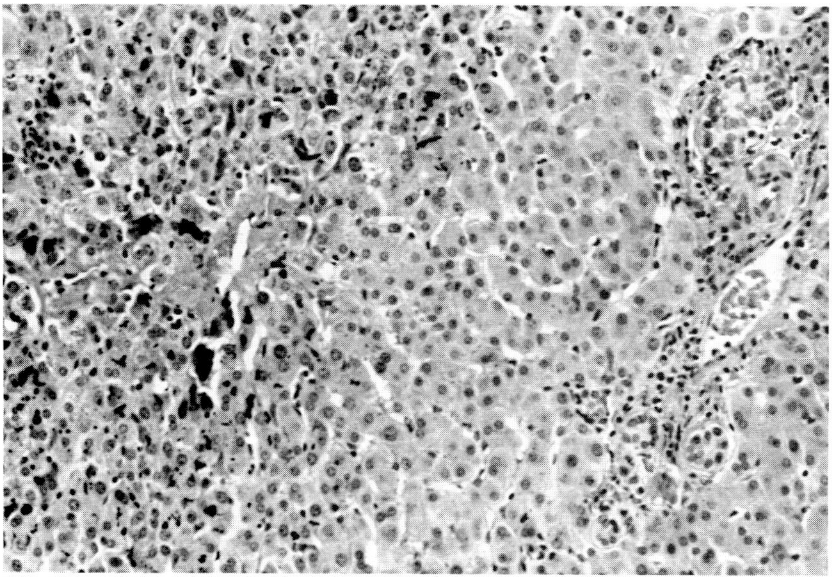

Fig. 3.9 There is much dark bile pigment in canaliculi situated around the central vein (left). Despite the absence of cholangitis this patient had extrahepatic obstruction due to pancreatic carcinoma (HE, × 195).

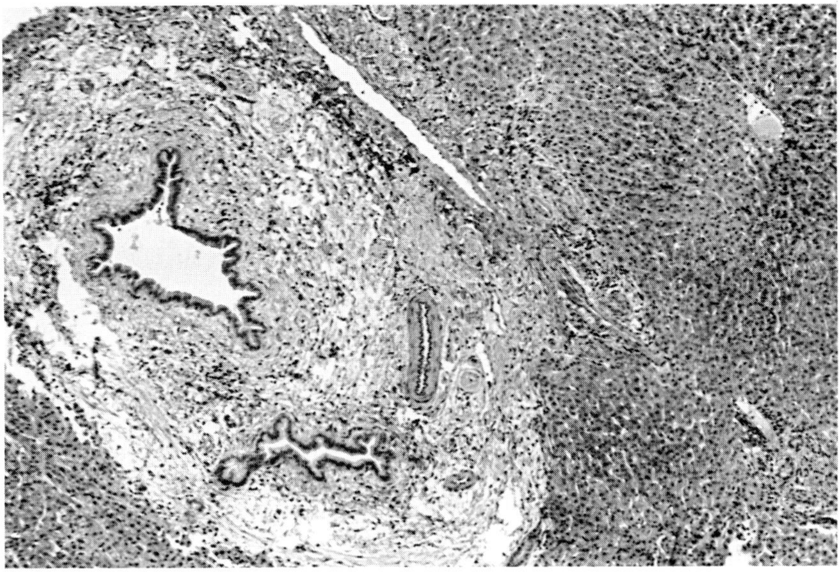

Fig. 3.10 Extrahepatic obstructive jaundice. There is considerable widening of the portal tract due to fibrosis and inflammatory oedema around dilated bile ducts (HE, × 72).

often less conspicuous compared with aggressive chronic active hepatitis. It is not surprising, however, that there may be biochemical evidence of liver cell damage as well as obstructive jaundice, and so the term 'cholangiohepatitis' is appropriate. Dilated ducts may contain inspissated bile (Fig. 3.11). There may be degeneration or hyperplastic polypoid proliferation of bile duct epithelium (Fig. 3.12), the latter being an effect of prolonged high biliary pressure. Depending on the virulence of infection pus cells may be present in and around these dilated ducts (Fig. 3.13), and large abscesses may form which can be the source of a severe or fatal septicaemia (Fig. 3.14), but occasionally no pathogenic organisms can be recovered (Ishak and Rogers, 1981). Duct dilatation is uncommon, and even without it the presence of acute inflammatory cells in the lumen and walls of ductules is very suspicious of an obstructive lesion (Fig. 3.15). Gallstones within dilated intrahepatic ducts and associated with chronic suppuration are rarely found; such lesions are commoner in the Orient where some are related to liver fluke (*Clonorchis sinensis*) infestation.

In less virulent infections macrophage activity is more prominent. In the portal tracts especially there may be granulomas (Fig. 3.16) or collections of foamy macrophages (xanthoma cells) (Fig. 3.17) which should not be confused with hepatocytes showing feathery degeneration (Fig. 3.18), but the distinction may be impossible. At these sites necrosis

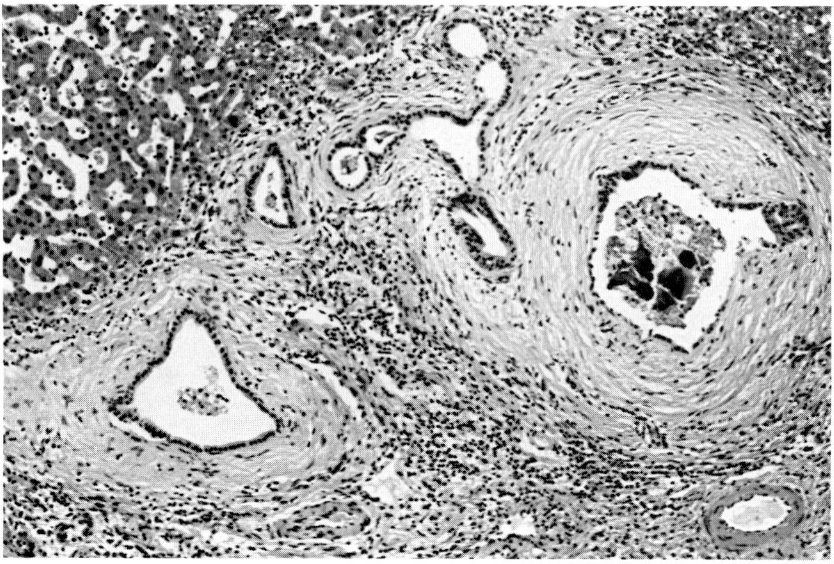

Fig. 3.11 Extrahepatic obstructive jaundice. There is concentric fibrosis around dilated ducts, one of which contains inspissated bile (HE, × 100).

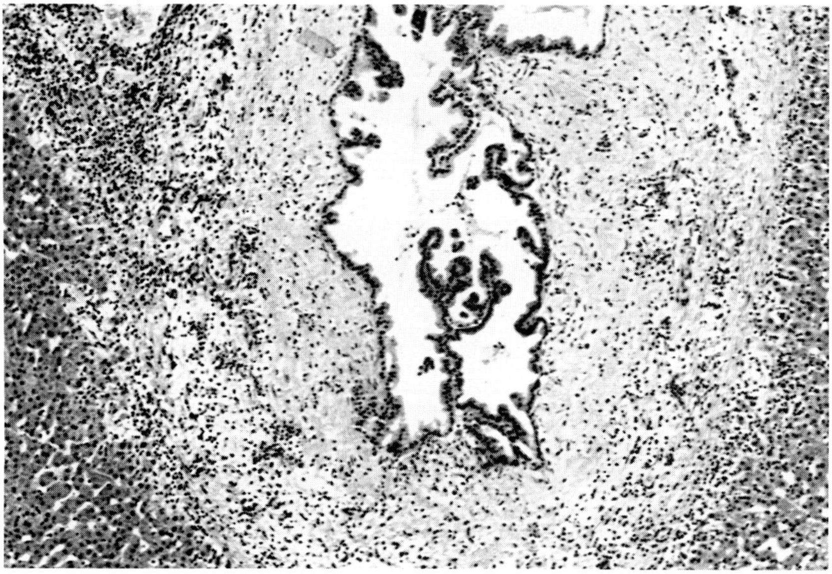

Fig. 3.12 Extrahepatic obstructive jaundice. There is epithelial papilliform hyperplasia of the bile duct, the wall of which shows inflammatory cell infiltration and oedema (HE, × 80).

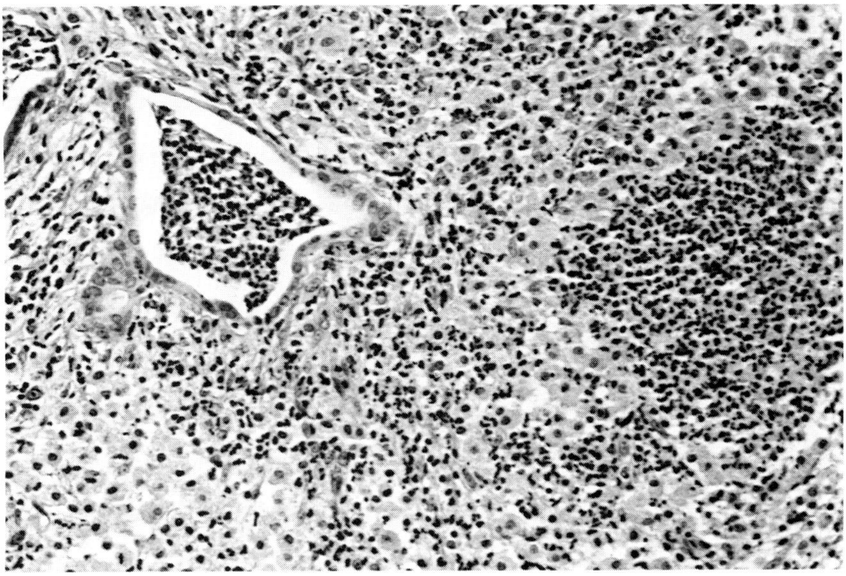

Fig. 3.13 Acute cholangiohepatitis in extrahepatic obstruction. There are many neutrophil polymorphs in portal and periportal tissues and within the lumen of a dilated bile duct lined by atrophic epithelium (HE, × 225).

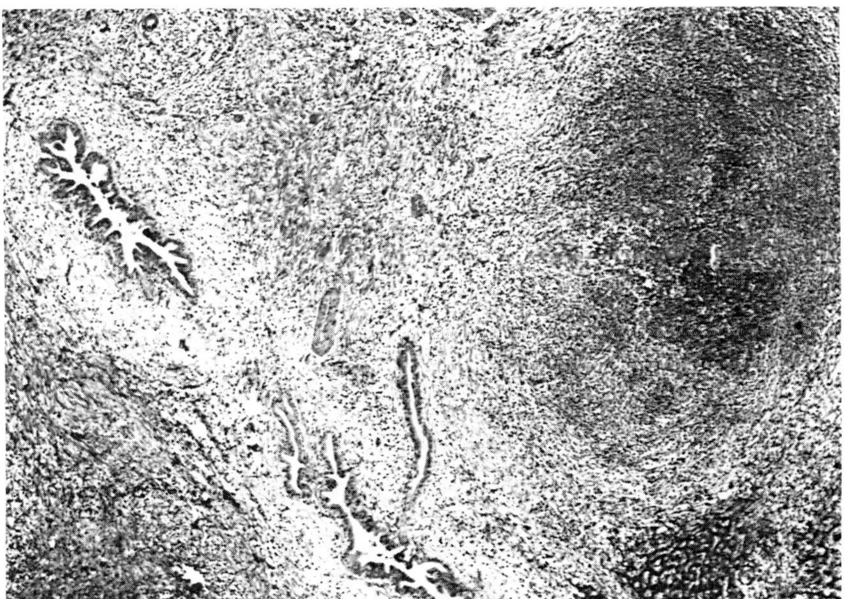

Fig. 3.14 Suppurative cholangiohepatitis. A large abscess is present (HE, × 45).

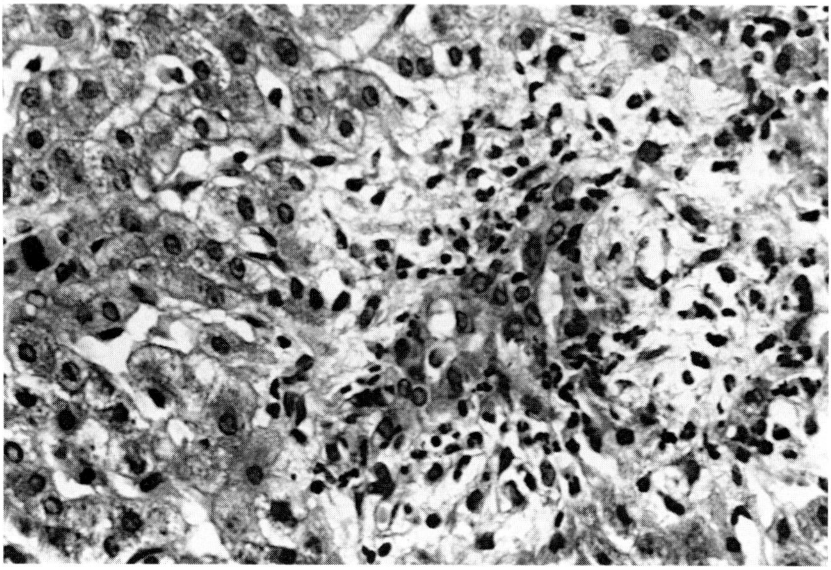

Fig. 3.15 Extrahepatic obstructive jaundice. There is acute ductular inflammation with destruction of epithelium. Dilatation of larger ducts was not a feature of this case (HE, × 180).

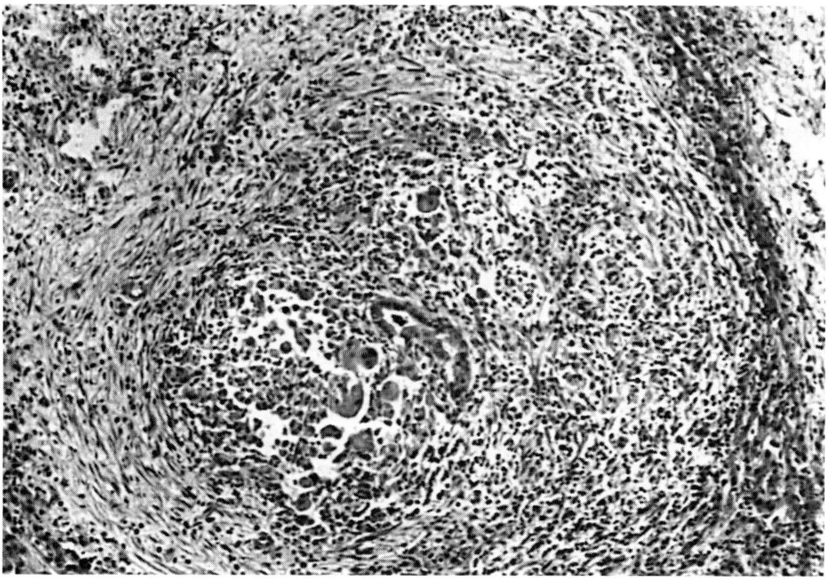

Fig. 3.16 Subacute 'cholangiohepatitis' in large duct obstruction. A granulomatous lesion has formed around a degenerate duct and extravasated bile pigment (HE, × 110).

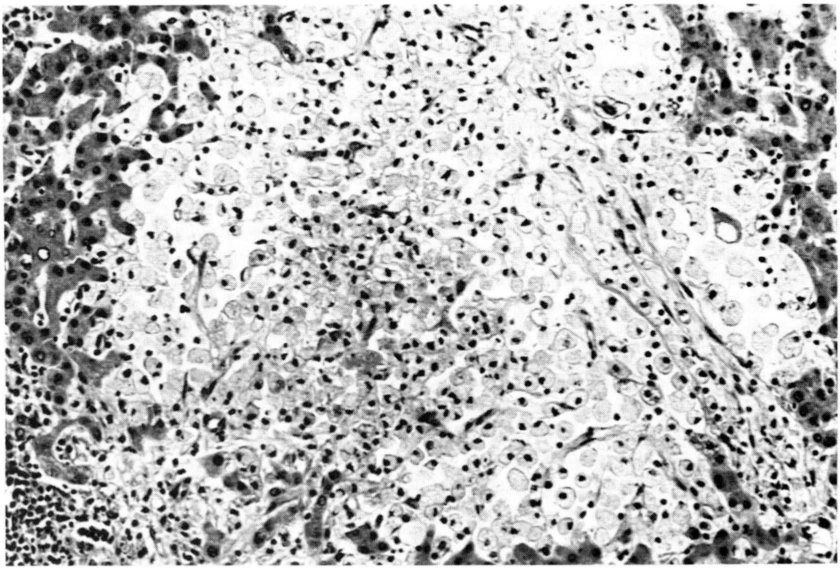

Fig. 3.17 Subacute 'cholangiohepatis'. Pale-staining foamy macrophages (xanthoma cells) are numerous in this granulomatous lesion (HE, × 180).

may supervene with the formation of 'bile infarcts' containing fibrin and bile-stained debris in place of parenchymal cells and stroma (Fig. 3.19). There may also be more prominent deposits of dark bile pigment derived from adjacent damaged ducts: these are surrounded by degenerate or necrotic cells and occasionally foreign-body-type giant cells and are referred to as 'bile lakes' (Fig. 3.20). It is usually stated that bile infarcts and bile lakes are diagnostic of large duct obstruction, but we have seen them occasionally in fatal primary biliary cirrhosis and fatal alcoholic cirrhosis with cholestasis.

These inflammatory, degenerative and necrotic lesions all have a portal and periportal location while the remainder of the parenchyma, apart from cholestasis and scanty small collections of foamy or feathery cells, often appears normal. Numerous ductules formed by proliferation of pre-existing channels (Fig. 3.21) or possibly by metaplasia of hepatocyte plates (Fig. 3.22) may be seen at the margins of portal tracts; occasionally this is very prominent and in our experience may be associated with obstruction due to bile duct carcinoma.

Failure to relieve obstruction may lead to acute exacerbations of cholangiohepatitis or the gradual progression of the chronic lesion with eventual coalescence of adjacent inflamed portal areas. Characteristically the centrilobular zones remain unaffected by fibrosis (Fig. 3.23). Apart

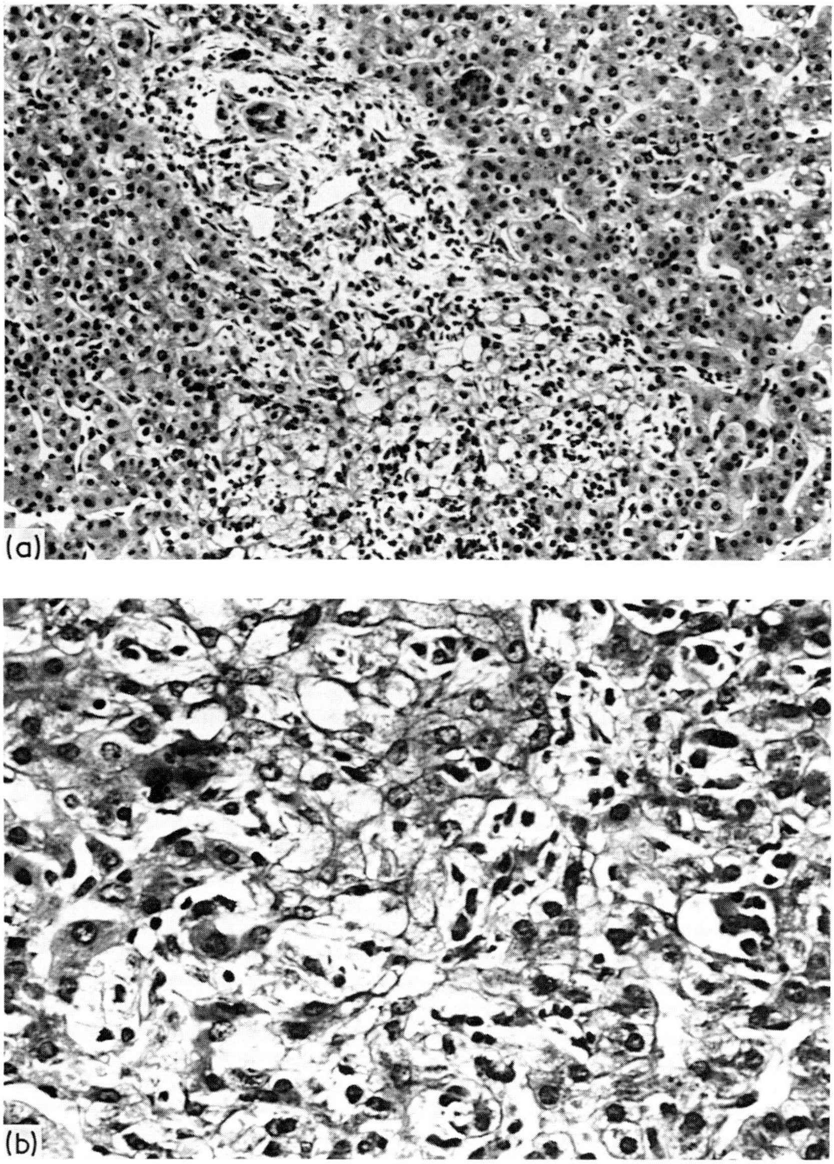

Fig. 3.18 (a) and (b), 'feathery degeneration' of hepatocytes. The affected cells are pale but not notably swollen. The cell margins are clearly defined and the cytoplasm faintly reticulated. There was bile staining of some of these cells (not apparent in the photographs). ((a) HE, × 180; (b) HE, × 450.)

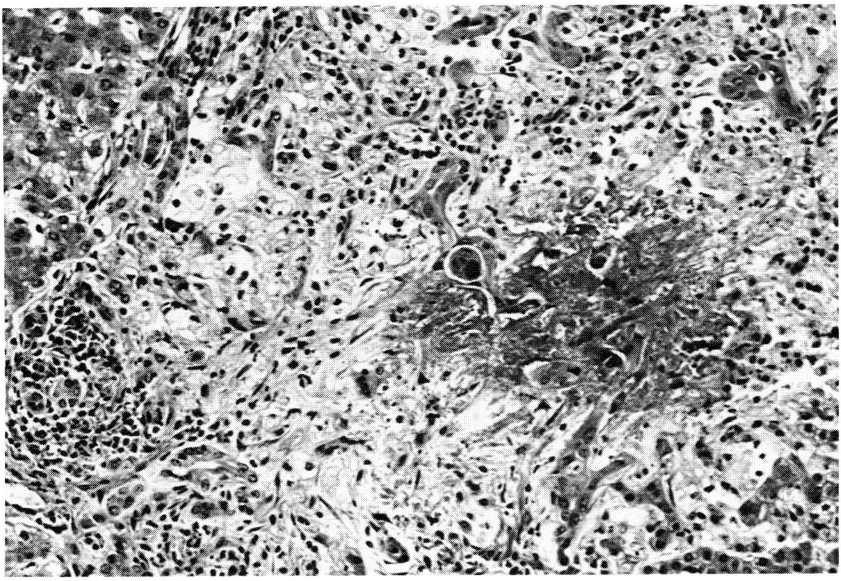

Fig. 3.19 'Bile infarct' (HE, × 180).

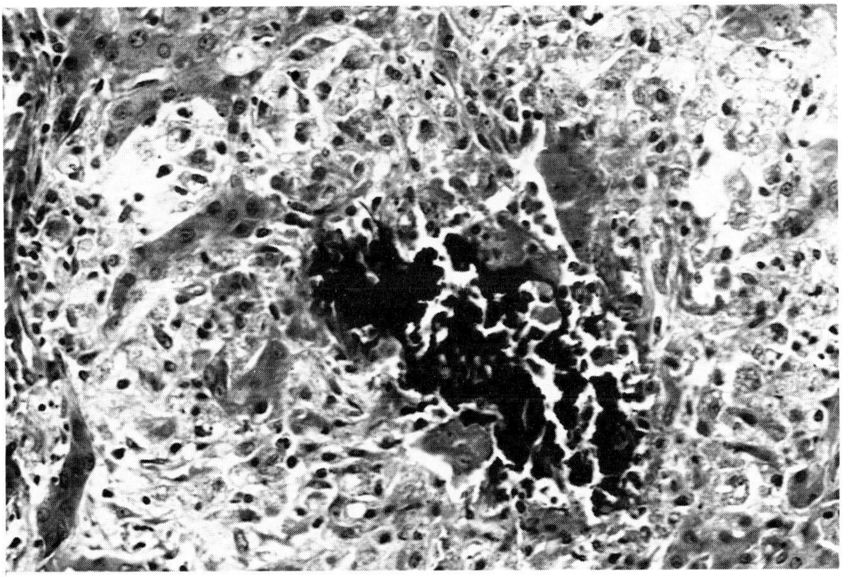

Fig. 3.20 'Bile lake' (HE, × 285).

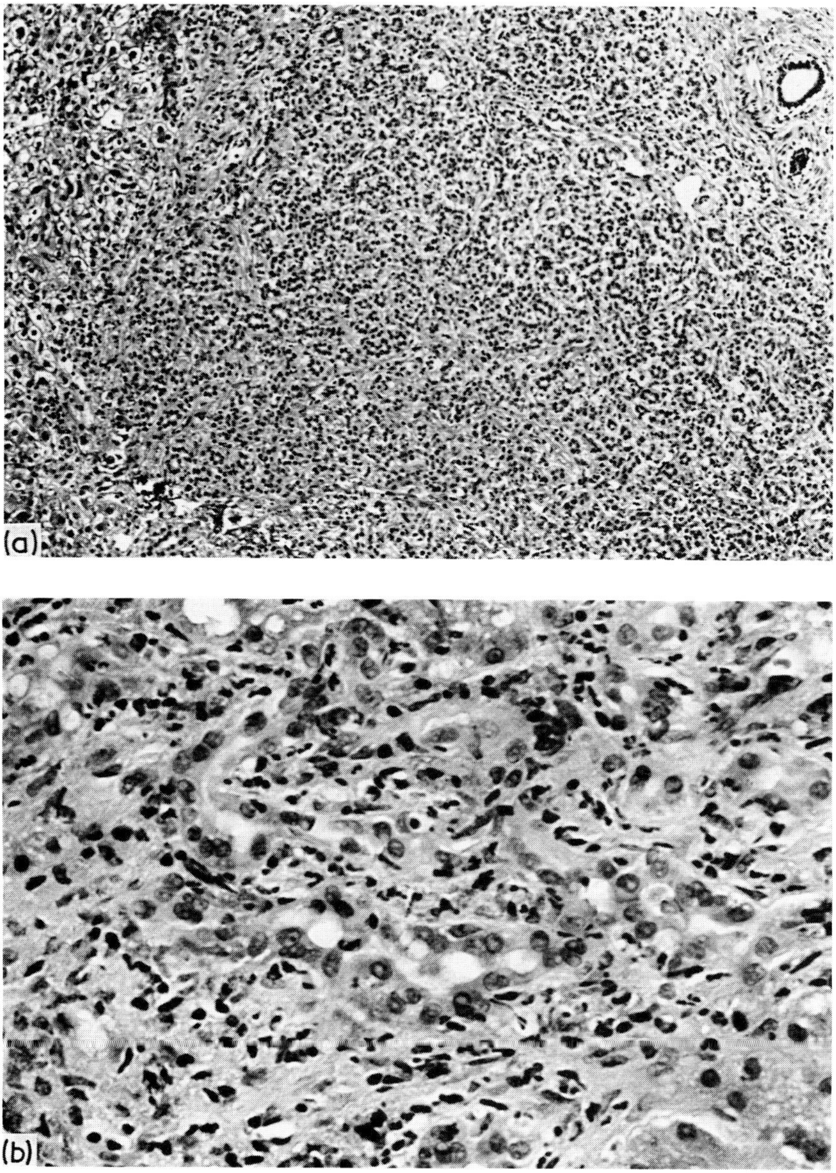

Fig. 3.21 (a) and (b), bile ductular proliferation in extrahepatic biliary obstruction ((a) HE, × 100; (b) HE, × 450.)

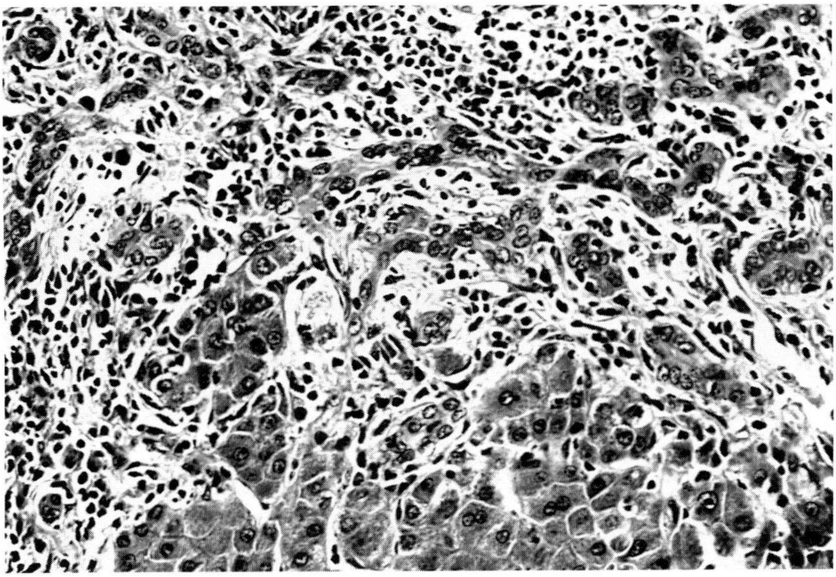

Fig. 3.22 Chronic 'cholangiohepatitis' in extrahepatic biliary obstruction. The numerous bile ductular structures in the inflamed portal area (above) appear to arise from hepatocyte proliferation (HE, × 280).

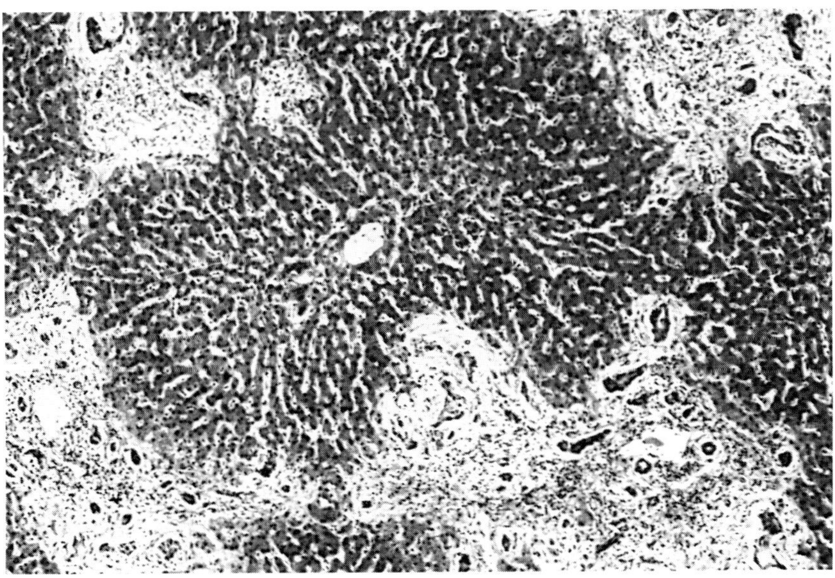

Fig. 3.23 Chronic 'cholangiohepatitis' in extrahepatic obstruction. The widened fibrotic portal tracts with ductular proliferation tend to coalesce while the centrilobular zone is unaffected (HE, × 90).

from extrahepatic bile duct atresia in infants in which these changes progress relatively quickly, it is now unusual to see advanced cases with loss of liver architecture and nodular regeneration of parenchyma ('secondary biliary cirrhosis'). The true nature of lesions falling short of this end-stage is suspected when surviving liver persists in irregular shapes rather like pieces of a jigsaw puzzle, each with a margin of hydropic hepatocytes and separated by fibrous connective tissue (Fig. 3.24). Advanced primary biliary cirrhosis has a similar histological appearance. With the final development of parenchymal nodules, porta-hepatic fibrovascular bridging lesions and the reduction in numbers of bile ducts and ductules, it can be impossible to determine the pathogenesis of such end-stage lesions from histopathology alone.

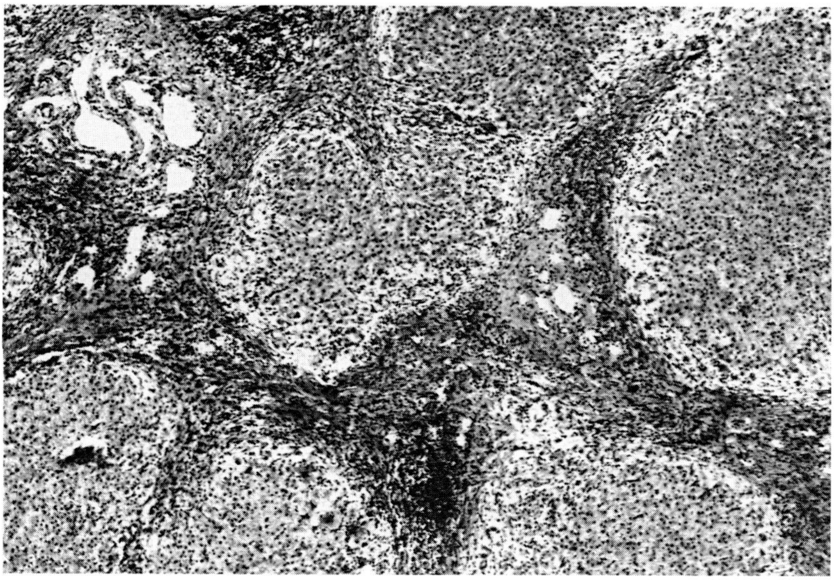

Fig. 3.24 Developing secondary biliary cirrhosis. There is a jigsaw pattern of surviving parenchyma. Each portion has a margin of pale-staining loose fibrous tissue in which there are small proliferating ductules (HE, × 72).

3.2 Special types of biliary disease

Apart from intrahepatic cholestasis complicating various types of liver disease which are described elsewhere, the following conditions require some comment here.

3.2.1 *Primary biliary cirrhosis*

This is a degenerative condition of interlobular bile duct epithelium of unknown aetiology and slow development which can terminate in biliary cirrhosis. Because of its importance in relation to other forms of chronic hepatitis it is described fully in Chapter 11.

3.2.2 *Sclerosing cholangitis and pericholangitis*

As the majority of these cases are associated with chronic inflammatory bowel disease they are discussed in Chapter 7.

3.2.3 *Liver damage and parenteral feeding*

The patients may suffer from cholestasis or cholestatic hepatitis. Adults can be affected but the majority are young children. It is dealt with in Section 3.4 and in Chapter 7.

3.2.4 *Liver damage in shock, sepsis and following surgery*

Jaundice may appear in a minority of patients within a day or two of major surgery even for conditions unrelated to hepatobiliary disease. Centrilobular cholestasis occurs and can persist for several weeks. Jaundice is usually more severe where there is sepsis and may be related to endotoxaemia. Shock from this cause and from circulatory failure could lead to centrilobular hepatocyte damage and thereby aggravate cholestasis. Cholestasis in shock may be seen also within proliferated bile ductules at portal tract margins rather than within canaliculi (Lefkowitch, 1982). These ducts appear to end blindly as cholestasis does not involve the normal interlobular channels (Fig. 3.25). Portal inflammation is usually minimal.

3.2.5 *Cholestasis in malignancy*

Intrahepatic cholestasis may complicate certain non-hepatobiliary cancers such as renal carcinoma, despite the absence of liver metastatic deposits. The cause is unknown but could be a humoral effect of some product of the primary tumour's cells, as it can regress following resection of the neoplasm. Likewise, biopsy may show no evidence of liver involvement in certain cases of Hodgkin's disease or patients with other types of lymphoma suffering from cholestatic jaundice.

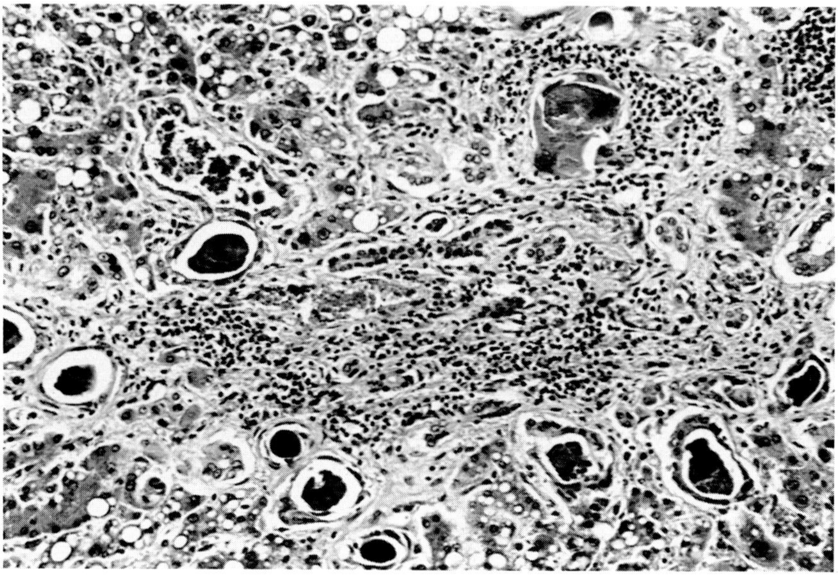

Fig. 3.25 Liver in shock. Ductular channels distended with bile pigment are arranged around a central portal tract which has a normal undistended interlobular bile duct and mild inflammatory cell infiltration (HE, × 180).

3.2.6 *Biliary damage in graft-versus-host disease* (Farthing *et al.*, 1982)

A minor degree of liver dysfunction is common in patients following bone marrow transplantation and becomes severe in some of the minority who develop graft-versus-host disease in skin and perhaps at other sites such as the gastrointestinal tract. The characteristic hepatic lesion is degeneration of the epithelial lining of smaller bile ducts, the larger channels being unaffected. Sheets of damaged cells are atrophic or swollen and pleomorphic with clear cytoplasm. The lesion has a patchy distribution throughout the liver but, unlike primary biliary cirrhosis, there is usually only mild lymphocytic infiltration of the affected portal tracts (Fig. 3.26). Bile capillary cholestasis is seen.

In addition, there can be a minor degree of focal hepatocyte necrosis and evidence of an opportunistic infection such as cytomegalic inclusion disease. Veno-occlusive disease has occurred in graft recipients not suffering from biliary damage and the relationship of this to chemotherapy and radiotherapy is uncertain (Berk *et al.*, 1979). Opportunistic infection is probably the main cause of liver damage in renal transplant recipients.

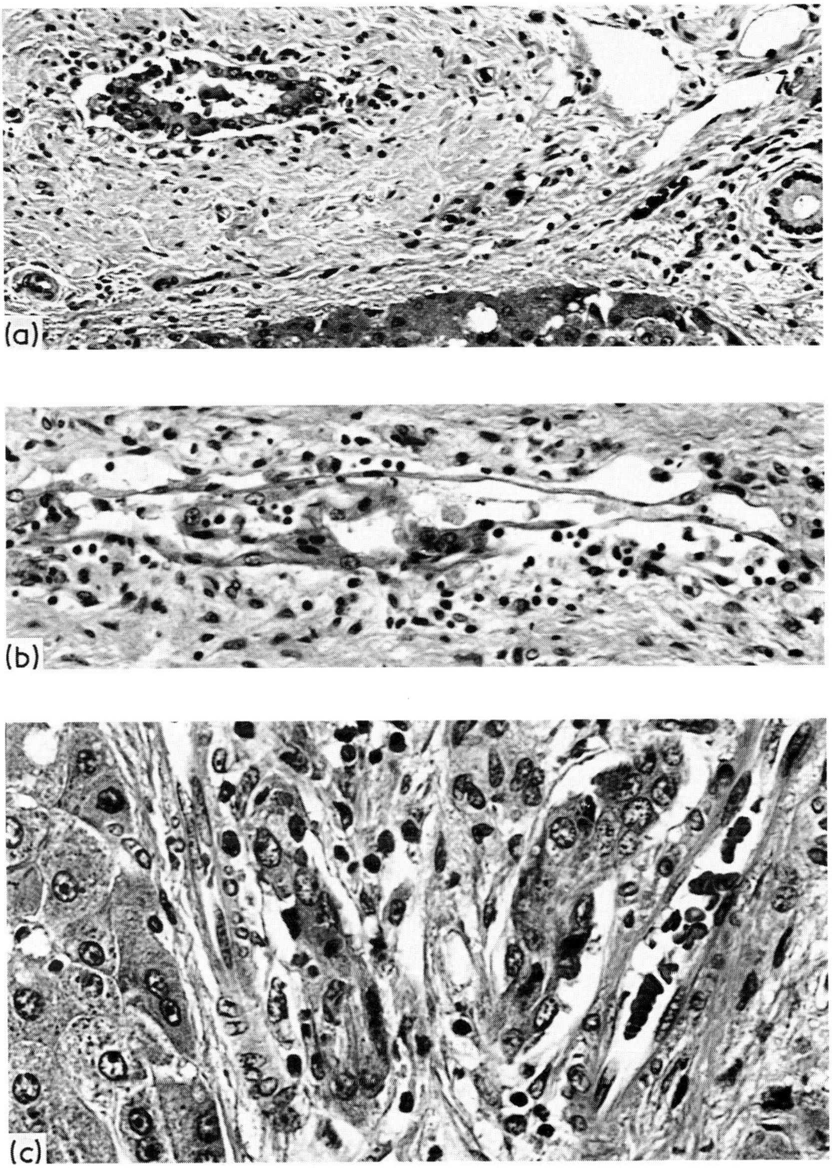

Fig. 3.26 Bile duct damage in graft-versus-host disease: (a) the larger damaged duct has a thick fibrous wall and a lining of degenerate epithelium compared with the smaller normal duct (right); (b) most of the bile duct epithelial cells have disappeared and the remainder are degenerate; (c) two small ducts have deeply staining degenerate epithelium with nuclear irregularity. In each specimen, there is relatively little inflammatory reaction. ((a) HE, × 180; (b) HE, × 225; (c) HE, × 560).

3.2.7 *Complications in liver transplantation* (Starzl *et al.*, 1976)

Biliary obstruction is a prominent feature of both early and late postoperative illness. Improvements in surgical technique have led to a noticeable reduction in cases which could be attributed to faults in biliary reconstruction and obstruction to the onward flow of bile. The problem has not been eliminated entirely and some cases have biliary obstruction from necrotic debris including degenerate collagen ('collagenous sludge') derived apparently from the walls of ducts damaged by ischaemia.

Biliary infection can occur also, often as a consequence of immunosuppressive therapy. It can be severe with suppurative cholangiohepatitis and be a cause of death.

Symptoms of such biliary complications can be difficult to distinguish from those due to graft rejection without the use of liver biopsy and cholangiography.

Allograft rejection, requiring a second transplant, occurs in about 16% of patients. Episodes of acute rejection which partially, completely or never resolve, occurs in 77% of cases and chronic rejection in about 10% of cases. The main structures attacked in rejection are bile ducts, central veins and hepatic arterioles. Normal liver cells do not express HLA class 1 antigens, unlike vascular endothelium and bile ducts, which may explain the focus of the immune attack; hepatocytes may, however, express HLA class 1 during rejection, and also in viral hepatitis (Barbatis *et al.*, 1981).

Hyperimmune rejection is uncommon in liver transplantation and is characterized by florid polymorph exudation in portal tracts, arteritis, fibrin deposition and extensive hepatic necrosis. Episodes of acute rejection, which is the commonest form, are evidenced by bile duct damage, mixed lymphocytic and polymorph infiltration of portal tracts which may be centred on bile ducts, and central venous endothelialitis; these features are diagnostic of acute rejection. The morphological features of chronic rejection and bad prognostic indicators (which may accompany evidence of acute rejection) are: arteritis; ballooning regeneration of central hepatocytes and hepatocyte necrosis; and paucity of bile ducts. The latter feature, in particular, may connote a bad outcome (Snover *et al.*, 1987). Those patients who survive have persistent cholestasis and extensive fibrosis or cirrhosis.

3.3 Differential diagnosis of cholestatic jaundice from liver biopsy

3.3.1 *There is no evidence of intrahepatic cholestasis*

This is possible when there is a hepatocyte defect in the secretion of organic anions other than bile salts. The serum contains elevated levels of conjugated bilirubin but normal amounts of bile salts, cholesterol and alkaline phosphatase. It is characteristic of Dubin–Johnson syndrome and Rotor syndrome and may occur during the early stage of viral hepatitis.

A second possibility is leakage of bile into the bloodstream through damaged bile duct epithelium in primary biliary cirrhosis. Cholestasis may ensue at a later stage of this disease when there is mechanical obstruction to bile flow caused by bile duct destruction and portal fibrosis.

3.3.2 *There is centrilobular intrahepatic cholestasis only*

This can be the only feature of intrahepatic or extrahepatic obstructive-type jaundice from various causes and is occasionally idiopathic. It is most characteristic, however, of the jaundice induced by anabolic steroids, especially 17-substituted testosterone compounds. Jaundice due to oral contraceptives is also of this type and these women are prone to cholestatic jaundice during the later months of pregnancy. The illness is mild and full recovery follows removal of the exciting cause.

Obstructive-type jaundice with intrahepatic cholestasis may occur in chronic venous congestion. As noted above it may complicate major surgery; septic conditions especially with endotoxaemia; shock and malignant disease without hepatic metastatic deposits. It can be prominent during regeneration after partial hepatectomy and in transplanted liver. Rarely cholestasis may occur in amyloid liver, and may be the only abnormality seen in biopsies from cases of viral hepatitis and alcoholism.

3.3.3 *There is peripheral intrahepatic cholestasis*

Bile canalicular cholestasis in this situation indicates mechanical obstruction caused by portal inflammation or by distortion of regenerated ducts. In the absence of centrilobular cholestasis it is characteristic of primary biliary cirrhosis, and may be seen occasionally in chronic active hepatitis. Prominent bile plugs in proliferated ductules are suggestive of shock. Concretions in larger ducts occur as a result of extrahepatic obstruction but are seen rarely in biopsies.

3.3.4 *There is intrahepatic cholestasis with hepatitis*

This is also difficult to assess as cholestasis from any cause may in itself induce hepatocyte damage and inflammatory changes in the liver. This form of hepatitis is confined to the areas of cholestasis, in contrast to the more diffuse inflammation in cholestatic viral and alcoholic hepatitis. Peripheral hepatocyte Mallory bodies are more typical of primary biliary disease than alcoholic disease but this distinction is not always clear;

peripheral Mallory bodies are common in alcoholic cirrhosis. Many cases with less severe hepatitis prove to be examples of drug-induced liver injury. Pericholangitis with cholestasis as a complication of colonic inflammation should not be overlooked.

Non-specific reactive-type hepatitis with cholestasis may be seen also during attacks of benign idiopathic recurrent intrahepatic cholestasis (Williams *et al.*, 1964). This is a rare familial disease with some clinical resemblance to Dubin–Johnson and Rotor syndrome; however, the liver biopsy appearances are different while liver function is normal between attacks (see Chapter 9).

3.3.5 *There is positive histological evidence of a large duct obstruction*

A large duct obstruction is likely if cholestasis is associated with inflammatory oedema of the portal tracts with a concentric arrangement of fibroblasts around dilated ducts; if these ducts contain inspissated bile or acute inflammatory cells; if acute inflammatory cells are prominent in the portal and periportal areas; if there are bile infarcts or bile lakes; or if there is well-defined proliferation of bile ductules at the portal tract margins.

3.3.6 *There is histological evidence of cholestasis in the absence of clinical jaundice*

This suggests that a portion of the liver only is affected, possibly from obstruction to a branch of an intrahepatic duct. Sometimes there is more widespread but irregularly distributed obstruction to a number of smaller ducts related to secondary carcinoma. On the other hand, some of these patients may be jaundiced if there is sufficient absorption of conjugated bilirubin which the unaffected parts of the liver are unable to re-excrete for reasons which are not understood (Tiesenga *et al.*, 1964). An irregular distribution of cholestasis without clinical jaundice can occur in certain readily recognizable liver diseases such as cirrhosis, congenital hepatic fibrosis and sclerosing cholangitis.

3.4 Cholestasis in children

The investigation of cholestatic jaundice, particularly in neonates and young infants, is an important part of paediatric practice and requires a separate discussion because the aetiology and pathology are usually different from cholestatic disorders in adults. Of particular importance is the distinction between extrahepatic biliary atresia and neonatal hepatitis, as correct surgical treatment of the former can be life-saving. Liver biopsy is of particular value here but interpretation may be difficult;

moreover, it seems possible that certain cases of atresia could arise as a consequence of hepatobiliary infection.

The following paragraphs refer to the more important aspects of the subject. Reference to special works on paediatric pathology and medicine is required for an exhaustive account of the numerous causes of cholestatic jaundice in infants and young children.

3.4.1 *Extrahepatic bile duct atresia*

There is segmental narrowing, obstruction or absence of extrahepatic ducts and the liver becomes enlarged and bile stained. The most important histological features are proliferation of small interlobular ducts and a rapid development of portal fibrosis (Fig. 3.27). Other less characteristic changes include bile ductular proliferation, cholestasis involving canaliculi and ductules, and portal and lobular inflammation. Bile infarcts and lakes are rarely seen and there is no duct dilatation.

These changes are not well developed during the first month or two of life, but biopsy should not be delayed much beyond this time as secondary biliary cirrhosis unresponsive to surgical relief can be established within 6 months. Biopsy should be supplemented by cholangiography and, if necessary, exploratory laparotomy.

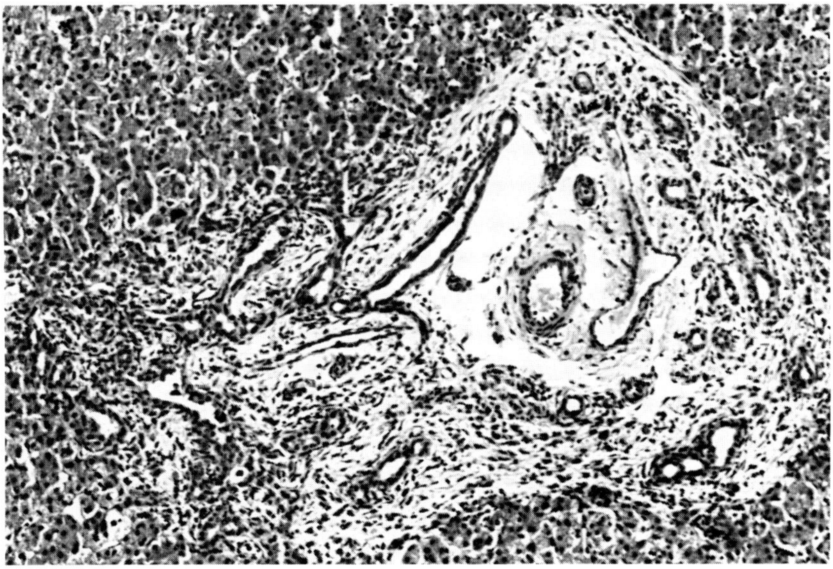

Fig. 3.27 Liver of a 6-week-old infant with extrahepatic biliary atresia. There is prominent proliferation of interlobular bile ducts and fibrosis of the expanded portal tract (HE, × 120).

Other causes of extrahepatic biliary obstruction in infants have the same effects on the liver but are very uncommon. These include choledochal cyst.

3.4.2 *Neonatal hepatitis*

This term tends to be used very loosely to include almost all primary non-neoplastic hepatic conditions in young infants characterized by cholestasis but with patency of extrahepatic bile ducts. Compared with the effects of extrahepatic obstruction, hepatitis including Kupffer cell hyperactivity is more severe and haematopoiesis in liver tends to persist. Hepatocyte degenerative changes are also more prominent; this includes giant cell transformation (see Fig. 2.17), but it must be noted that a similar feature can occur in biliary atresia and in other conditions. Cholestasis and bile ductular proliferation are present.

There are many causes of neonatal hepatitis including infections such as rubella, inherited metabolic disorders – especially alpha-1-antitrypsin deficiency – and various developmental anomalies. Cholestasis can complicate infection at sites outwith the liver, especially bowel and urinary tract, when endotoxaemia may be a factor. In many cases the discovery of the cause will depend on the clinical features and results of investigations other than liver biopsy, but biopsy can be diagnostic as in cytomegalic inclusion disease.

3.4.3 *Intrahepatic bile duct atresia*

Rarely older children suffer from cholestatic jaundice associated with extensive atrophy or complete absence of interlobular bile ducts and variable amounts of portal fibrosis (Fig. 3.28). It is possible that some of these cases may be a sequel to neonatal hepatitis. It has been noted also in combination with extrahepatic atresia when the condition can resolve if the normal flow of bile is established.

Intrahepatic atresia may be a component of certain rare developmental anomalies such as the 17–18 trisomy syndrome (Alpert *et al.*, 1969), syndromes which include cardiovascular and vertebral defects and hypogonadism (Alagille *et al.*, 1975; Kahn *et al.*, 1983), and in asociation with oedema of the lower limbs due to impaired lymphatic development (Sigstad *et al.*, 1970). It is a late manifestation of Byler's disease which is a familial cholestatic hepatitis appearing in older infants and causing fatal cirrhosis, usually within a few years. It is suggested that Byler's disease is part of a spectrum of familial cholestatic conditions ranging from benign recurrent cholestasis (see Chapter 9) to intrahepatic biliary atresia (Ballow *et al.*, 1973).

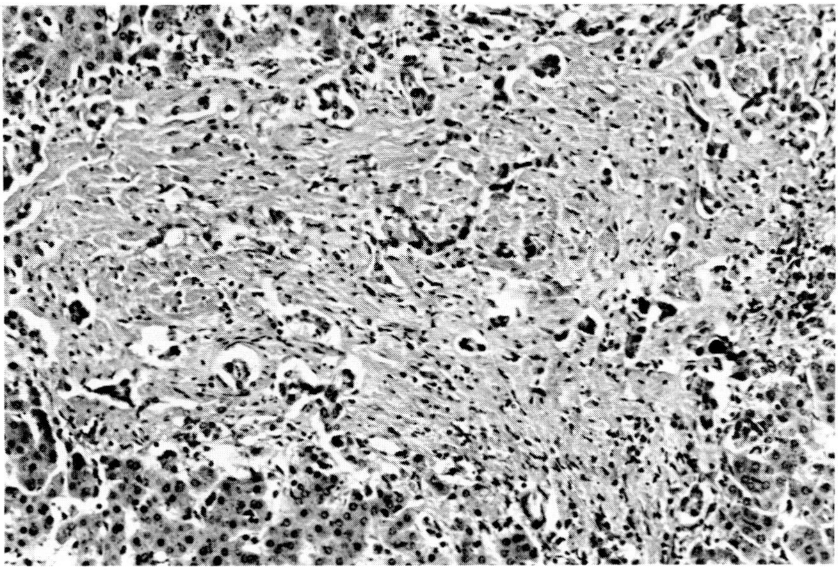

Fig. 3.28 Liver of a child aged 4 years with intrahepatic biliary atresia. The expanded fibrous portal tract contains a few tiny biliary channels but no normal bile duct. All portal tracts in this wedge biopsy were similar (HE, × 180).

A new syndrome which may form part of this spectrum has been described recently in adults, namely familial intrahepatic cholestatic cirrhosis in young adults (Haratake, 1985). In these cases of intrahepatic biliary hypoplasia there are no extrahepatic anomalies in contradistinction to Alagille's syndrome.

3.4.4 *Haemolytic disease of the newborn*

Hyperbilirubinaemia which is initially of the unconjugated type may later include conjugated pigment. This has been regarded as a manifestation of the 'inspissated bile syndrome' with intrahepatic cholestasis the result of bilirubin overload. In most cases, however, it indicates the impaired transport of bilirubin in damaged hepatocytes, as liver necrosis can be extensive.

3.4.5 *Parenteral feeding and cholestatic hepatitis*

This has been recorded rarely in very young infants, usually following major alimentary tract surgery. We have noted a similar condition in several older babies with severe infective gastroenteritis given intravenous casein hydrolysate and in one adolescent similarly treated after extensive small bowel resection. While liver damage in such cases could be multifactorial, the onset of jaundice usually follows closely on the introduction of parenteral feeding. Liver biopsy shows portal and periportal inflammation with hydropic degeneration of hepatocytes and centrilobular cholestasis. Parenteral feeding is referred to more fully on page 153.

3.4.6 *Inherited metabolic disorders*

Alpha-1-antitrypsin deficiency is the most important member of this group of conditions associated with cholestasis. In very young babies the characteristic PAS positive inclusions may not be evident but the retained protein can still be demonstrated immunohistochemically. Other conditions under this heading include Wilson's disease and galactosaemia in which plugs of bile are seen in abnormal hepatocyte acini (see Fig. 9.21). Impaired bilirubin conjugation within hepatocytes causes severe unconjugated hyperbilirubinaemia (Crigler–Najjar syndrome) but there are also intrahepatic cholestasis and hepatocyte damage. In older children mucoviscidosis may cause a patchy distribution of biliary cirrhosis from focal obstruction of small ducts within the liver by abnormal mucus (see Fig. 9.25).

References

Alagille, D., Odievre, M., Gautier, M. *et al.* (1975), Hepatic ductular hypoplasia associated with characteristic facies, vertebral malformations, retarded physical, mental and sexual development and cardiac murmur. *J. Pediatr.*, **86,** 63–71.

Alpert, L. I., Strauss, L. and Hirschhorn, K. (1969), Neonatal hepatitis and biliary atresia associated with trisomy 17–18 syndrome. *N. Eng. J. Med.*, **280,** 16–20.

Ballow, M., Margolis, C. Z., Schachtel, B. *et al.* (1973), Progressive familial intrahepatic cholestasis. *Pediatrics*, **51,** 998–1007.

Barbatis, C., Woods, J., Morton, J. A. *et al.* (1981), Immunohistochemical analysis of HLA (A,B,C) antigens in liver disease using a monoclonal antibody. *Gut*, **22,** 985–90.

Berk, P. D., Popper, H., Krueger, G. R. F. *et al.* (1979), Veno-occlusive disease of the liver after allogenenic bone marrow transplantation: possible association with graft-versus-host disease. *Ann. Intern. Med.*, **90,** 158–64.

Farthing, M. J. G., Clark, M. L., Sloane, J. P. *et al.* (1982), Liver disease after bone marrow transplantation. *Gut*, **23,** 465–74.

Haratake, J., Horie, A., Ishii, N. *et al.* (1985), Familial intrahepatic cholestatic cirrhosis in young adults. *Gastroenterology*, **89,** 202–9.

Ishak, K. G. and Rogers, W. A. (1981), Cryptogenic acute cholangitis: association with toxic shock syndrome. *Am. J. Clin. Pathol.*, **76,** 619–26.

Kahn, E. I., Daum, F., Markowitz, J. *et al.* (1983), Arterio-hepatic dysplasia: II. Hepatobiliary morphology. *Hepatology*, **3,** 77–84.

Lefkowitch, J. H. (1982), Bile ductular cholestasis: an ominous histopathologic sign related to sepsis and 'cholangitis lenta'. *Hum. Pathol.*, **13,** 19–24.

Morris, J. S., Gallo, G. A., Scheuer, P. J. *et al.* (1975), Percutaneous liver biopsy in patients with large bile duct obstruction. *Gastroenterology*, **68,** 750–4.

Sigstad, H., Aagenaes, O., Bjorn-Hansen, R. W. *et al.* (1970), Primary lymphoedema combined with hereditary recurrent intrahepatic cholestasis. *Acta Med. Scand.*, **188,** 213–9.

Snover, D. C., Freese, D. K., Sharp, H. L. E. *et al.* (1987), Liver allograft rejection. An analysis of the use of biopsy in determining outcome of rejection. *Am. J. Surg. Pathol.*, **11,** 1–10.

Starzl, T. E., Porter, K. A., Putnam, C. W. *et al.* (1976), Orthotopic liver transplantation in ninety-three patients. *Surg. Gynecol. Obstet.*, **142,** 487–505.

Tiesenga, M. F., Neal, R. H. and Hemwall, G. A. (1964), Jaundice due to right hepatic duct obstruction. *J.A.M.A.*, **187,** 367–8.

Williams, R., Carter, M. A., Sherlock, S. *et al.* (1964), Idiopathic recurrent cholestasis: a study of the functional and pathological lesions in four cases. *Q. J. Med.*, **131,** 387–99.

Further reading

Phillips, M. J., Poucell, S. and Oda, M. (1986), Mechanisms of cholestasis. *Lab. Invest.*, **54,** 593–608.

Popper, H. (1981), Cholestasis: the future of a past and present riddle. *Hepatology*, **1,** 187–91.

Review by an International Group (1983), Histopathology of the intrahepatic biliary tree. *Liver*, **3,** 161–75.

Rushton, D. I. (1981), Fetal and neonatal liver disease. *Diagn. Histopathol.*, **4,** 17–48.

Zimmermann, H. J. (1979), Intrahepatic cholestasis. *Arch. Intern. Med.*, **139,** 1038–45.

4 Viral hepatitis

Numerous viruses produce hepatitis. The term viral hepatitis, however, is restricted to those forms of hepatitis caused by hepatitis A virus (HAV), hepatitis B virus (HBV), hepatitis D (delta) virus (HDV) and by the putative aetiologic agent(s) of nonA, nonB (NANB) hepatitis. This chapter will deal mainly with hepatitis caused by these agents, all of which can produce similar acute disease. Consideration will then be given to other viral infections of the liver e.g. cytomegalovirus and Epstein–Barr virus (EBV) which cause quite different liver disease.

Viral A hepatitis (HAV, infectious, epidemic hepatitis) has a short incubation period of 2–6 weeks, occurs in epidemics or sporadically and is more common in children or young adults. Virus A is spread by the faecal–oral route and occasionally by blood products. The virus is present in faeces and serum late in the incubation period and usually disappears when symptoms appear. A viral antigen, HAAg, is demonstrable in hepatocyte and Kupffer cell cytoplasm in experimental animals during acute disease, unlike HBV markers. Immunoelectron microscopy, a technique in which viruses can be identified electron microscopically as aggregates after treatment with antibody, first positively identified virus A as a 27-nm particle and differentiated it from virus B. HAV has an RNA genome. Serum IgM antibody to HAV is indicative of recent infection while IgG anti-HAV correlates with past infection and protective immunity. HAV usually produces a benign self-limiting illness with occasional cases of fulminant hepatitis due to massive hepatic necrosis. Chronic infection and carrier states are rare or non-existent.

Viral B hepatitis (HBV hepatitis, serum hepatitis) by contrast has a longer incubation period of 2–6 months, rarely causes epidemics but sporadic cases are common. It is spread mainly by blood or blood products, and also by intimate sexual contact, particularly in male homosexuals; HBsAg and HBV have been found in the semen of chronic carriers. The Dane particle represents the entire B virion. It consists of a core of incomplete circular double-stranded DNA, DNA polymerase, core antigen (HBcAg) and HBeAg; and an envelope of coat protein, mainly HBsAg. The genome of hepatitis B virus consists of at least four genes, *S*, *P*, *C* and *X*. *S* codes for HBsAg; *P* for DNA polymerase; *C* for HBcAg and also HBeAg; the role of the *X* gene has not been fully worked out. The production of HBsAg and HBcAg is not coordinate and HBsAg is usually produced in excess of HBcAg. HBsAg appears in the serum late in the incubation period and early in the icteric phase in typical cases and then disappears. Serum HBeAg appears almost at the same time as HBsAg and disappears as the HB surface antigenaemia abates. Persistence of HBcAg, HBV DNA, and DNA polymerase connotes persistent viral replication. HBe antigen indicates high infectivity; antibody to e antigen usually indicates return to normal liver function and low infectivity, but relapse can occur (Koike *et al.*, 1987). On recovery from typical viral hepatitis, anti-HBcAg, anti-HBeAg and anti-HBsAg appear in serum with clearance of respective antigens. In chronic hepatitis antigenaemia persists and antibody production is delayed. HBsAg and HBcAg are demonstrable in chronic liver disease in paraffin sections. It is rare to find any HBV antigens in an acute infection but occasionally nuclear HBcAg is demonstrable in hepatocytes.

The delta antigen, which allowed the characterization of hepatitis D virus (HDV), was discovered by Rizzetto *et al.* in 1977. HDV is a defective RNA virus with a 1.7 Kb genome which encodes the delta antigen, a specific nucleoprotein. This virus only causes disease in patients who also have an HBV infection because HDV uses HBsAg as its coat. HDV occurs worldwide but does not parallel the distribution of HBV infection; it is rare in the Far East where about 75% of HBV carriers live. The highest incidence is in the Mediterranean and Middle East, particularly Kuwait. In endemic areas, HDV is transmitted by blood or blood products and is found almost exclusively in intravenous drug abusers, homosexuals and haemophiliacs, and in close family contacts. HDV modifies the hepatitis produced by HBV. In general, liver disease is more severe (p. 92). Serum delta antigen is found very fleetingly in acute infection and diagnosis is usually dependent on demonstrating anti-HDV in serum or demonstrating delta antigen in liver biopsy (for review see Rizzetto and Verme, 1985).

Studies on post-transfusional hepatitis have allowed the detection of a fourth form of viral hepatitis, nonA, nonB (NANB) hepatitis. There are no markers for NANB, hence this is a diagnosis by exclusion. There are two main types. The first is spread by blood or blood products and can be transmitted experimentally to chimpanzees. There may be at least two viruses involved (Bradley *et al.*, 1983). A short incubation (2–4 weeks) hepatitis has been described in haemophiliacs receiving factor VIII concentrates and in hypogammaglobulinaemic patients getting replacement therapy (Thompson *et al.*, 1987). A long incubation (6–10 weeks) hepatitis is transmitted mainly by blood transfusion.

Enteric NANB epidemics occur in India and North America. This is a mild self-limiting hepatitis which does not progress to chronic hepatitis. It is particulary severe in pregnant women.

Infection with these viruses may be subclinical or present with a spectrum of pathological and clinical abnormalities which are categorized as follows:

(I) Acute viral hepatitis.
- (1) Typical viral hepatitis.
- (2) Acute viral hepatitis with confluent necrosis.
- (3) Acute cholestatic viral hepatitis.

(II) Chronic viral infection.
- (1) Asymptomatic carriers.
- (2) Chronic hepatitis and cirrhosis.
- (3) Hepatocellular cancer.

4.1 Acute viral infection

4.1.1 Typical viral hepatitis

Infection may be subclinical. If symptomatic, the patient complains of a short period of vague ill-health with upper abdominal symptoms; nausea, dark urine and jaundice indicate liver disease. Symptoms, however, may be non-specific. Transaminases are characteristically very high and the prothrombin time gives a good overall idea of the extent of liver damage. Serum bilirubin parallels jaundice and alkaline phosphatase elevations are modest. In most cases diagnosis is apparent on clinical grounds. There is a group of patients, however, in which the best clinical diagnosis that can be arrived at is one of jaundice of hepatocellular or obstructive type. This is the type of case which usually comes to biopsy in the UK.

The light microscopic abnormalities produced by acute infection with HAV, HBV, HBV with HDV, and NANB are similar. The changes are qualitatively the same in icteric and non-icteric cases but are quantitatively more impressive in the former. There is no single morphological feature which is diagnostic of viral hepatitis, rather its pathological diagnosis depends on a combination of features.

The most characteristic abnormality on biopsy is lobular disarray (Fig. 4.1), which is due to a combination of hepatocyte degeneration and necrosis, hepatocyte regeneration, mesenchymal cell activation and inflammation. Throughout the lobules there is balloon degeneration of hepatocytes which is usually most prominent around central veins. Balloon cells are much larger than normal and have an empty appearance (Fig. 4.2). The cytoplasmic appearance is due to the presence of large clear

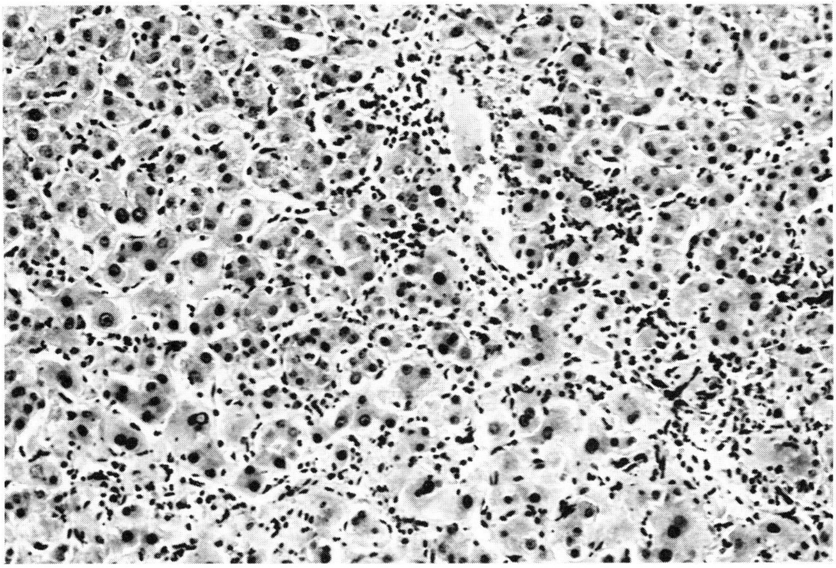

Fig. 4.1 Acute viral hepatitis. There is lobular disarray and increased cellularity due to inflammatory cell infiltration. Many hepatocytes are swollen and some are replaced by focal collection of inflammatory cells (HE, × 180).

spaces intersected by thin strips of cytoplasm which may also contain pigment granules. This is due to swelling and degranulation of the endoplasmic reticulum and loss of glycogen. These cells may contain more than one nucleus or alternatively the nucleus undergoes lysis. This form of degeneration, rather than acidophil change, is responsible for liver dysfunction. The other type of hepatocyte degeneration is acidophil body formation. This process starts by shrinkage of hepatocytes. The cytoplasm becomes intensely eosinophilic and the nucleus pyknotic (Fig. 4.3). As the process advances the hepatocyte is extruded from the liver cell plate and comes to lie in a sinusoid as a rounded mass of eosinophilic cytoplasm with or without a nucleus (Fig. 4.4). Acidophil bodies are also found in some cases of alcoholic liver injury and drug-induced liver disease (pp. 135–177). Ground-glass cells are not a feature of acute hepatitis and HBV antigens are rarely detectable in liver in the acute phase.

As these forms of cell degeneration progress, foci of necrosis appear in the lobule and these evoke a local exudation of lymphocytes, macrophages and plasma cells. The surrounding sinusoidal cells are enlarged and contain hepatocyte debris identified as PAS-positive diastase-resistant granules; these cells also give a positive Perls' reaction for iron (Fig. 4.5). Areas of focal necrosis also give rise to a local increase in the

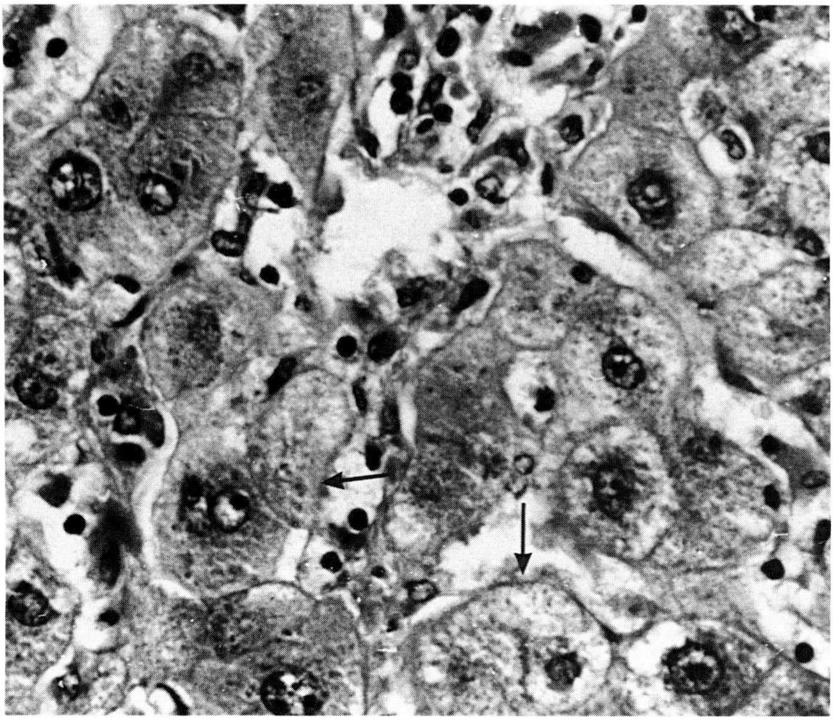

Fig. 4.2 Acute viral hepatitis. The hepatocytes show balloon degeneration – they are enlarged and their cytoplasm rarefied (arrows) (HE, × 600).

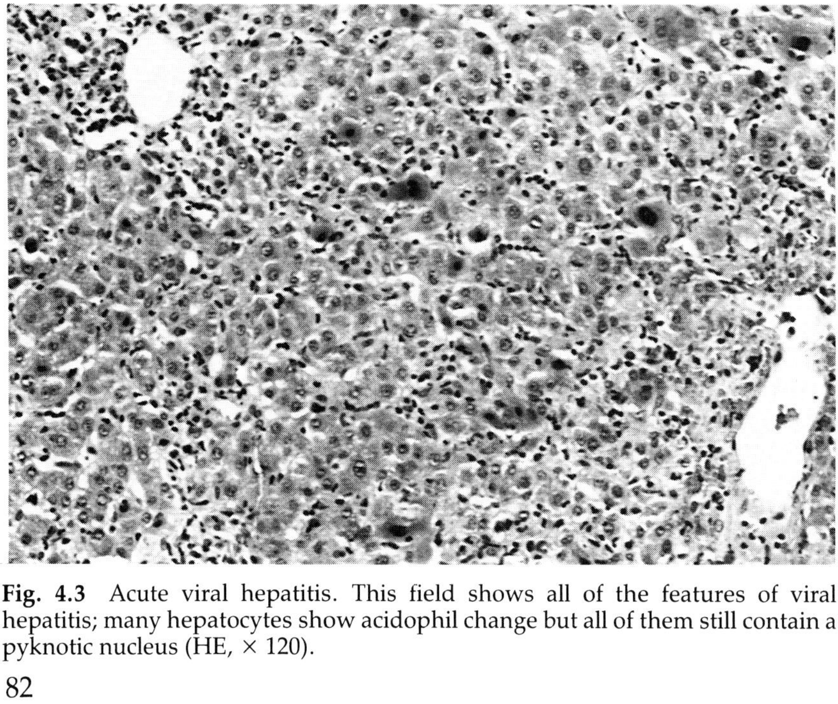

Fig. 4.3 Acute viral hepatitis. This field shows all of the features of viral hepatitis; many hepatocytes show acidophil change but all of them still contain a pyknotic nucleus (HE, × 120).

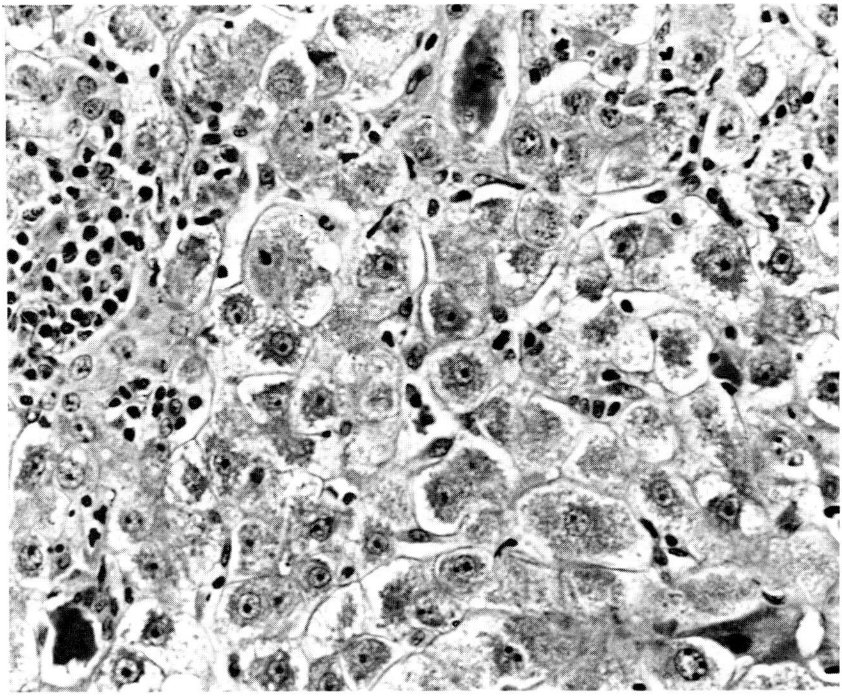

Fig. 4.4 Acute viral hepatitis. Three hepatocytes have undergone acidophilic necrosis (HE, × 450).

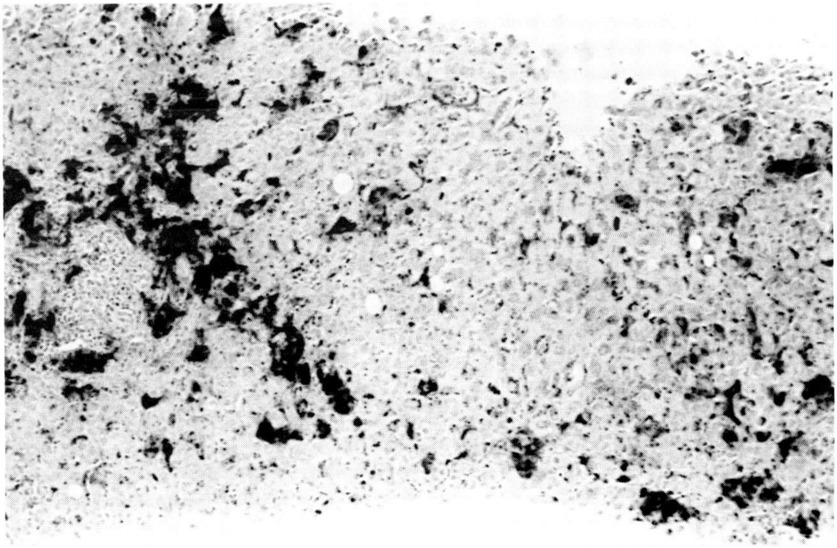

Fig. 4.5 Biopsy taken during recovery from acute viral hepatitis. Kupffer cells and portal tract macrophages have ingested debris containing haemosiderin pigment (Perls' stain, × 110).

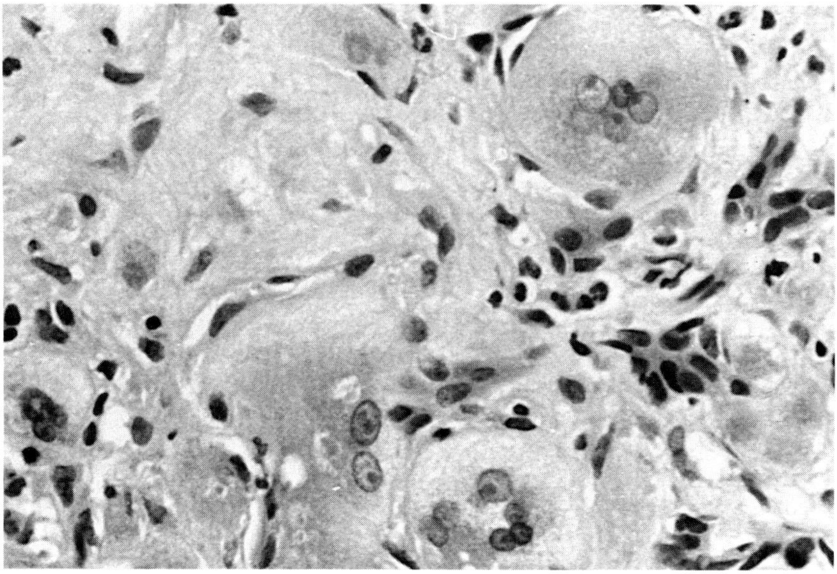

Fig. 4.6 Acute viral hepatitis. Many hepatocytes contain from two to six nuclei; this is usually interpreted as regenerative activity (HE, × 600).

reticulin stroma. A further response to hepatocyte death is regeneration, as evidenced by mitotic activity and the presence of two or more nuclei in hepatocytes (Fig. 4.6). Multinucleate hepatocytes, similar to those in neonatal hepatitis, are sometimes present. More extensive areas of zonal necrosis are sometimes found around central veins and these can form bridges between central venous radicles, and between central veins and portal tracts (see below).

Concomitant with these intralobular changes the portal tracts are invariably infiltrated by lymphocytes, mononuclear cells, plasma cells and an occasional polymorph (Fig. 4.7). The inflammatory infiltrate is usually diffuse but focal infiltrates of lymphocytes with or without germinal centres are sometimes found. These inflammatory cells usually spill over into the parenchyma but a diagnosis of viral-induced chronic active hepatitis should not be made unless there is good evidence that the disease has been present for 6 months. Cholangiolar proliferation is not a feature of typical viral hepatitis. Bile duct damage is sometimes evident as enlarged vacuolated epithelial ductal cells surrounded by lymphoid cells.

In the recovery phase all of the hepatocyte degenerative changes and much of the intralobular and portal tract inflammation disappear and the only residue seen in biopsies is the presence of hypertrophied sinusoidal cells which contain ceroid pigment; haemosiderin may also be prominent

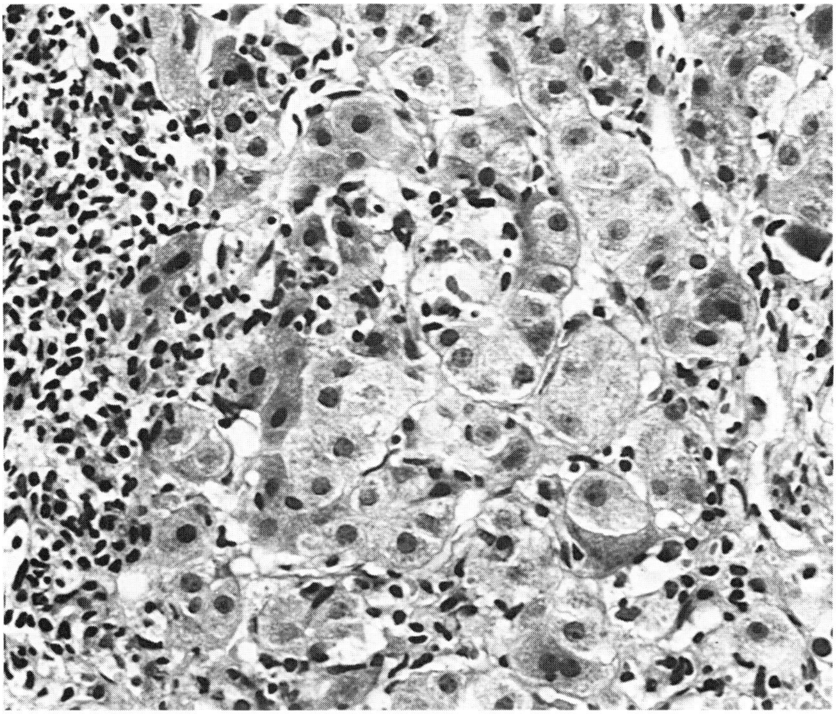

Fig. 4.7 Acute viral hepatitis. There is balloon degeneration and numerous acidophil cells. There is a prominent infiltration of lymphocytes and mononuclear cells in the portal tract (HE, × 280).

in the same cells. Macrophages containing similar pigment may be found in portal tracts.

4.1.2 Acute viral hepatitis with confluent necrosis

These patients usually have a more severe clinical illness with marked jaundice which may not resolve for months or, alternatively, be fatal. Subacute hepatitis or subacute hepatic necrosis are terms used to describe the latter condition. It is usually associated with massive necrosis or extensive bridging necrosis. The correlation between clinical presentation and morphology is good but not absolute. Some patients with fulminant hepatitis do not have massive necrosis.

Necrosis may extend throughout zone 3 of the acinus (Fig. 4.8) forming bridges between hepatic venules and portal tracts (Fig. 4.9). These lesions may be significant in the development of chronic hepatitis and cirrhosis but there is some disagreement regarding this (Schmid *et al.*, 1981;

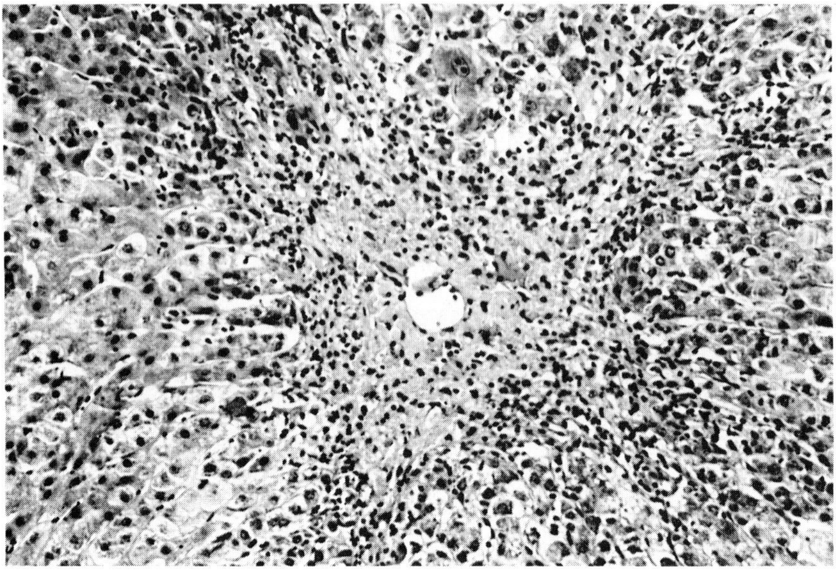

Fig. 4.8 Acute viral hepatitis. In this case there is prominent zonal necrosis around a central vein (HE, × 180).

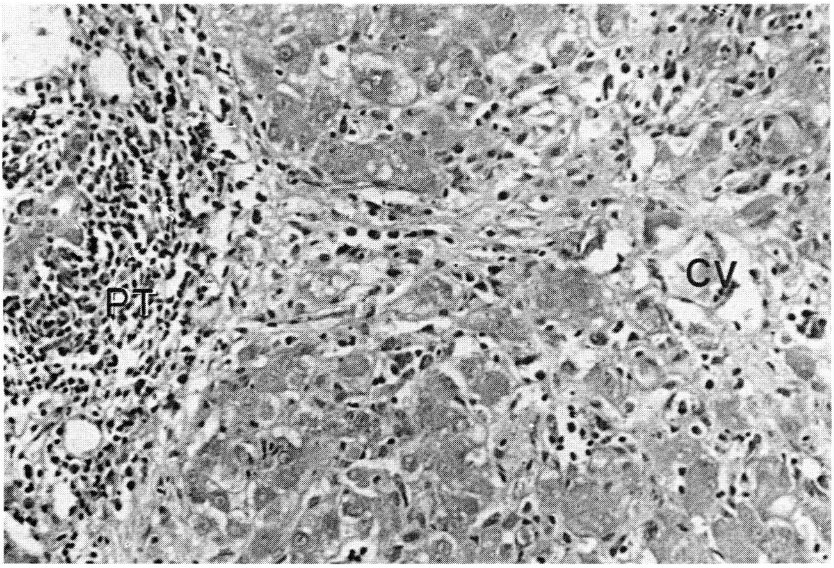

Fig. 4.9 Acute viral hepatitis with bridging necrosis. There is a tract of necrosis linking a central venous (CV) and portal tract area (PT) (HE, × 175).

Vanstapel *et al.*, 1983). Bridging between central vein and central vein or between portal tract and portal tract are probably unimportant in this respect. Zone 3 bridging (i.e. central–portal bridges) produces architectural disruption and sets the scene for subsequent chronic hepatitis and cirrhosis (see below).

Within the bridging areas there are lysed hepatocytes surrounded by lymphocytes, macrophages and plasma cells. When hepatocyte debris is resorbed, the residual loose connective tissue may thicken. There may occasionally be endophlebitis of the terminal hepatic venules. The surviving adjacent hepatocytes are ballooned and there is usually prominent cholestasis.

The presence of necrosis, collagen septa, regeneration, and florid inflammation in portal tracts with piecemeal necrosis may give the impression of chronic active hepatitis. The concomitant features of typical viral hepatitis usually differentiate the two disorders. The clinical data are also vitally important in making this distinction. A diagnosis of chronic active hepatitis should not be made in the absence of a 6-month history of liver disease.

When first described, bridging was thought to carry with it a high risk of developing chronic hepatitis and cirrhosis (Boyer and Klatskin, 1970). However, others have argued that bridging is not a predictor of chronicity (Nisman *et al.*, 1979). This dichotomy may be due to inclusion of different types of bridging (central–central, central–portal, portal–portal) under the same designation 'bridging necrosis'. Until the evidence has been firmly established, it is wise to consider zone 3 bridging as a serious abnormality which may predict conversion to chronicity.

The changes described above may progress to panacinar massive necrosis in which entire lobules undergo necrosis (Fig. 4.10). Clinically, this is manifest as fulminant hepatitis. Necrosis is accompanied by bile ductular proliferation in portal tracts and within the confines of former lobules. The striking finding in this lesion is the relative lack of inflammatory cell infiltration. Macrophages containing ceroid pigment are invariably present in necrotic areas. Necrosis can occur in HAV, HBV, HBV with superinfection by HDV and in NANB infections. About 80% of patients die within weeks or months. Remarkably, however, some patients make a complete recovery or go on to develop post-necrotic scarring or cirrhosis (Fig. 4.11). This type of scarring can also be found in the absence of a previous history of fulminant hepatitis. Conversely not all patients with severe liver dysfunction have massive necrosis. In these, the clinical features are probably explained by extensive hepatocyte degeneration and focal necrosis.

In some cases of typical acute hepatitis there is also prominent periportal necrosis and striking inflammation of portal tracts. There may be

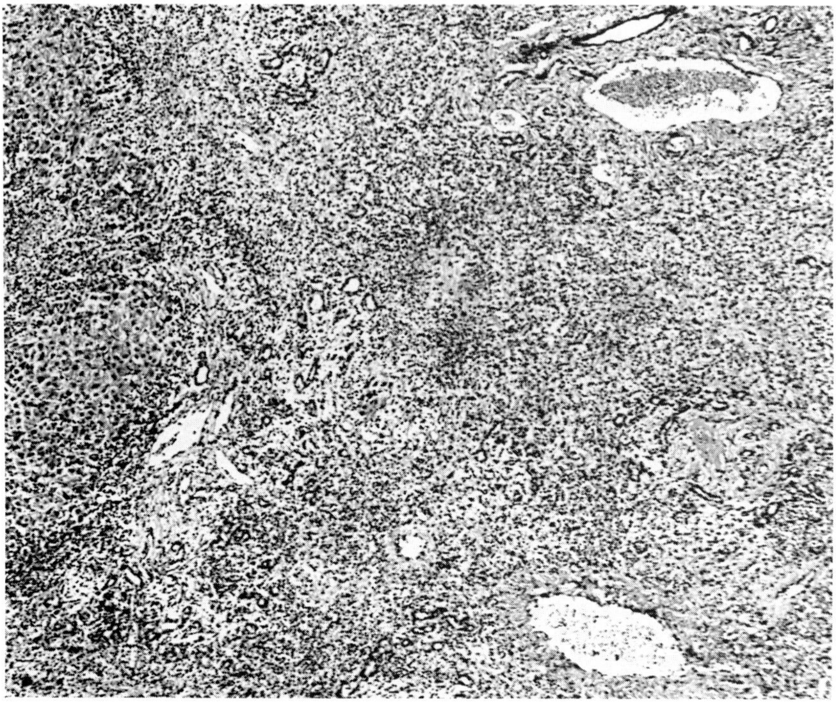

Fig. 4.10 Acute viral hepatitis with massive necrosis. In the right side of the field all of the hepatocytes in several lobules have undergone necrosis. There are numerous bile ductules and inflammatory cells in this area (HE, × 72).

bridging necrosis between adjacent portal tracts. Lymphocytes and plasma cells are most prominent in the inflammatory reaction and these spill over into the parenchyma as in piecemeal necrosis. Some authors attempt to distinguish periportal necrosis in viral hepatitis from piecemeal necrosis of chronic active hepatitis. In our opinion, this is impossible on morphological grounds. Superimposed on these changes are features of typical viral hepatitis. This lesion occurs in HAV infection which virtually never progresses to chronic hepatitis. Therefore, on its own this lesion has no predictive value. Viral type, serological viral markers and duration of clinical disease are more reliable indicators of outcome than morphology only. In some cases, liver dysfunction can persist for a year or more but recovery may occur.

4.1.3 *Acute cholestatic viral hepatitis*

This type of hepatitis although uncommon frequently comes to biopsy. Its main clinical features are intense jaundice, pale stools and pruritus.

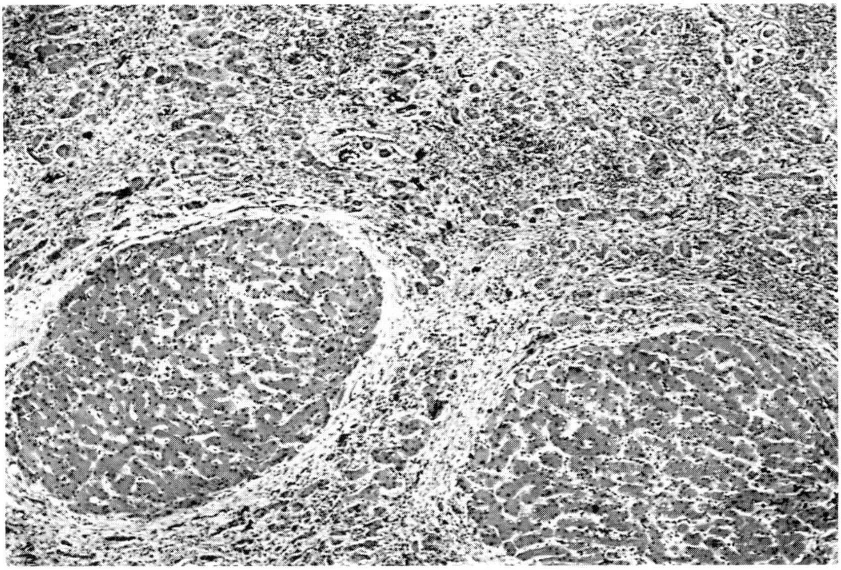

Fig. 4.11 Acute viral hepatitis with submassive necrosis. Two surviving islands of parenchyma are undergoing nodular regeneration which may lead to coarse post-hepatitis scarring or cirrhosis (HE, × 56).

The obvious clinical and pathological problem is to differentiate these cases from those with large duct obstruction (p. 94).

Biopsy examination of liver from these patients may show all the features of viral hepatitis: lobular disarray, ballooned cells, acidophil bodies and focal necrosis singly or in various combinations. The most prominent feature, however, is cholestasis. Bilirubin accumulates in the central zones as intracanalicular bile plugs or bile granules within hepatocytes (Fig. 4.12). Kupffer cells also may contain conspicuous bile pigment. Centrilobular necrosis is frequently present (Fig. 4.8). Two other features which distinguish this entity are the presence of bile ductular proliferation and polymorphs in portal tracts (Fig. 4.13); the cholangioles may also contain bile plugs. Polymorphs in portal tracts sometimes make it difficult to differentiate this form of viral hepatitis from large duct obstruction. In the former, however, balloon degeneration, viral markers in serum, and other features of viral hepatitis are helpful in differential diagnosis. In some cases, the issue is only resolved by endoscopic retrograde cholangiography.

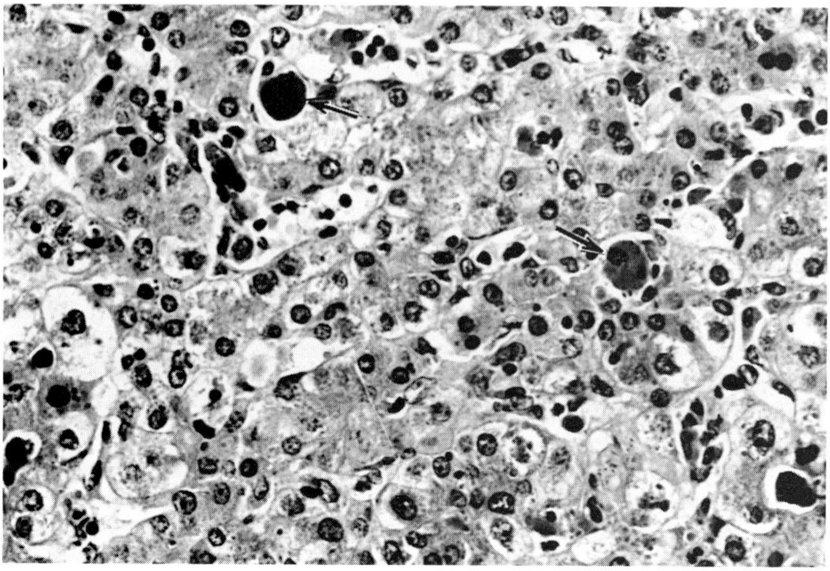

Fig. 4.12 Acute cholestatic viral hepatitis. In this field there are several dark bile plugs (arrows) as well as features of acute viral hepatitis (HE, × 450).

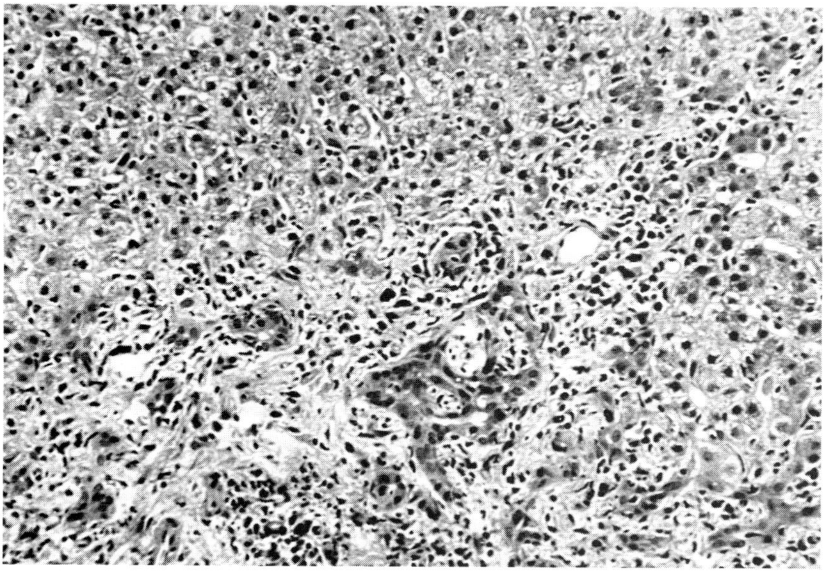

Fig. 4.13 Acute cholestatic viral hepatitis. There is bile ductular proliferation in a portal tract; the ductules are surrounded by inflammatory cells in which polymorphs are prominent (HE, × 180).

4.1.4 Morphological distinction of viral hepatitis types

The liver changes in viral hepatitis are similar irrespective of the causative virus. There are no absolute distinguishing routine morphological features for any of these viruses but there are patterns of injury which show some correlation with viral type. Demonstration of the specific viral antigen–antibody system in serum or antigen in the liver is the best evidence of aetiology. In the acute attack, however, viral antigen may not be demonstrable in the liver, especially in HBV hepatitis.

Hepatis A has been compared morphologically to Type B (Abe *et al.*, 1982; Okuno *et al.*, 1984). In general, the degree of periportal necrosis and inflammatory cell infiltration of periportal areas and portal tracts is more pronounced in HAV than HBV (Fig. 4.7). Perivenular liver damage and inflammation may be mild or absent in HAV but cholestasis may be prominent. Although piecemeal necrosis and plasma cell infiltration of portal tracts may be prominent, these cases do not progress to chronic active hepatitis and resolution is the rule. Serum IgM anti-HAAg indicates recent infection whereas IgG anti-HAAg connotes resolving/healed disease.

In *hepatitis B,* sinusoidal cell activation (i.e. increase in number and prominence of sinusoidal cells) is prominent, as is lymphocytic infiltration around ballooned hepatocytes. Hepatocellular damage is diffuse and more severe than in other types of viral hepatitis. Lymphocytes and macrophages are closely adherent and sometimes present within hepatocytes (Fig. 4.14), a phenomenon referred to as emperipolesis. There is still debate about the significance of piecemeal necrosis and bridging necrosis in HBV. Some claim that these features are more common in patients who develop chronic liver disease (Vanstapel *et al.*, 1983), while others have found no correlation (Schmid *et al.*, 1981).

In typical HBV which is self limiting, HBsAg, HBeAg, DNA polymerase and DNA, HBcAg and complete Dane particles are present in serum early in the illness before the immune response is established and indicate permissive replication of the virus. Anti-HBcAg (IgM) and anti-HBeAg become evident in serum as the corresponding antigens disappear, indicating the end of permissive viral replication; finally anti-HBsAg develops. In cases which become chronic, HBsAg, HBeAg, DNA polymerase, DNA and HBcAg persist and the corresponding antibodies are delayed. In the acute attack or preclinical stage of infection, there may be a very little HBeAg and membranous HBcAg in hepatocytes. However, these are usually negative because at the time of biopsy the tissue response has eliminated hepatocytes containing detectable levels of antigen. If liver antigens are found in a patient with established acute

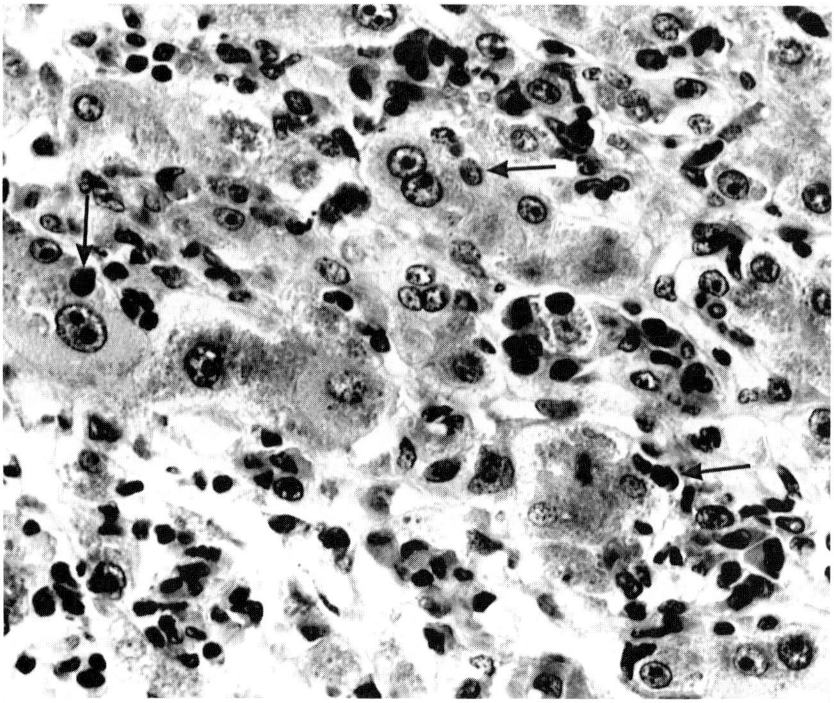

Fig. 4.14 Acute HBV infection. There is a heavy lymphocytic infiltration and some of these cells are in close apposition to degenerate hepatocytes (arrows) (HE, × 577).

disease, this implies an increased risk of chronicity. In an individual patient the outcome of the disease can be reasonably accurately predicted on the basis of antigen–antibody parameters. Most cases resolve, but some develop chronic liver disease, while a few die during the acute attack.

HDV infection modifies disease activity in HBV cases. Whether HDV is transmitted simultaneously (coinfection) or is superimposed (superinfection) on an existing HBV infection, the result is a more severe form of liver disease. The acute HBV illness is more severe and may be fulminant. In severe acute disease, there is fine vacuolation of hepatocytes which may be due to lipid accumulation and many hepatocyte nuclei contain HDV antigens. The latter disappear and then reappear as cirrhosis develops. In chronic carriers of HBV, superinfection with HDV results in a severe and sometimes fatal illness, or alternatively accelerates an already established chronic hepatitis (Lok *et al.*, 1985; Rizzetto and Verme, 1985). HDV infection in HBV carriers is manifest on biopsy by focal necrosis, inflammation and piecemeal necrosis. In acute HDV infection the delta antigen is present only fleetingly in serum very early in the illness. Thereafter, IgM and subsequently IgG anti-HDV appear. Repeat testing for antibody is advisable since IgM may be the only marker present during acute coinfection. Persistence of IgM beyond the period of clinical illness is associated with chronic disease and its disappearance indicates resolution (Aragona *et al.*, 1987).

NANB hepatitis (Dienes *et al.*, 1982) is the commonest cause of post-transfusion hepatitis and is frequent in haemophiliacs and drug abusers. The morphological features which suggest NANB hepatitis are prominent eosinophilic clumping of cytoplasm and acidophil degeneration of hepatocytes (Fig. 4.15). Microvesicular fat and patchy prominent sinusoidal activation are also present (Fig. 4.16). Sinusoidal cell activation may be the most prominent feature, with little liver cell damage. In some cases, lymphoid infiltration of portal tracts, with or without germinal centres, is striking. When this is associated with piecemeal necrosis, development of chronic liver disease should be suspected. NANB infection arises in patients after transfusion of blood or blood products and these patients have a high risk of developing chronic hepatitis. In some series, this is as high as 50%. The biopsy appearances are those of chronic hepatitis or cirrhosis; this is a particular hazard in patients with hypo-

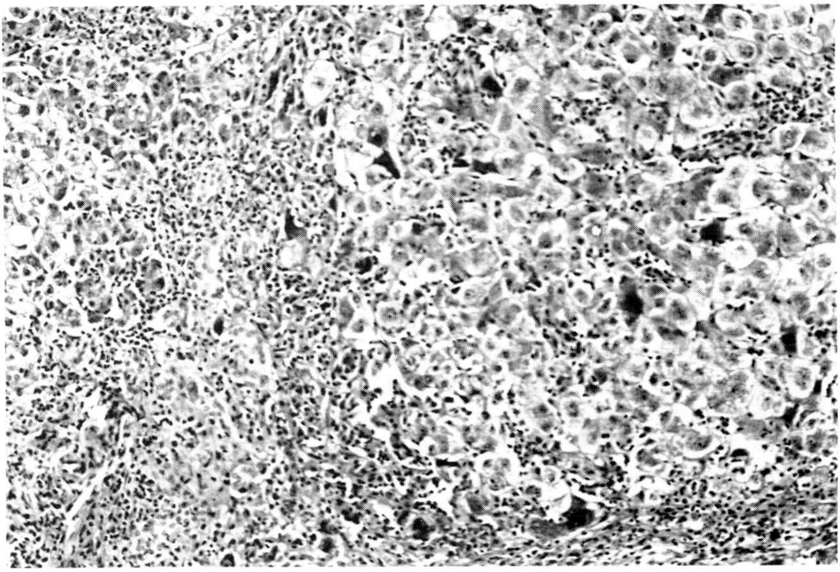

Fig. 4.15 Acute NANB infection. Acidophilic degeneration and necrosis of hepatocytes is a prominent feature (HE, × 120).

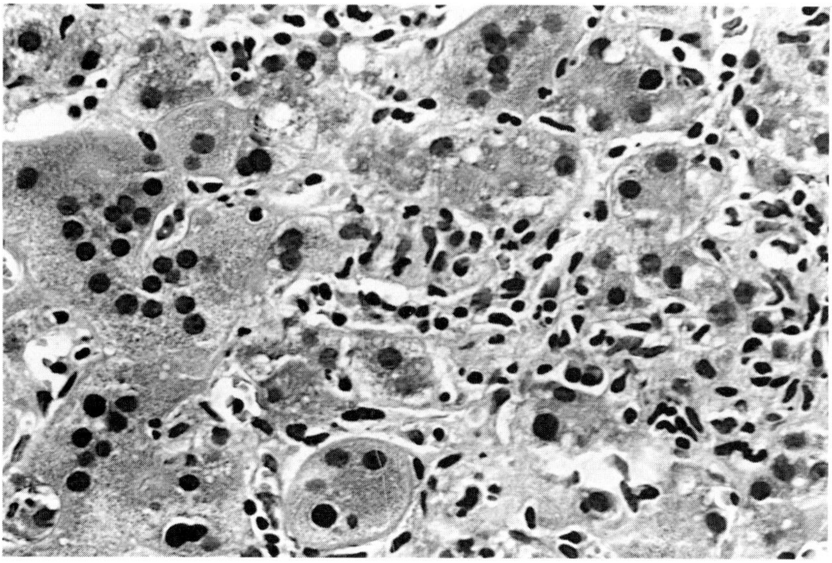

Fig. 4.16 Acute NANB infection. There is microvesicular fatty change and giant cell transformation of hepatocytes which is unusual in an adult (HE, × 450).

gammaglobulinaemia (Thompson *et al.*, 1987). Water-borne epidemics of NANB hepatitis in Northern India are common and have a high fatality rate during the acute attack due to massive necrosis.

4.1.5 *Differential diagnosis of viral hepatitis*

The diagnosis of acute viral hepatitis is usually straightforward. The most difficult entity to differentiate is drug-induced viral hepatitis-like disease. This occurs most frequently with α-methyldopa and isoniazid. The differentiation is made by obtaining a drug history, absence of viral markers in serum and recovery on withdrawal of the agent. Features which suggest a drug reaction are absence of portal tract inflammation, abundant neutrophils or eosinophils, granulomas and punched-out pericentral necrosis with few inflammatory cells. Alcoholic hepatitis may be confused with viral hepatitis when acidophil cells are prominent in the former. The presence of Mallory bodies and large globule fat are the most helpful features in establishing an alcoholic aetiology. Large globule fat is not a feature of viral hepatitis. Bile duct obstruction can cause a clinical picture similar to cholestatic viral hepatitis. The positive features in favour of biliary obstruction are prominent polymorph infiltration in portal tracts and within interlobular bile ducts; polymorphs within the parenchyma are common in bile duct obstruction and rare in viral hepatitis. Other viral infections, especially EBV infections, can also cause diagnostic difficulty. In EBV infection the single most prominent feature is sinusoidal activation, and hepatocyte ballooning is rare; acidophil cells are found in EBV. The bile duct lesion of viral hepatitis is distinguished from primary biliary cirrhosis (PBC) by the concomitant lobular disarray and clinical and serological features. Granulomas are a feature of PBC but not viral hepatitis. The granulomas found in drug abusers with viral hepatitis do not usually cause difficulty in correct diagnosis.

4.2 Chronic viral infection

The availability of screening tests for the specific aetiological agent of viral hepatitis makes it clear that some of these viruses can give rise to a chronic carrier state or the development of chronic persistent hepatitis (CPH), chronic active hepatitis (CAH), cirrhosis and primary hepatic cancer.

4.2.1 HBsAg carriers

The incidence of HBsAg carriers varies widely throughout the world; in South East Asia 20% of the population are carriers while in Europe and the USA the rate is about 0.1–0.5% (Blumberg, 1976). The factors which predispose to the development of the carrier state are not properly understood. HBsAg carriers may be clinically and biochemically normal, or alternatively they may develop the various forms of chronic liver disease described below (Anderson and Murray-Lyon, 1985).

The biopsy hallmark of the carrier state is the ground-glass cell. In HE sections, ground-glass hepatocytes are characterized by areas of cytoplasm having a lightly eosinophilic ground-glass appearance. This change may affect the entire cell or part of the cell, or form an arc around the nucleus (Fig. 4.17). The number and distribution of these cells vary depending on the associated pathology in the liver (see below), but in asymptomatic carriers they are found in large numbers diffusely throughout the lobule (Ishak, 1976). Ground-glass cells can be differentiated from other forms of hepatocyte abnormality by staining routine paraffin sections with aldehyde fuchsin (Fig. 4.18), orcein or thionine, in which the ground-glass areas of cytoplasm are stained purple, brown or blue respectively. The ultrastructural correlate of this cytoplasmic change is proliferation of smooth endoplasmic reticulum which contains circular and tubular profiles of HBsAg. The antigen is easily and specifically demonstrated on routine paraffin sections by the immunoperoxidase procedure using antibody to HBsAg (Fig. 4.19). HBcAg may be found in carriers in the absence of HBsAg (Anderson and Murray-Lyon, 1985). HBcAg, in the absence of HBsAg, can also confer a ground-glass

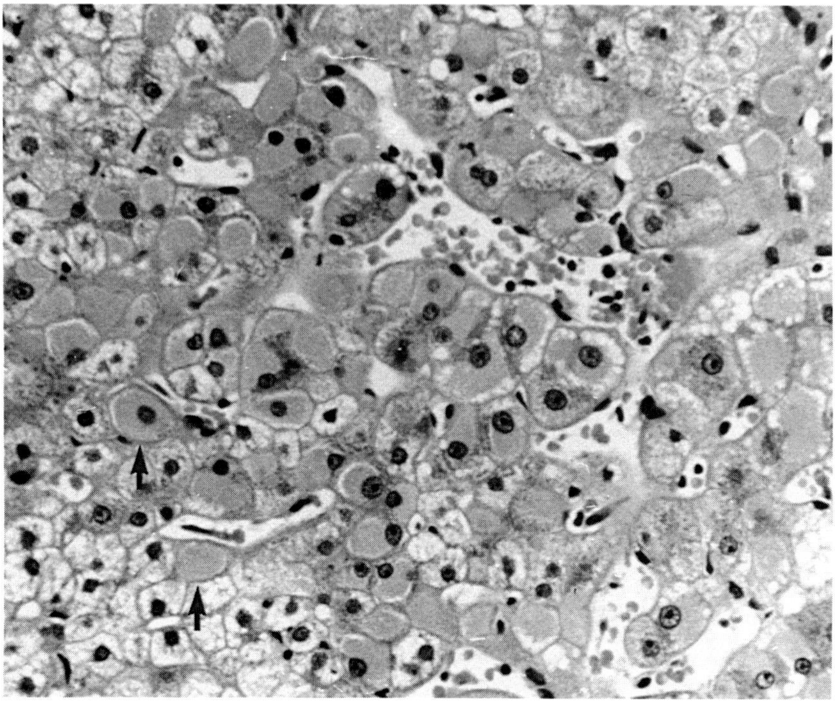

Fig. 4.17 HBsAg carrier. There are numerous ground-glass cells (arrows) (HE, × 320).

appearance in hepatocytes (Facchetti *et al.*, 1986). HBcAg is usually found in the nuclei where it gives rise to 'sanded nuclei', particularly in babies or immunosuppressed patients. Delta agent is sometimes found in nuclei and cytoplasm in Italian carriers (Facchetti *et al.*, 1986).

Cells somewhat similar to ground-glass cells also occur after long-term drug therapy (chlorpromazine and barbiturates) but do not stain with orcein or antibody to HBsAg (p. 171).

4.2.2 *Chronic persistent hepatitis*

CPH may follow viral B and NANB infections but is commoner after the latter. The symptoms and signs are generally mild and include anorexia, fatiguability and slight jaundice. Portal tracts are infiltrated by lymphocytes, plasma cells and occasional eosinophils; portal tract fibrosis and piecemeal necrosis are absent. Acidophil bodies, and foci of necrosis infiltrated by lymphocytes and plasma cells are almost always found. Balloon degeneration is usually absent except during relapse. In CPH

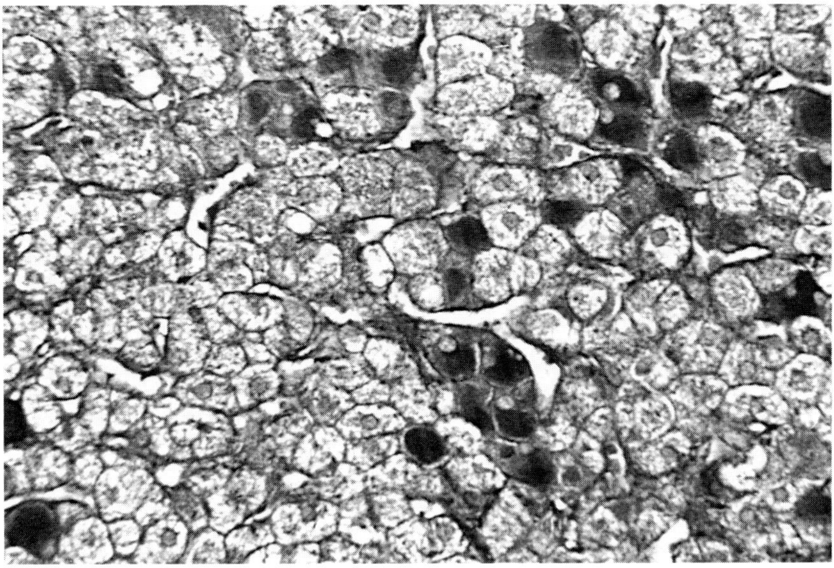

Fig. 4.18 HBsAg carrier. Ground-glass cells are intensely stained with aldehyde fuchsin (× 320).

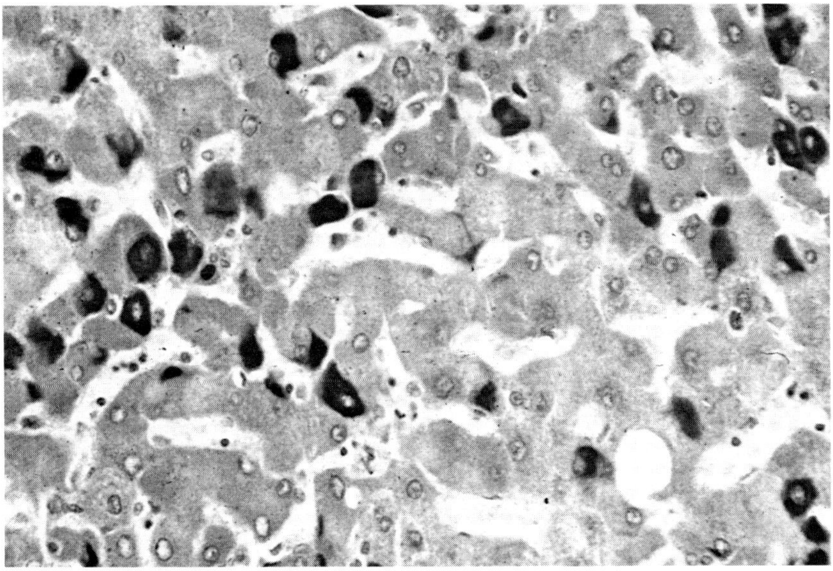

Fig. 4.19 HBsAg carrier. The surface viral antigen in ground-glass cells is specifically demonstrated in this section by the immunoperoxidase procedure. Note in this carrier that the antigen is present diffusely in the cytoplasm (× 320).

associated with viral B infection HBsAg may be demonstrable in large numbers of hepatocytes diffusely throughout the lobule and correspond to ground-glass cells; abundant core antigen may also be present in nuclei (Gudat *et al.*, 1975). The course of post-viral CPH may last for months or years but complete healing is to be anticipated in most cases.

In drug addicts additional features may be present. These include birefringent material in portal tract macrophages, granulomas and a prominent infiltrate of eosinophils in the portal areas (Ishak, 1976).

4.2.3 *Chronic active hepatitis*

A small proportion of patients with acute viral B hepatitis progress to CAH or present initially with the CAH syndrome. Concomitant HDV infection increases the risk of chronic hepatitis. NANB infection is a frequent cause of CAH. The histological features of CAH are inflammatory cell infiltration of portal tracts, piecemeal necrosis and fibrosis. During relapse of CAH all of the features of acute viral hepatitis may be present. However, the best morphological features which reflect a viral aetiology are ground-glass cells and the presence of viral antigens in hepatocytes. HBsAg, HBcAg and HDV have all been demonstrated in chronic active viral hepatitis. The surface antigen is usually found in small numbers of hepatocytes distributed in a spotty fashion throughout the lobule. Of considerable pathogenic interest is the finding that this antigen in CAH is frequently present on or close to the hepatocyte membrane (Ray *et al.*, 1976) (Fig. 4.20). Viral antigen may modify the hepatocyte cell surface so that it becomes accessible to attack by the immune system. Other factors which may contribute to the development of post-viral CAH are the development of bridging necrosis during acute hepatitis (p. 87). HBcAg and HDV are most often nuclear but both have now been described in hepatocyte cytoplasm (Facchetti *et al.*, 1986). Seroconversion of HBeAg to anti-HBe is associated with termination of HBV replication and return to normality of transaminases and liver biopsy. Seroconversion however may be associated with exacerbation of disease, presumably due to clearing of remaining infected hepatocytes by the immune system.

It is important to be aware that HBV markers may be present only in the liver but absent from serum. Therefore, it is useful to look for viral markers in chronic hepatitis even when serum markers are not detectable (Cuccurullo *et al.*, 1987).

4.2.4 *Post-hepatitic cirrhosis*

Cirrhosis occasionally results from a preceding viral B, HBV plus HDV, and NANB infection. The cirrhosis is of macronodular or mixed type (Fig.

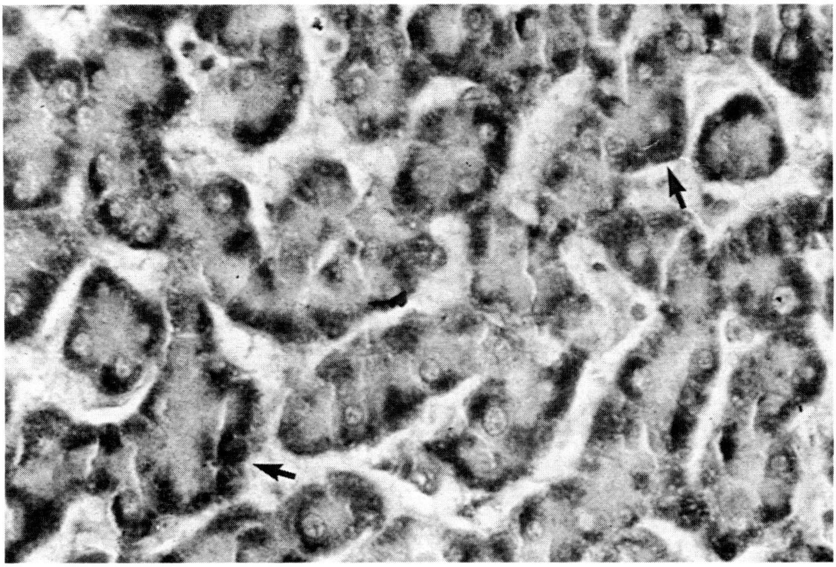

Fig. 4.20 In this case of viral-induced chronic active hepatitis HBsAg is present close to the cell membrane of all the hepatocytes (arrows). Immunoperoxidase procedure using antibody to HBsAg (× 480).

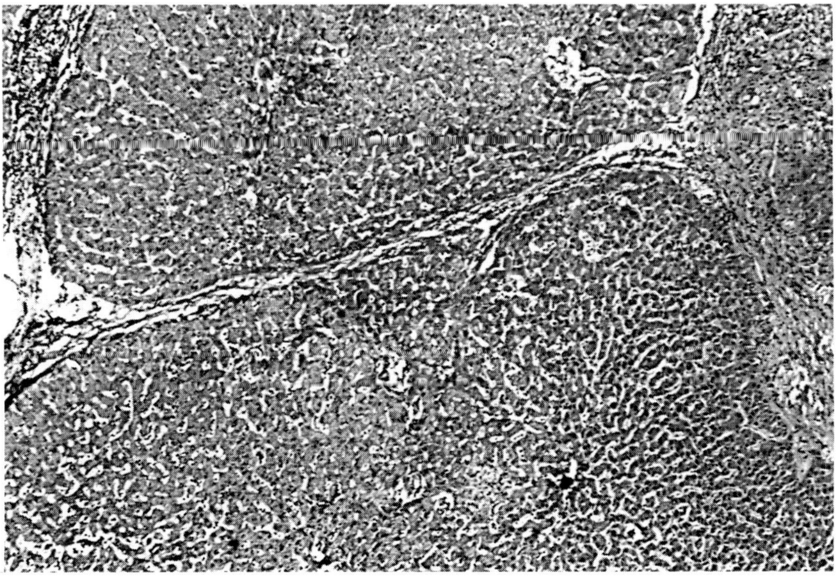

Fig. 4.21 Cirrhosis following viral B hepatitis. The parenchyma is intersected by fine bands of connective tissue which delineate large nodules of parenchyma (HE, × 58).

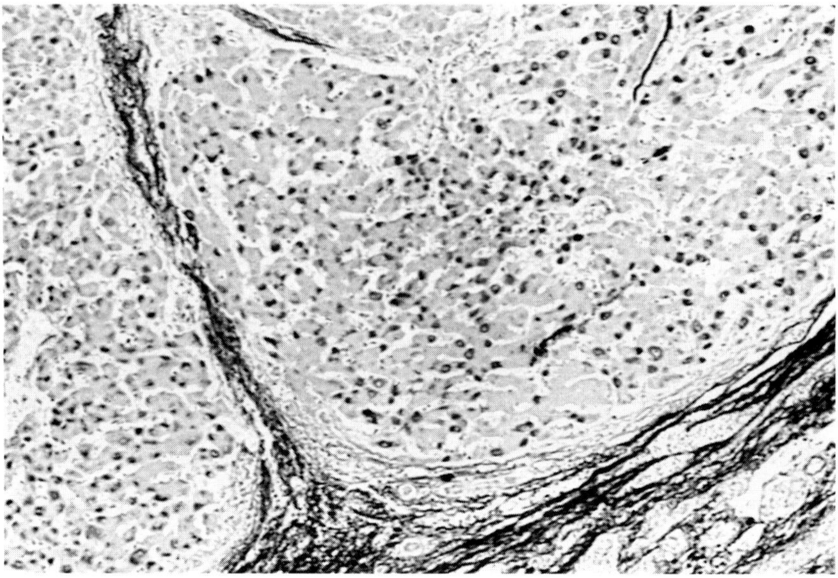

Fig. 4.22 Post-hepatitis cirrhosis. Some hepatocytes contain HBsAg (orcein, × 152).

4.21) and in most instances there are no morphological features which testify to its viral association except for the presence of ground-glass cells and viral antigens in hepatocytes (Fig. 4.22). Some claim that the features of acute hepatitis (balloon cells, acidophil bodies, etc.) may persist in the cirrhotic liver thereby indicating its viral aetiology, while others attribute these features to continuing activity of the cirrhotic process. In inactive post-hepatitis cirrhosis, HBsAg is usually present homogeneously in the cytoplasm and the positive cells are scattered diffusely throughout the lobule, whereas in active cirrhosis the antigen occurs in fewer cells and is distributed on or near the cell membrane (Ray *et al.*, 1976), as in Fig. 4.20.

Post-necrotic scarring results from any viral hepatitis which has produced massive necrosis.

4.2.5 Hepatocellular cancer

Since viral hepatitis may be complicated by the development of cirrhosis, it is to be expected that it would also lead indirectly to hepatic cancer. HBs antigenaemia occurs in patients with hepatocellular cancer and its incidence varies from 4.5% in Hong Kong to 40% in Uganda. Ground-glass cells and HBsAg can be demonstrated in the cirrhotic liver adjacent to primary hepatic cancer and occasionally in the tumour cells themselves (Turbitt *et al.*, 1977). The presence of HBcAg in cancer nuclei and the integration of the HBV genome indicates a possible oncogenic role for virus B in hepatic cancer. HDV infection does not appear to predispose to hepatic cancer, presumably because it shortens the time to the development of chronic liver failure. There is no good evidence that HAV or NANB cause hepatic cancer.

4.3 Other viral infections

4.3.1 Adenoviruses

These DNA viruses rarely involve the liver but when they do they produce a characteristic picture of sharply circumscribed areas of coagulation necrosis. The adjacent hepatocytes contain intranuclear inclusions, sometimes surrounded by a clear halo reminiscent of cytomegalovirus (CMV) infection. Similar inclusions have also been documented in bile ducts.

4.3.2 Cytomegalovirus (Snover and Horwitz, 1984)

Disseminated CMV infection in neonates present with ascites, jaundice, haemolytic anaemia and thrombocytopenic purpura. Typically, the liver shows focal necrosis, giant cell transformation of hepatocytes, cholestasis and polymorph infiltration of portal tracts. The pathognomonic feature is the presence of basophilic intranuclear inclusions surrounded by a clear halo (Fig. 4.23). Inclusions occur in hepatocytes, sinusoidal cells and bile duct epithelium (Fig. 4.24), but often a careful search has to be made because they may be present in small numbers. In doubtful cases, virus isolation from the urine or saliva may be necessary for definitive diagnosis.

Disseminated fatal CMV infections are rare in adults and occur in grossly debilitated or immunosuppressed patients. The commonest pathology is a necrotizing pneumonia and adrenal necrosis and the liver pathology is similar to that in infants. Healthy adults may develop a severe hepatitis after an infected blood transfusion without developing typical inclusions in hepatocytes. On biopsy, there may be a liver lesion similar to acute viral hepatitis, subacute hepatic necrosis, or granulomatous hepatitis (Ten Napel *et al.*, 1984). Positive diagnosis depends on demonstrating viral antibodies in serum.

4.3.3 Herpes simplex

Generalized herpetic infections are limited to neonates and immunosuppressed adults. In neonates, confluent hepatocellular necrosis has been

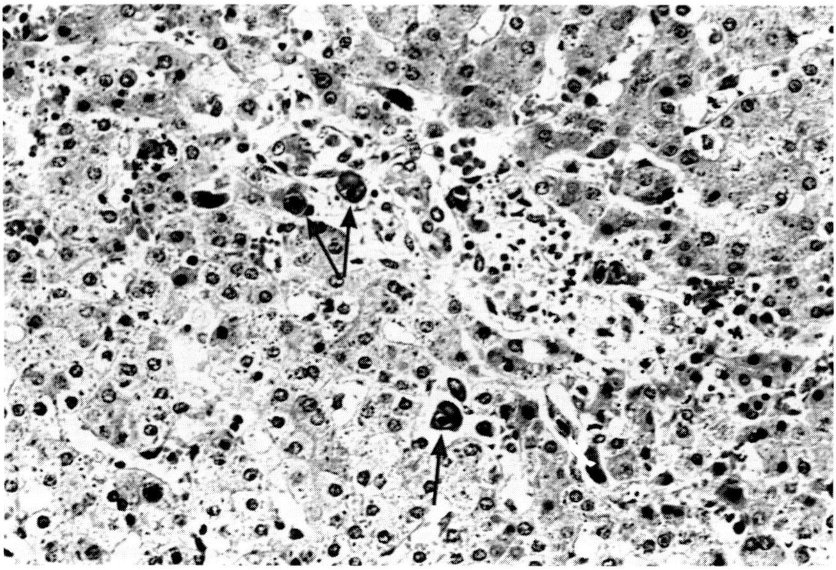

Fig. 4.23 Cytomegalovirus hepatitis. There is focal necrosis and an accumulation of cells with intranuclear viral inclusions (arrows) producing an 'owl's eye' appearance (HE, × 247).

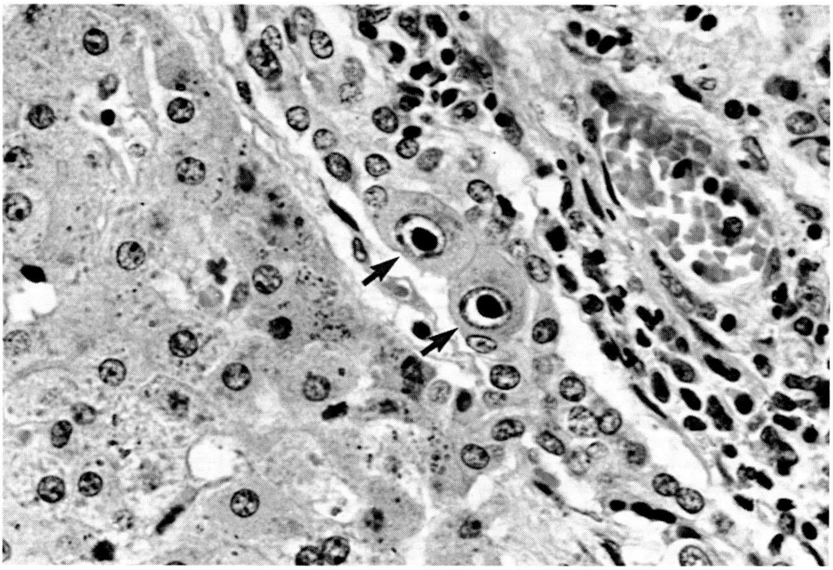

Fig. 4.24 Cytomegalovirus hepatitis. The typical intranuclear inclusions in this case were more prominent in bile duct epithelium (arrows) than in hepatocytes (HE, × 450).

described which may be central or midzonal and associated with giant cell transformation of hepatocytes (p. 31). Definitive diagnosis can be made by demonstrating intranuclear eosinophilic inclusions in hepatocytes which are purple in HE sections and are separated from the nuclear envelope by a pale halo.

4.3.4 *Infectious mononucleosis*

Some degree of hepatic structural abnormality is almost invariable in infectious mononucleosis. Clinically, however, jaundice and hepatomegaly occur in only 5–20% of cases. The disease usually presents in adults with pharnygitis and lymphadenopathy. There is diffuse infiltration of sinusoids and portal tracts by cells of the lymphoid series (including plasma cells) and large monocytes and occasionally polymorphs (Fig. 4.25). Compared with viral hepatitis, hepatocyte damage is inconspicuous in the majority of cases. Sinusoidal cell swelling, together with the infiltrating mononuclear cells, sometimes gives rise to small granulomas. Acidophil cells are usually found in small numbers and, as in many systemic infections, mitotic activity in hepatocytes is common. In severe cases endophlebitis of central veins has been described. Infectious mononucleosis does not give rise to chronic liver disease.

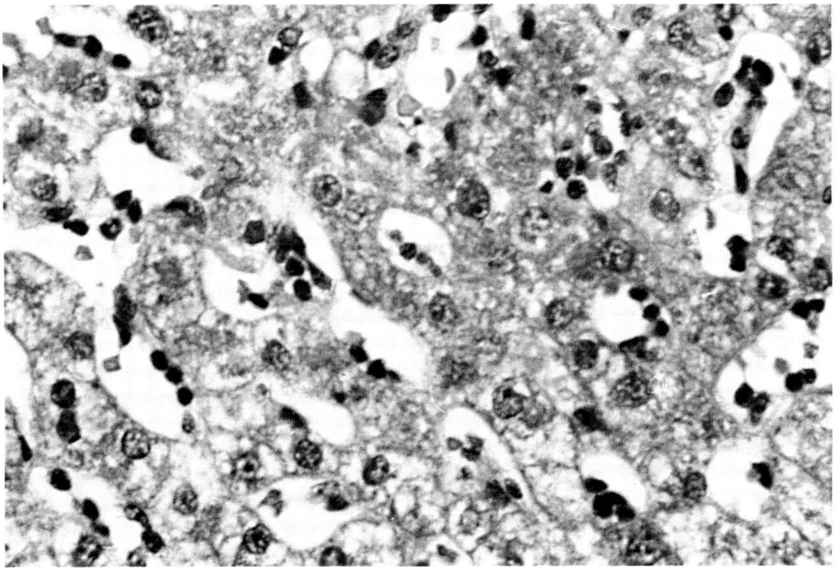

Fig. 4.25 The liver in infectious mononucleosis. Lymphoid cells are prominent in all of the sinusoids (HE, × 360).

4.3.5 *Rubella*

Some children with the congenital rubella syndrome (cataracts, deafness, congenital anomalies of the heart and mental retardation) develop hepatosplenomegaly which usually subsides in a few weeks. In some cases, however, jaundice develops with variable hepatic pathology. This may take the form of a fatal giant cell hepatitis or the biopsy findings may be indistinguishable from biliary atresia. An uncharacterized green pigment is sometimes found in Kupffer cells. Recovery from rubella hepatitis has been documented; progression to cirrhosis is rare.

4.3.6 *Yellow fever*

This is caused by infection with a group B arbovirus and may produce no clinical symptoms or a rapidly fatal illness characterized by jaundice, vomiting, haemorrhagic manifestations, oliguria and coma. In the liver there is fatty change, midzonal necrosis, scattered acidophil (Councilman) bodies and little inflammatory reaction. Councilman bodies are said to differ from acidophil cells in viral hepatitis; they have a distinct outer membrane and contain ceroid and lipid pigment (Vieira *et al.*, 1983). Massive necrosis may also be found. The virus has been demonstrated in hepatocytes and sinusoidal cells.

4.3.7 *Coxsackie viruses*

In infancy, Coxsackie B infection produces a clinical picture similar to herpes simplex. Myocarditis is the dominant feature of the illness but jaundice and hepatic pathology occur in some cases. The liver usually shows focal hepatitis or centrilobular necrosis, but in contrast to herpetic liver disease intranuclear inclusions are absent. In adults, Coxsackie B is associated with aseptic meningitis, pleurodynia, myocarditis and rarely hepatitis in which the portal tracts and veins are infiltrated by mononuclear cells and a few polymorphs. Centrilobular cholestasis and swelling of hepatocytes may be found.

Coxsackie A viruses may occasionally cause hepatitis in children and adults.

4.3.8 *ECHO viruses*

These viruses rarely produce neonatal hepatic necrosis similar to that produced by herpes simplex except that inclusions are not found in ECHO infections. Hepatic necrosis is usually central but may extend to become massive. In adults this virus has been associated with the clinical manifestations of an acute hepatitis.

4.3.9 *'Marburg virus'* (Bechtelsheimer *et al.*, 1972)

In Germany and Yugoslavia there were epidemics of hepatitis in laboratory workers who had handled tissues from vervet monkeys. The organism responsible for these outbreaks has not been characterized and may be a rickettsia. In these cases the liver showed acidophilic necrosis involving hepatocytes and sinusoidal cells, portal tract inflammation and intranuclear inclusions.

4.3.10 *Varicella*

In fatal varicella infections there are areas of focal coagulation necrosis which in our experience are most commonly related to portal tracts. These may be infiltrated by polymorphs. Acidophilic intranuclear inclusions in hepatocytes, bile ducts and sinusoidal cells are characteristic.

References

Abe, H., Benniger, P. R., Ikejiri, N. *et al.* (1982), Light microscopic findings of liver biopsy specimens from patients with hepatitis type A and comparison with type B. *Gastroenterology*, **82,** 938–47.

Anderson, M. G. and Murray-Lyon, I. M. (1985), Natural history of the HBsAg carrier. *Gut*, **26,** 848–60.

Aragona, M., Macaguo, S., Caredda, F. *et al.* (1987), Serological response to the hepatitis delta virus in hepatitis D. *Lancet*, **i,** 478–80 .

Bechtelsheimer, H., Korb, G. and Gedigk, P. (1972), The morphology and pathogenesis of 'Marburg Virus' hepatitis. *Hum. Pathol.*, **3,** 255–64.

Blumberg, B. S. (1976), Bioethical questions related to hepatitis B antigen (1976). *Am. J. Clin. Pathol.*, **65,** 848.

Boyer, J. L. and Klatskin, G. (1970), The pattern of necrosis in acute viral hepatitis: prognostic significance of bridging (subacute hepatic necrosis). *N. Engl. J. Med.*, **283,** 1063–71.

Bradley, D. W., Maynard, J. E., Popper, H. *et al.* (1983), Post-transfusion non-A, non-B hepatitis: physiochemical properties of two distinct agents. *J. Infect. Dis.*, **148,** 254–65.

Cuccurullo, G., Rambaldi, M., Iaquinto, G. *et al.* (1987), Importance of showing HBsAg and HBcAg positivity in the liver for better aetiological definition of chronic liver disease. *J. Clin. Pathol.*, **40,** 167–71.

Dienes, H. P., Popper, H., Arnold, W. *et al.* (1982), Histologic observations in human hepatitis non-A, non-B. *Hepatology*, **2,** 562–71.

Facchetti, F., Tardamio, R., Bonetti, M. F. *et al.* (1986), HBsAg, HBcAg and delta-Ag in liver tissue: simultaneous visualisation in a single section by triple immunostaining. *Histol. Histopathol*, **1,** 181–5.

Gudat, F., Bianchi, L., Sonnabend, W. *et al.* (1975), Pattern of core and surface expression in liver tissue reflects state of specific response to hepatitis B. *Lab. Invest.*, **32,** 1–9.

Ishak, K. G. (1976), Light microscopic morphology of viral hepatitis. *Am. J. Clin. Pathol.*, **65,** 787–93.

Koike, K., Shiro, I., Kurai, K. *et al.* (1987), IgM anti-HBc in anti-HBc positive chronic type B hepatitis with acute exacerbations. *Hepatology*, **7,** 573–6.

Lok, A. S. F., Lindsay, I., Scheuer, P. J. and Thomas, H. C. (1985), Clinical and histological features of delta infection in chronic hepatitis B virus carriers. *J. Clin. Pathol.*, **38,** 530–3.

Nisman, R. M., Ganderson, A. P., Vlahelric, R. and Gregory, D. H. (1979), Acute viral hepatitis with bridging necrosis. An overview. *Arch. Intern. Med.*, **139,** 1289–91.

Okuno, T., Sano, A., Seguchi, T. *et al.* (1984), Pathology of acute hepatitis A in humans. Comparison with acute hepatitis B. *Am. J. Clin. Pathol.*, **81,** 162–9.

Ray, M. B., Desmet, V. J., Fevery, J. *et al.* (1976), Distribution patterns of hepatitis B surface antigen (HBsAg) in the liver of hepatitis patients. *J. Clin. Pathol.*, **29,** 94.

Rizzetto, M., Canese, M. G., Arico, S. *et al.* (1977), Immunofluorescence detection of a new antigen–antibody system (S/anti-S) associated to hepatitis B virus in liver and in serum of HBsAg carriers. *Gut*, **18,** 997–1003.

Rizzetto, M. and Verme, G. (1985), Delta hepatitis – present status. *J. Hepatol.*, **1,** 187–93.

Schmid, M., Pirovino, M., Altorfer, J. *et al.* (1981), Acute viral hepatitis B with bridging necrosis: a follow-up study. *Liver*, **1,** 222–9.

Snover, D. C. and Horwitz, C. A. (1984), Liver disease in cytomegalovirus mononucleosis: a light microscopic and immunoperoxidase study of six cases. *Histopathology*, **4,** 408–16.

Ten Napel, H. H., Houthoff, H. J. and The, T. H. (1984), Cytomegalovirus hepatitis in normal and immunocompromised hosts. *Liver*, **4,** 184–94.

Thompson, B. J., Doran, M., Rever, A. M. L. and Webster, A. D. B. (1987), Alpha interpheron therapy for non-A, non-B hepatitis transmitted by gammaglobulin replacement therapy. *Lancet*, **i,** 539–41.

Turbitt, M. L., Patrick, R. S., Goudie, R. B. *et al.* (1977), Incidence in South West Scotland of hepatitis B surface antigen in the liver of patients with hepatocellular carcinoma. *J. Clin. Pathol.*, **30,** 1124–7.

Vanstapel, M. J., van Stenbergen, W., De Wolfe-Peeters, C. *et al.* (1983), Prognostic significance of piecemeal necrosis in acute viral hepatitis. *Liver*, **3,** 46–57.

Vieira, W. T., Gayotto, L. C., de Lina, C. O. *et al.* (1983), Histopathology of the human liver in yellow fever with special emphasis on the diagnostic role of the Councilman body. *Histopathology*, **7,** 195–208.

5 Hepatitis associated with non-viral infections

Liver dysfunction may accompany any severe localized or disseminated infection. Since the primary site of infection dominates the clinical presentation the pathogenesis of hepatic malfunction is usually not actively investigated and is therefore obscure in many instances. The infections considered in this Chapter are those which generally produce a structural abnormality and may be investigated by liver biopsy.

5.1 Bacterial and mycobacterial infections

5.1.1 Pyogenic abscess

In the pre-antibiotic era pyogenic abscesses were apparently relatively common as a sequel to acute appendicitis. They are now rare. Liver abscesses may result from drainage of bacteria via the portal vein from a septic focus in the abdomen such as appendicitis, proctitis, diverticulitis and any ulcerative lesion in the alimentary canal and umbilical sepsis in infants. Immunocompromised patients are more susceptible. A more common route of infection is local spread from the gall bladder or as a sequel to cholangitis. Hepatic arterial dissemination of organisms from conditions such as bacterial endocarditis is uncommon. Abscess formation may follow penetrating or non-penetrating hepatic trauma. About half of all hepatic abscesses are idiopathic in that a pre-existing extra-hepatic focus of sepsis cannot be demonstrated. In all cases the hepatic pathology is similar: the lesions are usually single and have the morphological features of abscesses elsewhere. Clinically and grossly, pyogenic and amoebic abscesses are similar. It is important to differentiate these conditions because of the different therapy required for each.

Focal collections of polymorphs in sinusoids, resembling microabscesses, are a frequent finding in wedge biopsies. These are due to surgical handling of the liver and not bacterial infection.

5.1.2 *Brucellosis*

Although any bacterium can theoretically colonize the liver, this is uncommon in most systemic infections. However, *Brucella abortus* is an exception as it frequently produces granulomas; *B. suis* and *melitensis* rarely involve the liver. The granulomas are distributed at random in the liver lobule and are composed of epithelioid cells, a few lymphocytes, plasma cells and giant cells (Fig. 5.1); these granulomas resemble those in tuberculosis but differ in that caseation is not a feature. Small microgranulomas composed of about ten macrophages unaccompanied by other cell types have been attributed to brucella infection.

Serial biopsy studies have shown that brucellosis is rarely associated with the development of cirrhosis in which granulomas have disappeared.

5.1.3 *Leprosy*

Although about half of the cases of lepromatous leprosy that come to autopsy show miliary granulomas in the liver, the liver is rarely biopsied in this condition. In biopsy material, granulomas may be found in portal tracts or in the lobules. They are composed of aggregates of epithelioid cells some of which are markedly vacuolated (Virchow cells) and an

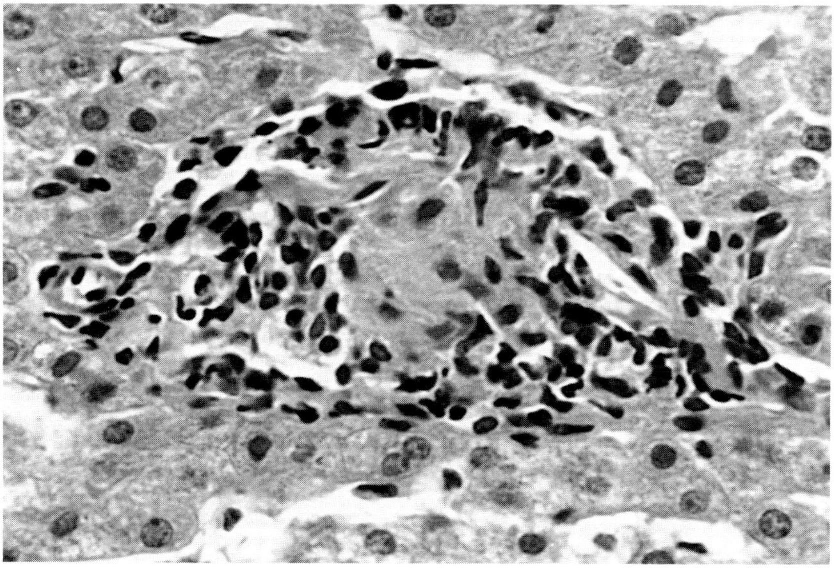

Fig. 5.1 Brucellosis of liver. There is a small granuloma composed of large histiocytes and surrounded by a cuff of lymphocytes (HE, × 720).

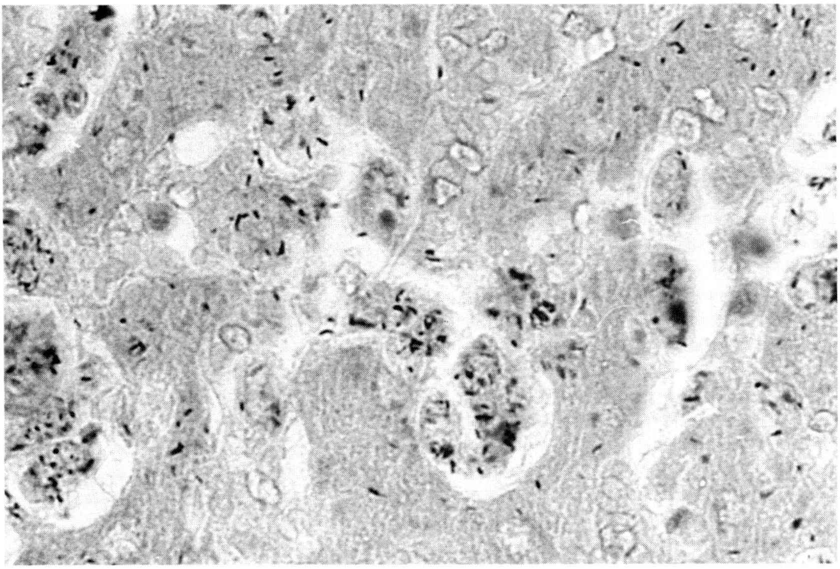

Fig. 5.2 Leprosy bacilli in the liver. The bacilli have been engulfed by sinusoidal cells which are grossly enlarged (Fite stain, × 680).

occasional Langhans-type giant cell. The lepra bacillus is readily demonstrated in these lesions with the Fite stain (Fig. 5.2).

5.1.4 *Tuberculosis*

Tuberculous involvement of the liver seldom produces evidence of dysfunction. Instead liver biopsy is normally undertaken when the patient presents with hepatomegaly and pyrexia of undefined cause or as part of the investigation of lung lesion. In positive cases there are granulomas in the liver which may be in portal tracts or anywhere in the lobule. Tubercles in the liver show the same features as in all other organs, except that in the liver caseation within follicles is unusual (Fig. 5.3), and the demonstration of acid-fast bacilli, even in proven tuberculous cases, is extremely uncommon in our experience. However, it is useful to examine serial sections stained by the Ziehl–Neelson method. Other non-specific complications of tuberculous infection are fatty change and amyloidosis. Even when granulomas are non-caseating and negative for acid-fast bacilli and other obvious causes of granulomas have been excluded, a course of antituberculous therapy is advisable.

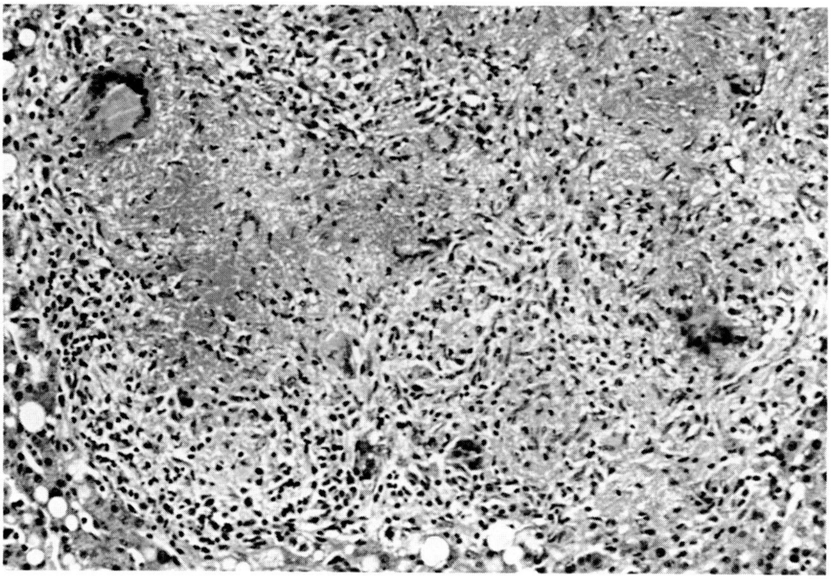

Fig. 5.3 Tubercles in the liver. There is a large granulomatous lesion with central caseation. At the periphery are epitheloid cells, prominent Langhans-type giant cells and a few lymphocytes (HE, × 180).

5.2 Spirochaetal infections

5.2.1 Relapsing fever

Borrelia recurrentis infection, spread by lice and ticks, frequently produces hepatomegaly, jaundice and fever. The two latter features disappear simultaneously when the acute phase of the illness is over. In fatal cases sinusoidal cells contain ingested red cells and degenerate spirochaetes, and there may be portal tract inflammation.

5.2.2 Syphilis

Luetic infection of the liver before birth may result in stillbirth or congenital syphilis. In the latter, there is diffuse fibrosis in which almost every liver cell is enclosed in fibrous tissue which derives from hepatic sinusoids (Fig. 5.4). In this lesion Levaditi-positive spirochaetes are abundant (Fig. 5.5). The findings in secondary syphilis in the liver are so protean as to be useless diagnostically. In late syphilis the characteristic lesions are gummas which are usually multiple: these consist of a central area of necrosis surrounded by a capsule of connective tissue in which

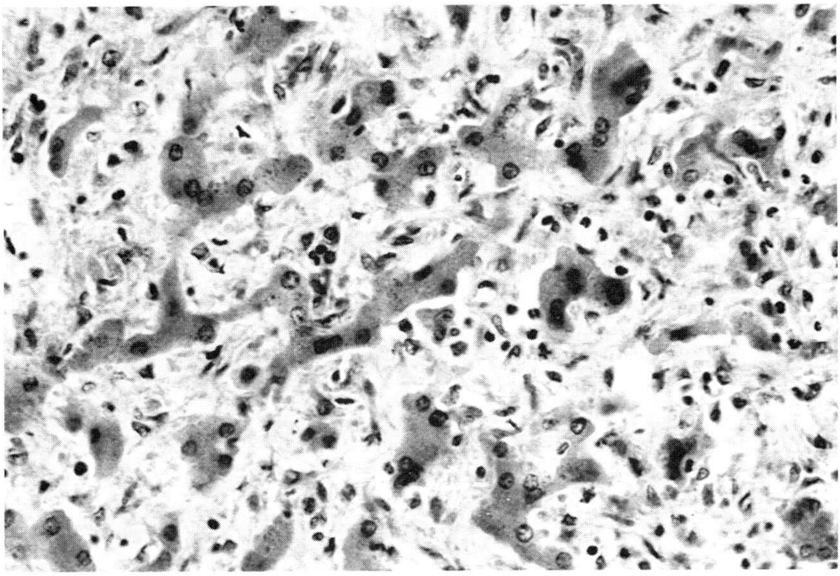

Fig. 5.4 Congenital syphilis. The hepatocytes which are atrophic are widely separated by massive quantities of connective tissue and occasional inflammatory cells (HE, × 640).

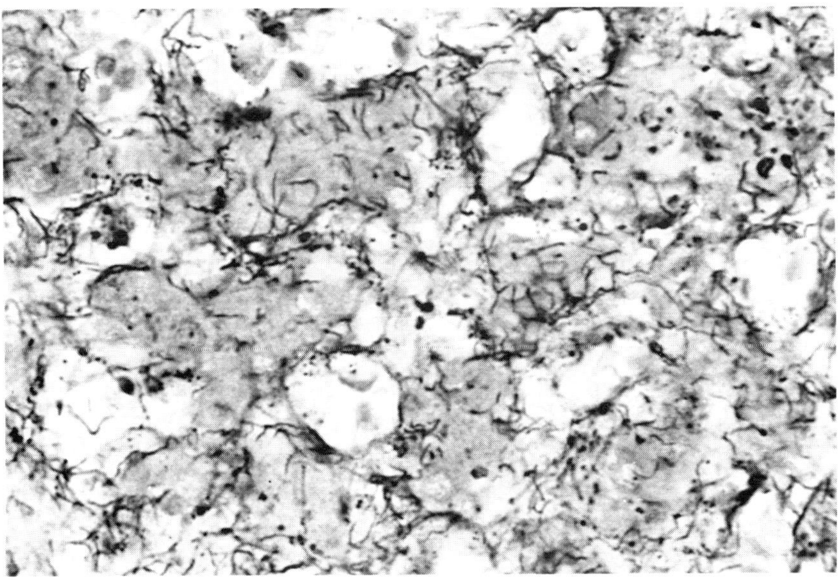

Fig. 5.5 Congenital syphilis. There are abundant spirochaetes in hepatocytes and in the sinusoidal connective tissue (Levaditi stain, × 720).

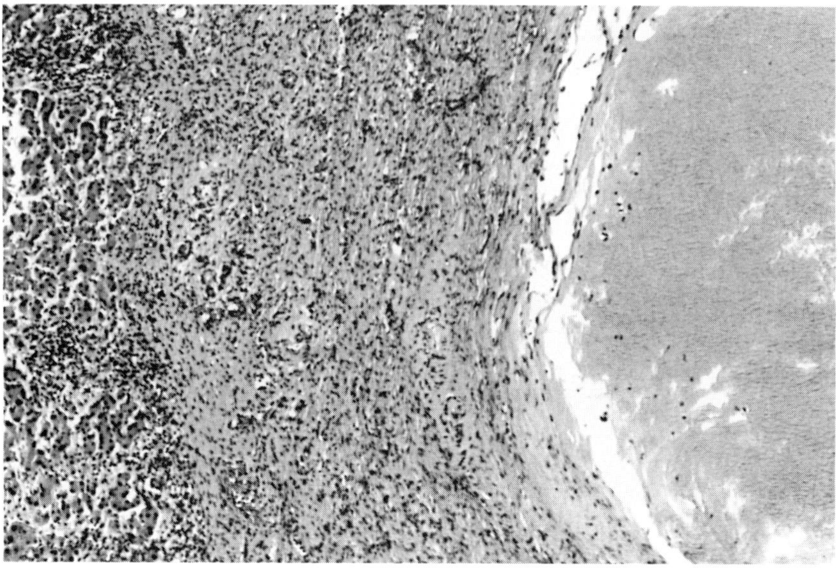

Fig. 5.6 Gumma of liver. On the right is a large area of acellular necrosis separated from hepatic tissue on the left by a thick band of fibrous tissue infiltrated with chronic inflammatory cells (HE, × 90).

giant cells may be found (Fig. 5.6). The sequelae of this process is stellate scarring and rarely nowadays hepar lobatum, a condition which superficially resembles macronodular cirrhosis but differs from it in that there is coarse scarring of the liver while the intervening hepatic parenchyma has a normal architecture.

5.2.3 Weil's disease

Cases of Weil's disease in this country seldom, if ever, come to biopsy. Therefore, the main experience is derived from autopsy material. In spite of the very dramatic hepatic dysfunction that may occur in leptospirosis, the liver pathology is usually unimpressive. The latter consists mainly of dissociation of hepatocytes, evidence of hepatic regeneration as evidenced by mitotic activity and sometimes centrilobular cholestasis. In about half of the fatal cases leptospirae can be demonstrated in the liver. Focal necrosis is described but is unusual.

5.3 Mycotic infections

5.3.1 *Actinomycosis*

Actinomycetic infection of the liver is almost always secondary to spread of the organism from the gut or rarely the gall bladder. Initially it causes a granulation tissue reaction in the liver with much fibrosis. The centre of these lesions then undergoes necrosis producing a multilocular abscess (Fig. 5.7(a)). The organisms are readily identified in tissue sections, where they are surrounded by numerous polymorphs (Fig. 5.7(b)), and in aspirates of the abscess cavity.

5.3.2 *Pseudomycosis*

Pseudomycosis (botryomycosis, granuloma pyogenicum) resembles actinomycosis grossly and 'sulphur granules' may be present also. This condition is due to infection with non-filamentous organisms, e.g. streptococci and staphylococci have been grown from these lesions (Bagby and Gunning, 1978).

5.3.3 *Candidiasis*

The liver is often involved in disseminated candidiasis but the diagnosis is rarely made ante mortem because of more serious manifestations elsewhere. The liver contains granulomas or abscesses (Fig. 5.8) and a positive diagnosis is made by demonstrating Gram-positive mycelia and yeast-like forms of the fungus (Lewis *et al.*, 1982).

5.3.4 *Blastomycosis*

Blastomyces dermatitides, and *Paracoccidioides brasiliensis* the organism responsible for Brazilian blastomycosis, only involve the liver after dissemination from the primary site of infection in the skin and lung, or mouth respectively. These small spherical/oval yeasts usually localize in portal tracts producing a granulomatous reaction in which eosinophils and multinucleate giant cells are characteristic.

5.3.5 *Coccidioidomycosis*

Coccidioides immitis gives rise to an acute self-limiting pulmonary illness with no liver involvement or a chronic fatal granulomatous disease in which liver pathology is frequent. In the latter granulomas are found in the liver and are indistinguishable from those of sarcoidosis except that

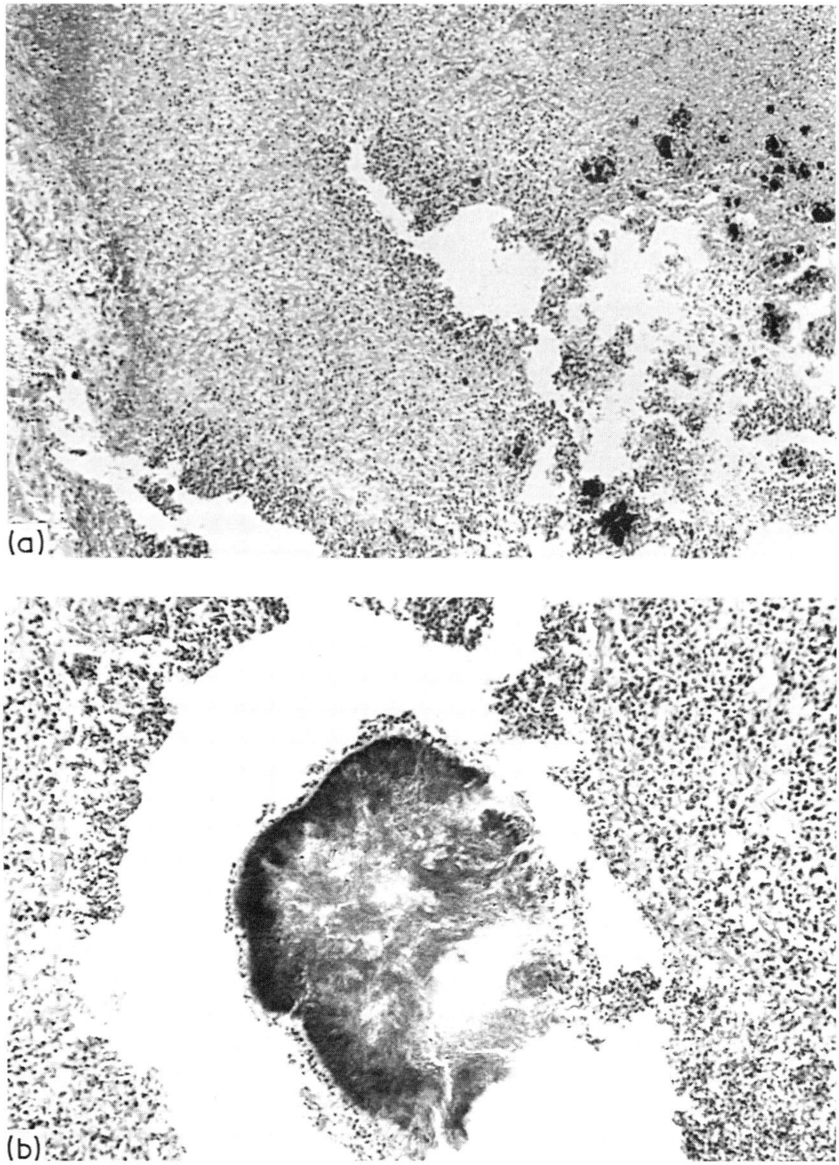

Fig. 5.7 (a) The liver in actinomycosis. Much of the field shows part of a large abscess with degenerate hepatic tissue on the left. The lesion consists of necrotic tissue infiltrated by acute inflammatory cells. There are numerous small dark clumps of *Actinomyces israeli,* the causative organism. (b) The central clump of actinomycetes is surrounded by numerous polymorphs and a layer of granulation tissue containing many chronic inflammatory cells. ((a) PAS and H, × 90; (b) HE, × 200.)

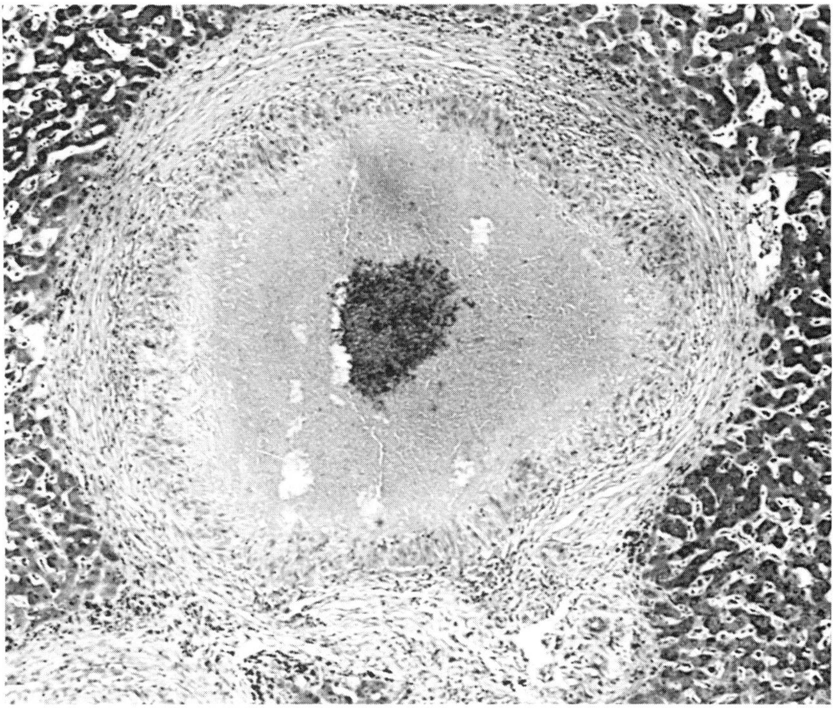

Fig. 5.8 Liver in candidiasis. A clump of *Candida* is surrounded by an area of necrosis and a palisade of macrophages and connective tissue (HE, × 102).

the organism may be demonstrable as PAS-positive spherules in macrophages.

5.3.6 *Cryptococcosis*

Inhalation of the yeast *Cryptococcus neoformans* produces benign granulomatous pneumonitis which rarely disseminates to produce granulomas in the liver. Small spheroidal yeasts are present in these granulomas and particularly adjacent to areas of necrosis.

5.3.7 *Histoplasmosis*

Histoplasma capsulatum infections usually present with benign pulmonary lesions and skin sensitivity to histoplasmin. Rarely, however, the organism is disseminated widely and in these cases the liver becomes enlarged; it contains granulomas composed of epithelioid cells and giant cells and the organism can usually be demonstrated as PAS-positive spheroidal spores in sinusoidal cells and epithelioid cells. The granulomas of histoplasmosis are similar to those found in sarcoidosis and may contain Schaumann bodies and asteroids. Occasionally, these granulomas show central necrosis and so resemble tuberculosis. Histoplasmosis of the liver heals by resolution or calcification.

5.4 Protozoal infections (Marcial-Rojas and Ramirez-Randa, 1982)

5.4.1 *Amoebiasis* (Aikat *et al.*, 1979)

Entamoeba histolytica involves the liver in 1–8% of cases where intestinal infection is present. The amoebae gain access to the liver by passage up the portal vein. The characteristic pathology is usually a single abscess in the liver at the site of the hepatic flexure of the colon or the dome of the right lobe. The abscess consists of a central mass of necrotic debris surrounded by a zone of stromal cells in which the PAS-positive amoebae are present together with degenerate hepatocytes (Fig. 5.9). The organism, which also stains with methenamine silver, is sometimes seen in the adjacent normal liver tissue.

A clinical syndrome, associated with *Entamoeba* infection, characterized by abdominal discomfort and a systemic illness has been ascribed to chronic amoebic hepatitis. Liver pathology in this syndrome is inconsistent and many doubt its existence.

5.4.2 *Leishmaniasis (kala-azar)*

Visceral leishmaniasis caused by *Leishmania donovani* produces hepatomegaly and splenomegaly. In this form of infection large numbers of organisms are present in sinusoidal cells and this is associated with fatty change and infiltration of portal tracts by plasma cells and lymphocytes (Fig. 5.10). A fine intralobular and perilobular fibrosis may also occur. The organisms can be identified in paraffin-fixed material by Giemsa staining but are more readily seen by the same method in unfixed touch preparations.

5.4.3 *Malaria*

In chronic malarial infection the liver becomes enlarged. This is due to infiltration of the sinusoids by mononuclear cells and lymphocytes; sinusoidal cell hyperplasia is also a feature. The hepatocytes may show mitotic activity and areas of focal necrosis. A characteristic feature is the presence of black 'malarial' pigment in sinusoidal cells (Fig. 5.11). This pigment is birefringent in polarized light and soluble in alcoholic picric acid. There is usually some degree of portal tract fibrosis that may become

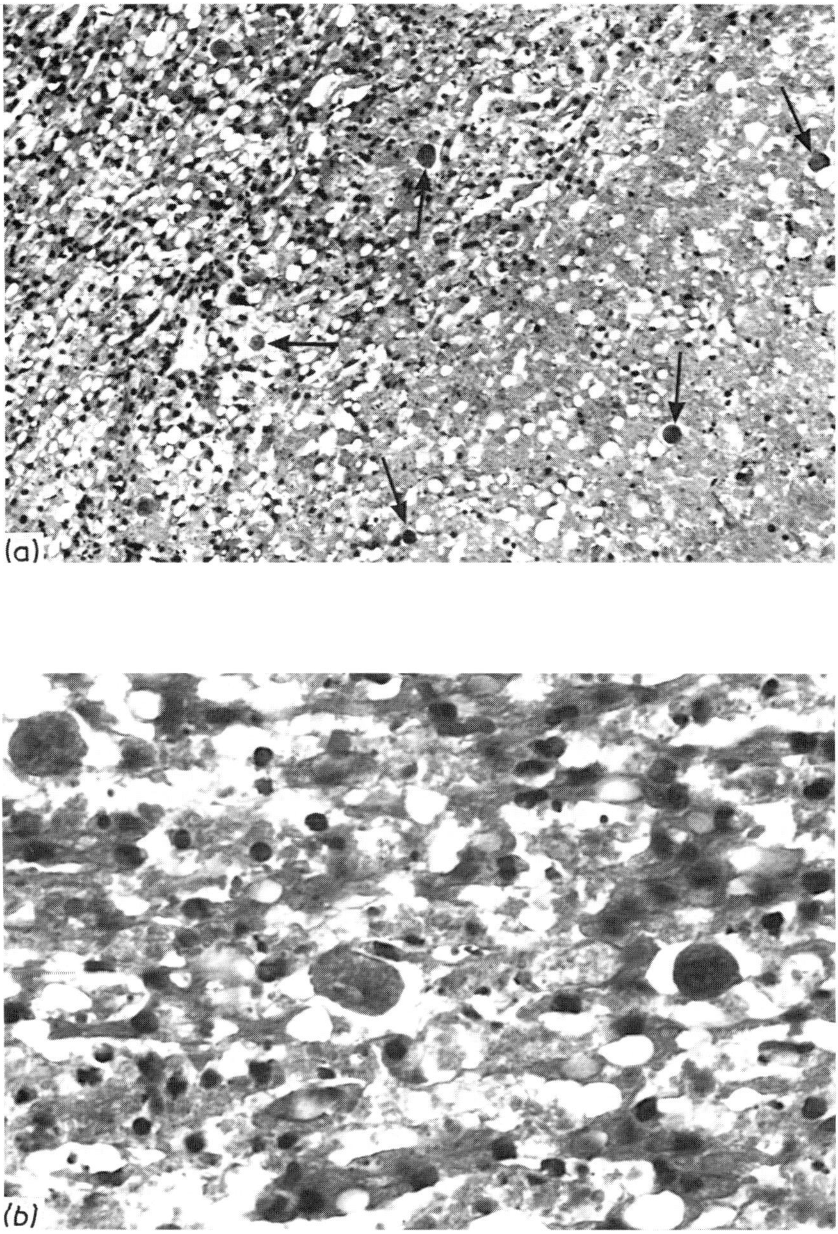

Fig. 5.9 (a) Amoebic abscess in liver. The centre of the abscess consists of necrotic debris surrounded by small numbers of lymphocytes. The amoebae are indicated by arrows. (b) A higher magnification showing several round or oval amoebae within degenerate hepatic tissue. ((a) HE, × 180; (b) HE, × 785.)

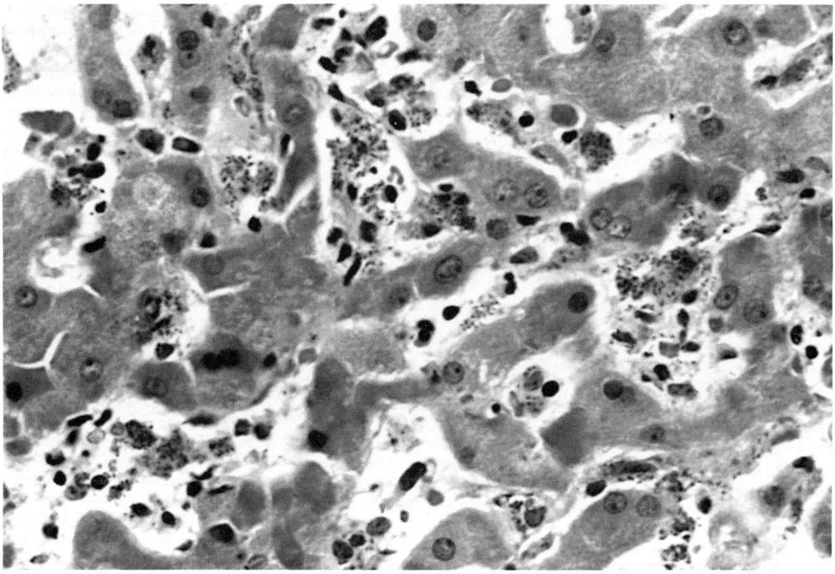

Fig. 5.10 Leishmaniasis (kala-azar) of liver. The protozoa have been engulfed by sinusoidal cells; there is no associated inflammation (HE, × 527).

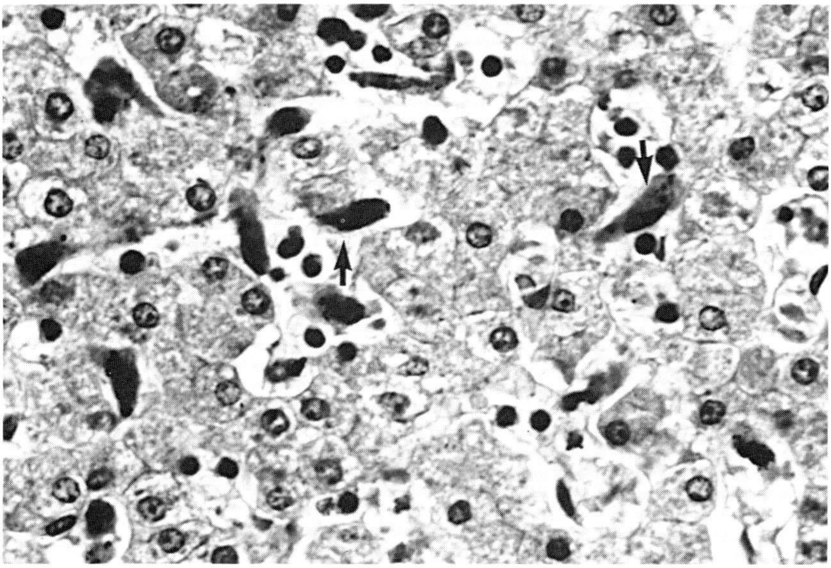

Fig. 5.11 The liver in malaria. Malarial pigment has been phagocytosed by sinusoidal cells (arrows). (HE, × 520).

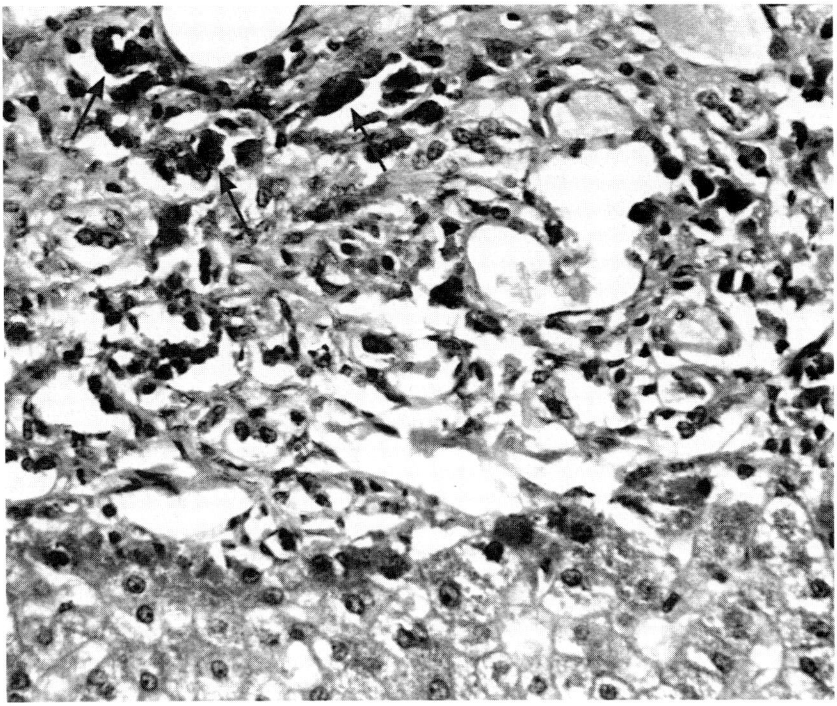

Fig. 5.12 The liver in malaria. When immunity is established malarial pigment is found in macrophages (arrows) in portal tracts (HE, × 440).

more diffuse. As immunity to the organism develops the sinusoidal pigment is redistributed and is then apparent in portal tract connective tissue (Fig. 5.12). Centrilobular necrosis and atrophy have also been described in chronic cases.

Tropical splenomegaly ('big spleen' disease) is associated with malarial infection. In this condition the hepatic sinusoids are packed with numerous small lymphocytes and monocytes (Fig. 5.13). This disease may be due to an abnormal immunological reaction to persistent malarial infection and sometimes responds to prolonged antimalarial therapy. Hepatic pigmentation is not a feature of this condition.

5.4.4 Toxoplasmosis

The infantile form of toxoplasmosis, inherited from the mother *in utero*, typically presents with encephalomyelitis, hydrocephalus and chorioretinitis. Many cases develop hepatomegaly and jaundice but morphological changes in the liver are rare; hence there is no satisfactory explanation of the functional abnormality.

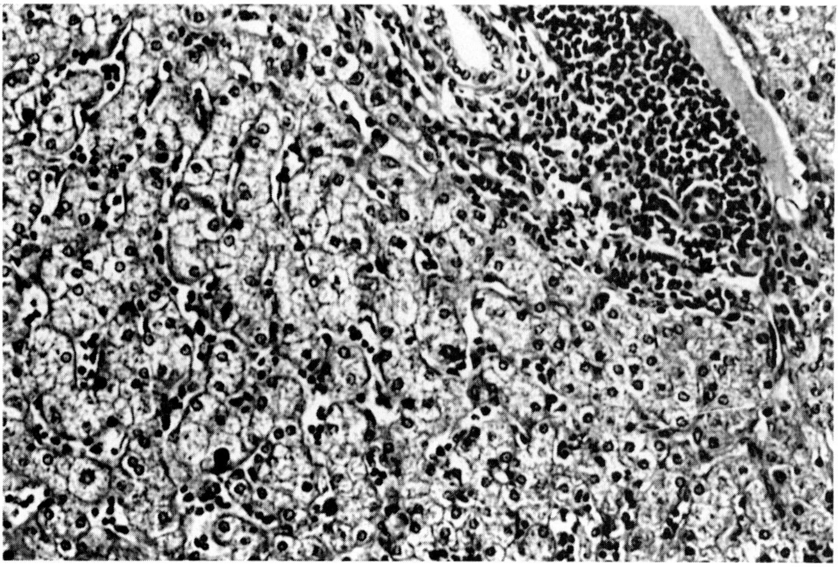

Fig. 5.13 'Big spleen' disease. There are numerous immunocytes in the hepatic sinusoids and in a portal tract (HE, × 180).

Liver involvement is rare in toxoplasmosis in adults. Clinically it resembles viral hepatitis, or infectious mononucleosis, heterophile antibodies being present in the latter. On liver biopsy, there is focal necrosis, diffuse infiltration of sinusoids with large mononuclears containing large lobulated nuclei and, occasionally, the organism can be seen in damaged hepatocytes and sinusoidal cells in Giemsa preparations (Weitberg *et al.*, 1979). Granulomas also occur but *Toxoplasma gondii* are not present in these lesions.

5.5 Helminthic infestations (Marcial-Rojas and Ramirez-Randa, 1982)

5.5.1 Ascariasis

In the Far East particularly, *Ascaris lumbricoides* may occasionally migrate and lodge in the common bile duct where it produces obstructive jaundice and ascending cholangitis. The larvae and ova may also gain access to the liver parenchyma eliciting a granulomatous reaction in which eosinophils are prominent.

The ascarides, *Toxocara canis* and *cati* are rarely responsible for liver disease in man. The larvae may produce a granulomatous hepatitis.

5.5.2 *Hydatid disease* (DeBakey and Jordan, 1982)

Echinococcus granulosus infection of the liver is, in a universal context, the most common cause of hepatic cysts. The dog is the definitive host with man, sheep and cattle being occasional intermediate hosts. Ingestion of the ova is commonest in children but hepatic cysts may not develop until adulthood. Hydatid cysts in the liver are usually single, unilocular and large (10–15 cm in diameter) and are composed of an outer layer of connective tissue, a middle acellular laminated layer and an inner germinal layer from which the worm grows into a central cavity containing degradation products of the helminth (Fig. 5.14). Rupture of these cysts into the biliary system or peritoneum are recognized complications.

Echinococcus multilocularis, which is common in Russia, produces a very different pathology. The helminth forms multiple small cysts surrounded by fibrous tissue. These lesions infiltrate into the liver substance and frequently masquerade as malignant tumours. When they occur around hepatic veins they cause portal hypertension.

5.5.3 *Liver flukes*

Millions of people in China, Vietnam and Hong Kong are infected by *Clonorchis sinensis*. *Opisthorchis viverrini* infections occur most commonly in Thailand. Both flukes colonize the biliary system resulting in hyperplasia of the biliary epithelium and periductal fibrosis infiltrated by eosinophils. Heavy infestation leads to obstructive jaundice, secondary bacterial cholangitis and abscess formation. A common sequel to the adenomatous hyperplasia of bile duct epithelium is cholangiocarcinoma.

Fasciola hepatica has a world-wide distribution and is chiefly a parasite of cattle and sheep. When ingested by man, the larvae migrate in the liver parenchyma producing areas of coagulation necrosis surrounded by eosinophils and polymorphs. Wedge biopsy is better at detecting the superficial larval tracks which radiate from pale necrotic nodules on the liver surface (Acosta-Ferreira *et al.*, 1979). The larvae eventually reach the biliary system where they produce obstructive jaundice and periductal fibrosis. Cases have occurred in the South of England from eating contaminated watercress.

5.5.4 *Schistosomiasis (bilharziasis)*

The liver frequently becomes involved in *Schistosoma japonicum* and *mansoni* infections and much less frequently with *haematobium* infestation. Although dead adult worms occasionally produce structural damage in the liver, the characteristic pathology is due to the presence of

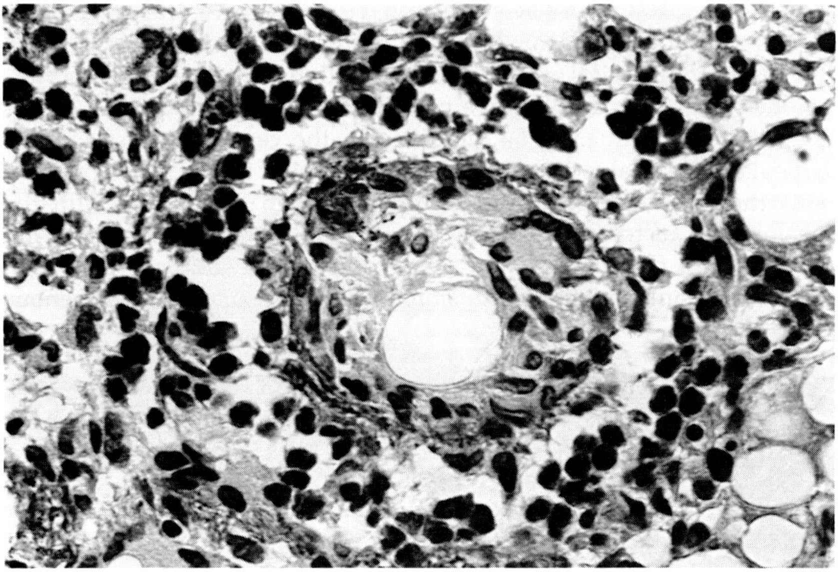

Fig. 5.16 Liver in Q fever. The small granuloma contains a central space ('doughnut' granuloma) and is delineated in part with thin wavy fibres of fibrin. The lesion has an outer cuff of inflammatory cells and the liver in this case is fatty (MSB, × 720).

multinucleate giant cells outside of which there is a fibrin ring (Fig. 5.16). The granulomas are associated with areas of focal necrosis. In some cases the necrotic cells elicit a polymorph and mononuclear response rather than a granulomatous reaction. Fibrin ring granulomas also occur in allopurinol-induced liver disease and occasionally in Hodgkin's disease (van der Stigel *et al.*, 1986).

5.7 Granulomatous hepatitis

Many of the infections discussed in this Chapter can produce a granulomatous hepatitis. This is characterized by the focal accumulation of epithelioid cells and multinucleate giant cells in the hepatic parenchyma or in portal tracts. The commonest causes in this country are tuberculosis, sarcoidosis, drug reactions and primary biliary cirrhosis. The causes of granulomatous hepatitis are listed in Table 5.1 Structural features helpful in differentiating these causes are listed in this Chapter. In about 30% of cases, it is impossible to discern the aetiological agent. In a case where granulomatous hepatitis is suspected, or in one which is rebiopsied, it is useful to send a piece of tissue for microbiological analysis.

Table 5.1 Causes of granulomatous hepatitis.*

Infections	Immunological disorders
Actinomycosis	Aids (atypical mycobacteria, TB, etc.)
Amoebiasis	Drugs (allopurinol, aspirin, cephalexin, diazepam, halothane, isoniazid, methyldopa, nitrofurantoin, oxyphenylbutazone, penicillin, phenytoin, quinidine, sulphonamides)
Ascaris	
Aspergillosis	
Blastomycosis	
Candidiasis	
Capillariasis	Granulomatous disease of childhood
Coccidioidomycosis	Polyarteritis nodosa
Cryptococcosis	Temporal arteritis
Cytomegalovirus	Wegner's granulomatosis
Fascioliasis	
Granuloma inguinale	
Giardiasis	Neoplasms
Histoplasmosis	
Hydatid disease (granulosa)	Hodgkin's disease
Infectious mononucleosis	Liver reaction to extrahepatic neoplasms
Kala-azar	Non-Hodgkin's lymphoma
Leprosy	
Listeriosis	
Melioidosis	Foreign material
Mucormycosis	
Nocardiosis	Barium, beryllium, intravenous drug abuse, mineral oil, silica, starch, suture material, talc, thorotrast
Opisthorchiasis	
Paracoccidioidomycosis	
Psittacosis	
Pyogenic infections	
Q fever	Others
Schistosomiasis	
Strongyloidiasis	Biliary obstruction
Syphilis (secondary and tertiary)	Crohn's disease
	Jejuno-ileal bypass
Toxoplasmosis	Lipogranulomas (alcohol-induced)
Toxocariasis	Primary biliary cirrhosis
Tuberculosis (and atypical mycobacteria)	Sarcoidosis
	Whipple's disease
Tularaemia	
Typhoid	

* See also Guckian and Perry (1966) and Ishak (1983).

In sarcoidosis, hepatic granulomas are found in as many as 66% of cases but liver dysfunction is rare. The granulomas may be found in any location within the liver but are commonest in portal tracts and around central veins. They are composed of epithelioid cells and multinucleate giant cells (Fig. 5.17). Schaumann bodies and asteroids are rare findings in sarcoidosis of the liver. Typically the granulomas do not undergo

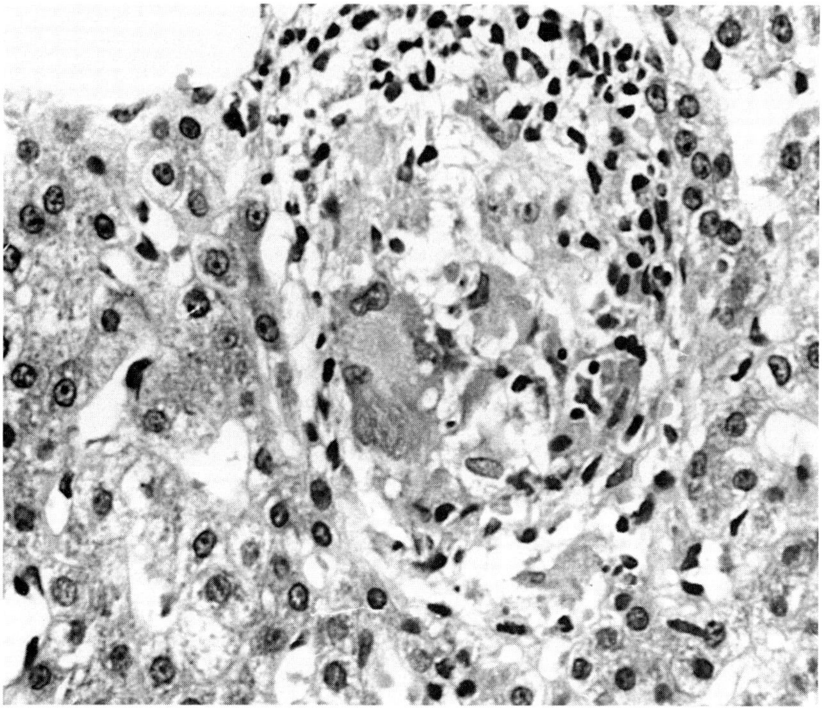

Fig. 5.17 Sarcoidosis of liver. This granuloma is typical of sarcoidosis in being composed of large histiocytes, a multinucleate giant cell and a peripheral rim of lymphocytes (HE, × 360).

necrosis but small foci of fibrinoid at their centres sometimes occur. On morphological grounds it is frequently impossible to differentiate the granulomas found in tuberculosis, primary biliary cirrhosis and other lesions which cause granulomatous hepatitis (see p. 125).

References

Acosta-Ferreira, W., Vercelli-Retta, J. and Falloni, L. M. (1979), *Fasciola hepatica* human infection. *Virchows Arch.* [*Pathol. Anat.*], **383,** 319–27.

Aikat, B. K., Bhusnvurmath, S. R., Pal, A. K. *et al.* (1979), The pathology and pathogenesis of fatal hepatic amoebiasis: a study based on 79 autopsy cases. *Trans. R. Soc. Trop. Med. Hyg.*, **73,** 188–92.

Bagby, G. C. and Gunning, J. J. (1978), Botryomycosis as an obstructive hepatic disease. *Arch. Intern. Med.*, **138,** 472–3.

DeBakey, M. E. and Jordan, G. L. (1982), Surgery of the liver, in *Diseases of the Liver*, 5th edn. (eds L. Schiff and E. R. Schiff), Lippincott, Philadelphia, pp. 1199–220.

Guckian, J. C. and Perry, J. E. (1966) Granulomatous hepatitis. Analysis of 63 cases and review of the literature. *Ann. Intern. Med.*, **65,** 1081–100.

Ishak, K. G. (1983) Granulomas of the liver, in *Pathology of Granulomas* (ed. H. L. Ioachim), Raven Press, New York, p. 307–69.

Lewis, J. H., Patec, H. C. and Zimmerman, H. J. (1982), The spectrum of hepatic candidiasis. *Hepatology*, **2,** 479–87.

Marcial-Rojas, R. A. and Ramirez-Randa, C. H. (1982), Parasitic diseases of the liver, in *Diseases of the Liver*, 5th edn (eds. L. Schiff and E. R. Schiff), Lippincott, Philadelphia, pp. 1165–98.

Pellegrin, M., Delsol, G. Auvergnat, J. C. *et al.* (1980), Granulomatous hepatitis in Q fever. *Hum. Pathol.*, **11,** 51–7.

van der Stigel, M., Safrani, E. S., Lejong, J. L. *et al.* (1986), Allopurinol hypersensitivity syndrome as a cause of hepatic fibrin-ring granulomas. *Gastroenterology*, **90,** 186–90.

Weitbert, A. B., Alper, J. C., Diamond, I. *et al.* (1979), Acute granulomatous hepatitis in the course of acquired toxoplasmosis. *N. Engl. J. Med.*, **300,** 1093–6.

6 Alcoholic liver disease

The incidence of chronic liver disease in a community bears a direct relationship to the prevalence of excessive alcohol consumption. In recent years there has been a considerable increase in this consumption in many Western countries and the pathologist can now expect to see examples of alcohol-induced hepatic damage frequently in hospital practice. Clinical histories from such patients particularly with regard to their addiction are notoriously unreliable and the initial suspicion of the disease may be expressed by the pathologist rather than the clinician. On the other hand one must be careful to avoid both the assumption that the changes described below invariably have this cause, and the temptation to disregard an apparently conflicting history.

There is no clear relationship between the severity of alcoholism and severity of liver damage and it is well known that many alcoholics, possibly the majority, have little or no hepatic dysfunction or structural abnormality in their livers; on the other hand, severe alcoholic liver damage unsuspected during life has been found at autopsy, and liver biopsy would appear to be the only reliable method for its detection and assessment of severity (Bruguera *et al.*, 1977). This may be explained by the multiplicity of pathogenic factors in alcoholic liver disease, such as an inherited susceptibility to liver fibrosis and cirrhosis in those with certain HLA types (Bell and Nordhagen, 1978) and a variable ability of hepatocytes to oxidize ethanol and its toxic metabolic product acetaldehyde (Thomas *et al.*, 1982).

In those alcoholic subjects with liver damage the characteristic histopathological features are fatty change, Mallory bodies and loss of hepatocytes (which tends to be maximal in centrilobular zones), infiltration by inflammatory cells including neutrophil polymorphs and, in advanced cases, cirrhosis which is usually but not invariably of the micronodular type. It is customary to recognize the broad classification of (a) alcoholic fatty liver without hepatitis, (b) alcoholic hepatitis with or without fatty change and (c) alcoholic cirrhosis.

It has been customary also to recognize clinically acute and chronic alcoholic liver disease. The chronic illness affects certain persons who have been imbibing alcohol over a period of many years and in amounts which have an energy equivalent to that of one-third or more of the total food intake but females may be susceptible to less than this. Fatty change is variable and regarded as a consequence of oxidation of alcohol within liver cells which leads to accumulation of both fatty acids and glycerol. More severe cellular damage and the inflammatory response which it evokes are probably caused by a toxic action of alcohol, and more particularly acetaldehyde, on hepatocyte membranes, although this is not always readily reproduced in experimental animals. It is now generally agreed that nutritional deficiencies are not essential to the pathogenesis of liver damage, although many chronic alcoholics are undernourished and show clinical improvement with dietary therapy. Many present with signs of portal hypertension which can occur in the absence of established cirrhosis.

Acute alcoholic liver disease is a severe illness with fever, jaundice and painful hepatomegaly which follows an episode of heavy drinking, usually as an acute exacerbation of the chronic condition. Mallory bodies and hepatitis are prominent and there may be centrilobular or submassive necrosis, possibly aggravated in severe or fatal cases by hypotensive or hypoglycaemic shock. In some acute cases severe fatty change is the only liver abnormality found, and a fatal outcome could be due to a variety of causes such as fat embolism, physical injuries resulting from inebriation or the effect of alcohol on the nervous or cardiovascular systems (Randall, 1980). Attacks of severe liver failure may hasten the development of chronic disease in those who survive them, but they are by no means essential for the development of extensive fibrosis or cirrhosis.

There is experimental evidence that hepatic steatosis in acute ethanol poisoning follows mobilization of fat from adipose tissue, but this has not been thoroughly investigated in human cases (Lieber, 1984). Chemical analysis of hepatic lipid helps to differentiate the fatty liver of acute from that of chronic alcoholism but histological examination is not helpful in this respect.

In some acute cases of alcoholic hepatitis there may be hyperlipidaemia and haemolytic anaemia associated with pancreatic injury (Zieve's syndrome) (Zieve, 1958); this diagnosis does not rest on biopsy appearances. In general there is an association between alcoholic liver disease and chronic, but not acute, pancreatitis (Dutta *et al.*, 1978).

Ethanol stimulates the hypertrophy of agranular endoplasmic reticulum (AER) and this could be a factor in hepatomegaly due mainly to steatosis and protein retention. Various drugs have a similar effect on AER (see Chapter 8), and hepatocytes not affected by steatosis or other

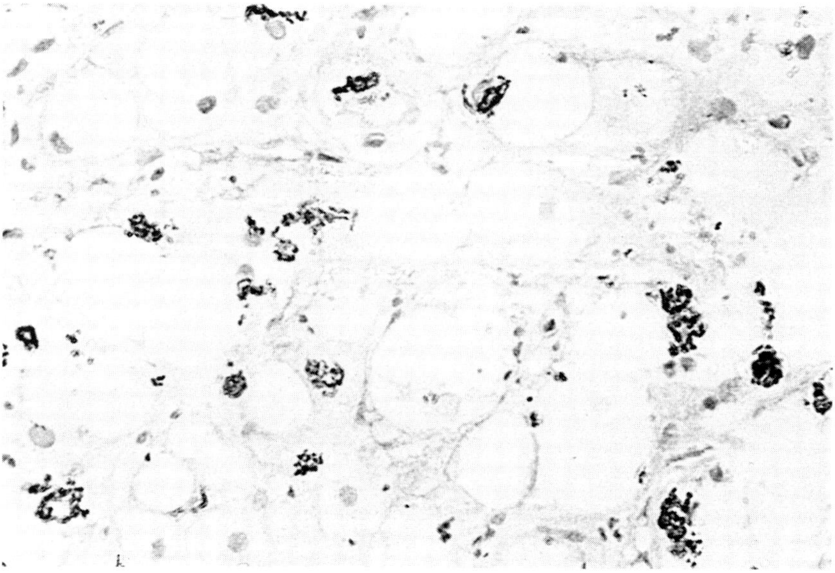

Fig. 6.9 Alcoholic hyalin stained by the immunoperoxidase procedure with antibody prepared from isolated Mallory bodies. This provides a more accurate assessment of the extent of this degenerative change compared with more conventional staining methods (× 400).

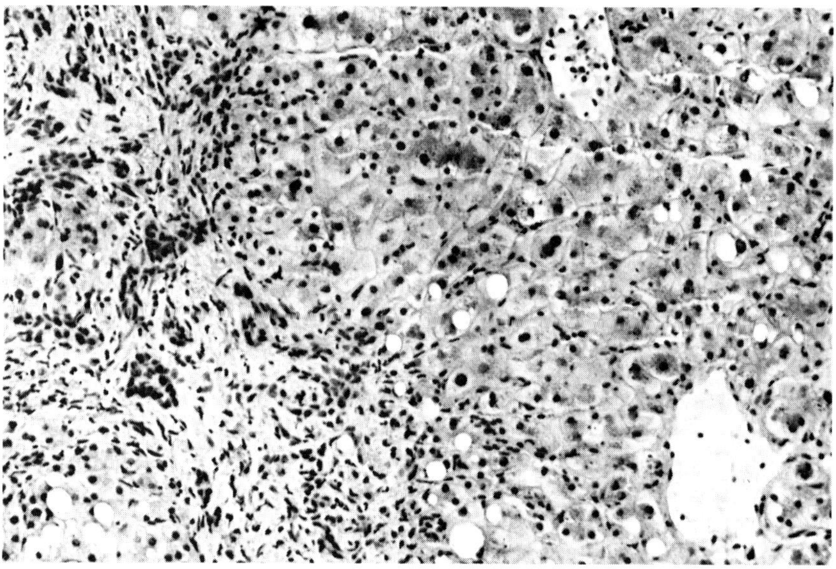

Fig. 6.10 Alcoholic hepatitis. Neutrophil polymorphs are included in the portal and periportal infiltrate. There are some swollen hydropic hepatocytes especially around centrilobular veins (HE, × 180).

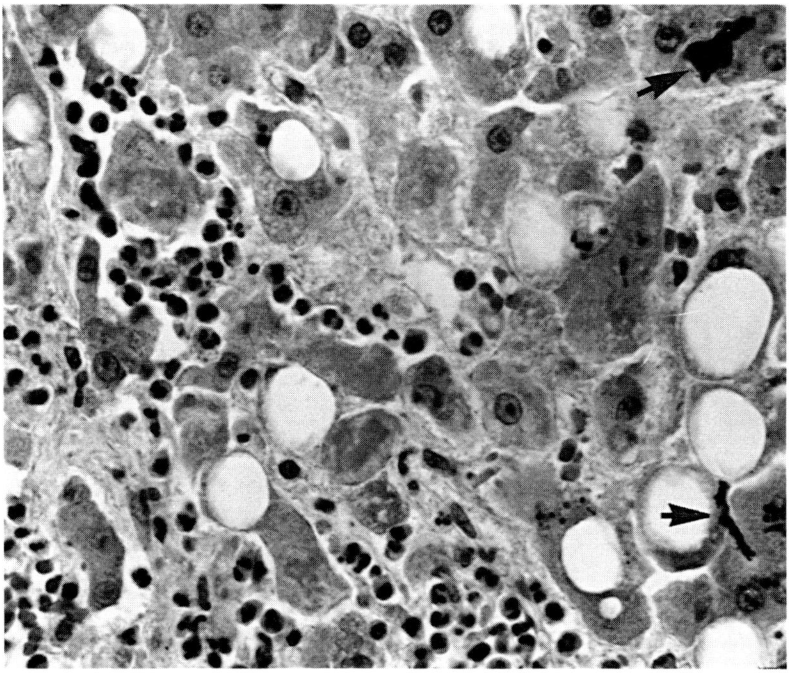

Fig. 6.11 Alcoholic hepatitis. There is infiltration by acute inflammatory cells adjacent to necrotic hepatocytes in which alcoholic hyalin is seen. Cholestasis is also present (arrow) (HE, × 560).

polymorphs and these can be numerous, especially around hepatocytes showing hyaline degeneration (Fig. 6.11). This suggests some positive chemotactic property of alcoholic hyalin, but the response is by no means present around all cells with this form of degeneration. It is difficult to relate intensity of inflammatory cells to intensity of liver damage, as it may be minimal in severe cases because of bone marrow depression.

Some degree of fibrosis is usually present and is indeed more common in alcoholic compared with other forms of hepatitis. It is present in both periportal and centrilobular zones, with development of bridging lesions between these two parts of the liver lobule (Fig. 6.12). There is often a diffuse involvement of sinusoidal walls (Fig. 6.13) that is well seen in sections stained for collagen, while Alcian blue preparations show excessive amounts of acidic ground substance.

Other less constant histological features of alcoholic hepatitis require additional description:

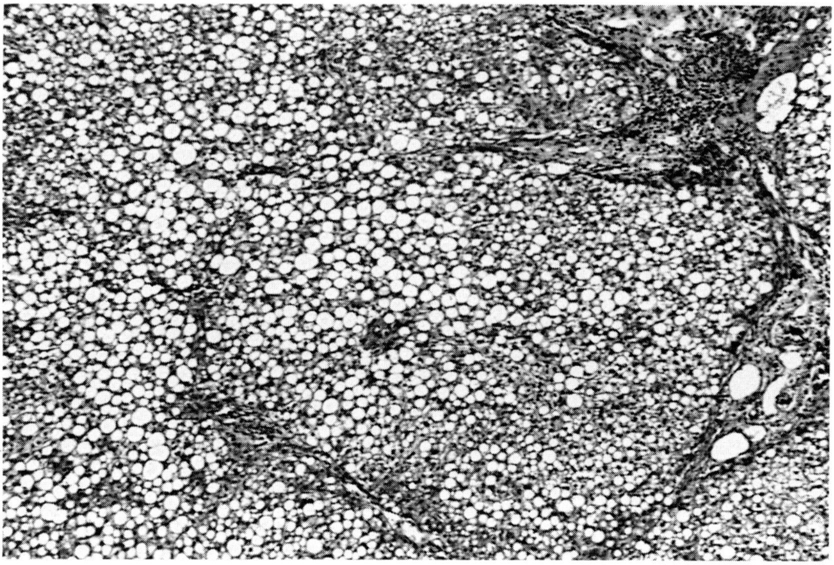

Fig. 6.12 Alcoholic hepatitis. Portal fibrosis and fatty change are prominent (HE, × 72).

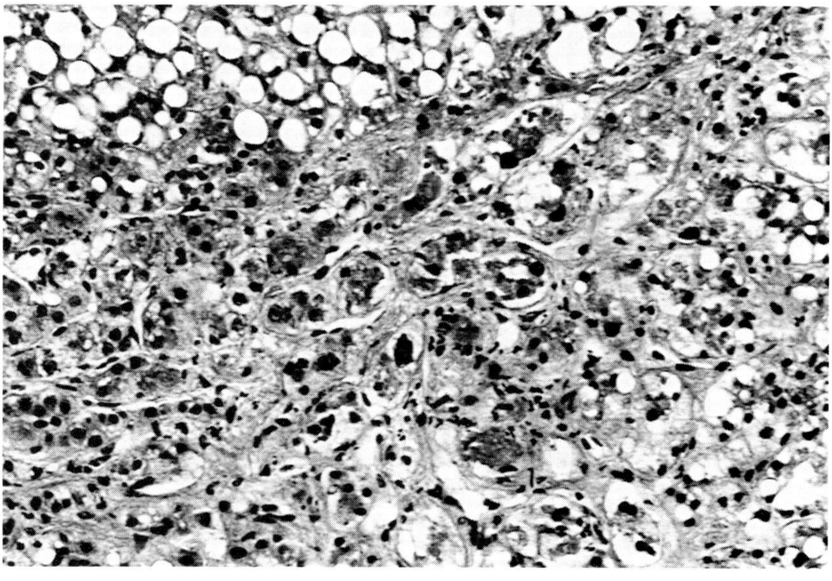

Fig. 6.13 Alcoholic hepatitis. This shows diffuse fibrous thickening of sinusoidal walls and pericellular fibrosis (HE, × 225).

(a) Central sclerosing hyaline necrosis

In some acute cases there are numerous Mallory bodies in centrilobular hepatocytes and sclerosis with constriction or obliteration of centrilobular veins and adjacent sinusoids (Fig. 6.14). Probably because of continuous heavy drinking there is little or no regenerative activity of surviving liver parenchyma, so that true cirrhosis need not be present; nevertheless, these patients suffer from severe ascites and other manifestations of portal hypertension. There is some resemblance to veno-occlusive disease of liver and we have occasionally observed subintimal proliferation of areolar tissue in hepatic veins rather similar to that which is characteristic of the vascular disease (Fig. 6.15).

(b) Chronic active hepatitis in alcoholism

In some cases of alcoholic hepatitis portal and periportal chronic inflammation with piecemeal loss of parenchyma are prominent. Moreover, some examples of CAH without the additional characteristic histological features of alcoholic hepatitis are found in alcoholics. There is evidence for immunological hyperactivity associated with antibody production to an alcoholic hyalin antigen and the development of lymphocyte toxicity to autologous liver cells which may be important in the progressive destruction of hepatic tissue in chronic alcoholism (Kakumu and Leevy, 1977). The T-cell population of liver containing Mallory bodies is increased, and these cells can be demonstrated in close apposition to necrotic hepatocytes (French *et al.*, 1979). Coexistence of other causes of liver disease such as persistent viral hepatitis must, of course, be considered. Carriers of HBsAg have raised susceptibility to alcohol-induced liver disease.

(c) Cholestasis in alcoholic hepatitis

Minor degrees of bile canalicular cholestasis with bile ductular proliferation are not infrequent (Fig. 6.11). Occasionally cholestasis is prominent and associated with clinical evidence of obstructive-type jaundice. Alcoholic pancreatitis may be a factor in causing common bile duct constriction.

(d) Liver haemosiderin in alcoholism

Haemosiderin deposition in hepatocytes is not uncommon in alcoholic hepatitis (Fig. 6.16). All degrees of severity can be seen, up to a condition indistinguishable from haemochromatosis; indeed it has been suggested

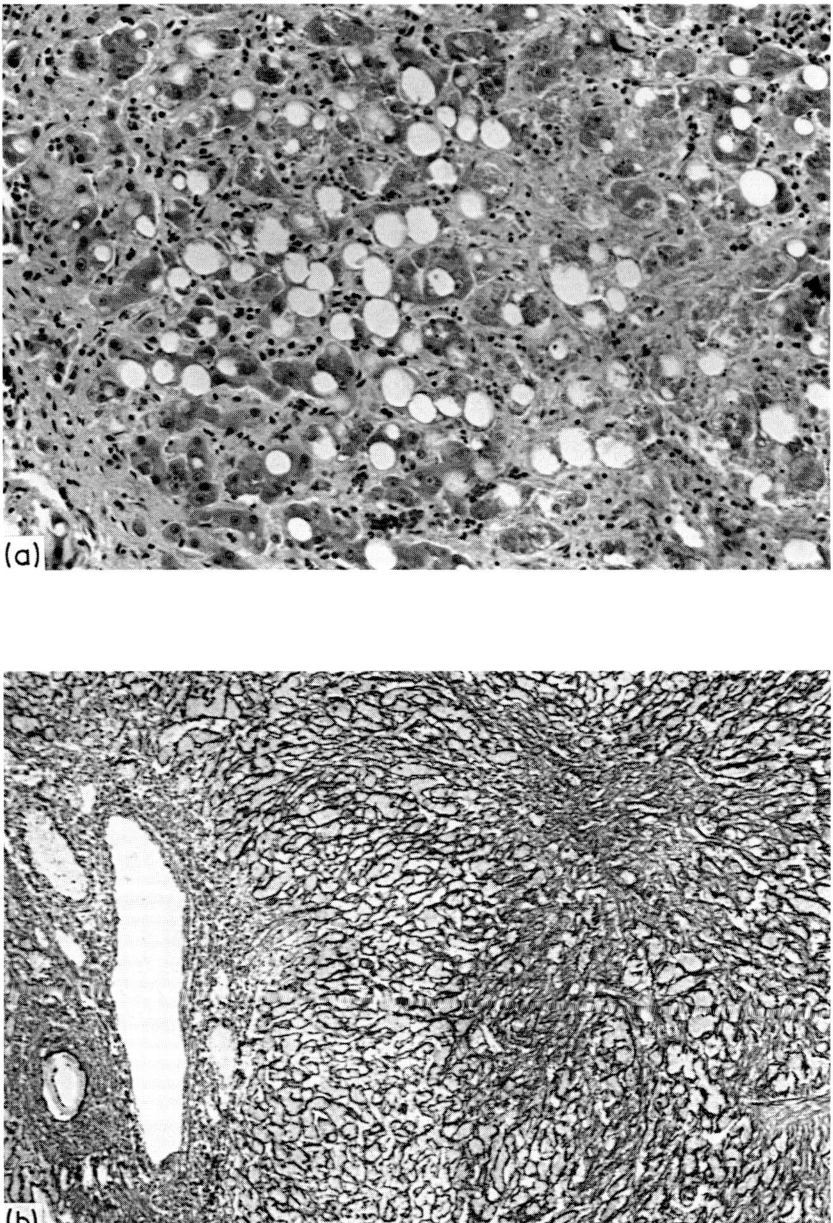

Fig. 6.14 Sclerosing hyaline necrosis: (a) There is hyaline degeneration and loss of hepatocytes in the centrilobular zone (right) where diffuse fibrosis completely obliterates the central vein. (b) A stellate scar occupies the centrilobular zone. ((a) HE, × 180; (b) reticulin stain, × 72.)

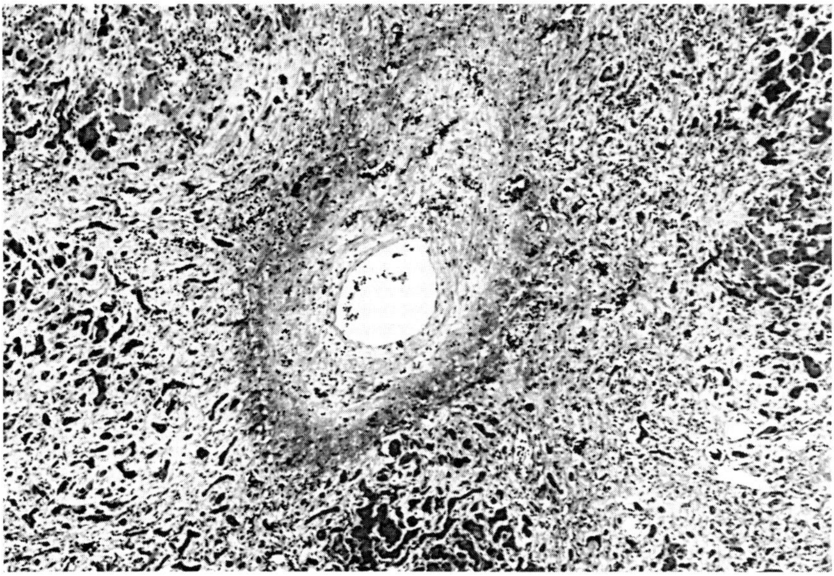

Fig. 6.15 Sclerosing hyaline necrosis. A large centrilobular vein shows subintimal fibrosis as in veno-occlusive disease (HE, × 72).

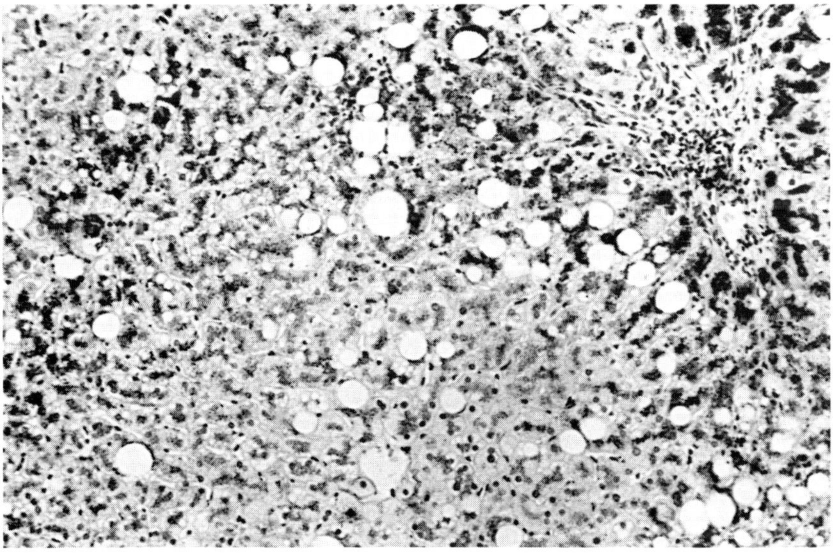

Fig. 6.16 Alcoholic hepatitis. The periportal cells contain dark deposits of haemosiderin (Perls' stain, × 140).

that haemochromatosis is really a form of alcoholic liver disease caused by consumption of beverages rich in iron, such as red wine. Considerable controversy still exists on this topic. Haemosiderosis is not essential to the development of fibrosis and cirrhosis in alcoholism.

In porphyria cutanea tarda, biopsy may show abundant haemosiderin in the liver as well as subacute hepatitis or cirrhosis. These changes may be due at least in part to alcoholism, which is commonly associated with this disease.

Occasionally haemosiderin pigment is seen mainly or entirely in Kupffer cells. We have noted this in biopsies from cases of Zieve's syndrome although the haemolytic component of this disease is not severe, the patient's jaundice being attributed to hepatocyte damage and cholestasis. Lipid may also be conspicuous in Kupffer cells in these cases.

6.1.3 Alcoholic cirrhosis

This is typically of the micronodular (regular) type, with persistence of fatty change (Fig. 6.17) which can vary from case to case and even between different nodules in the same specimen (Fig. 6.18). It should not be confused with fatty change in cirrhosis caused by ischaemia, when the central part of the nodule is most severely affected. The features of

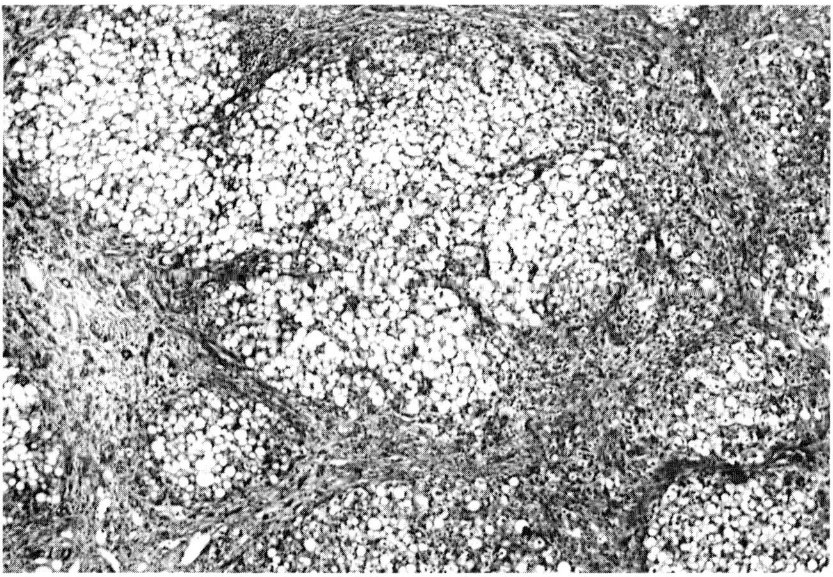

Fig. 6.17 Alcoholic cirrhosis. Fatty micronodular cirrhosis with persistent hepatitis and bile ductular proliferation (HE, × 56).

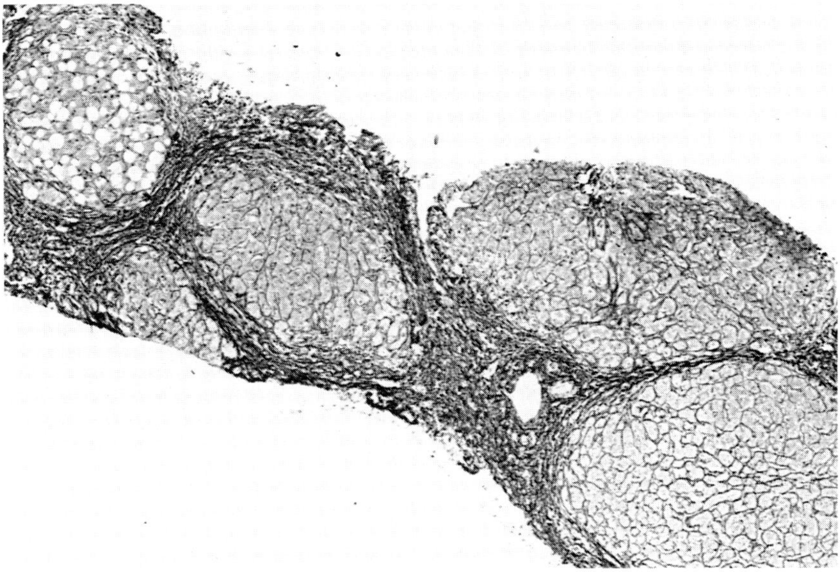

Fig. 6.18 Alcoholic cirrhosis seen in needle biopsy. Fatty change is absent apart from one regenerating nodule (reticulin stain, × 56).

hepatitis as already described may persist in alcoholic cirrhosis and fat granulomas may be present. Copper deposits can be detected in hepatocytes at the periphery of nodules; this feature is not related to cholestasis and is not present in alcoholic hepatitis or steatosis (Berresford *et al.*, 1980). The parenchymal nodules are small and contain no recognizable portal tracts.

With the passage of time fibrous septa become thicker and fat tends to disappear from the parenchymal nodules which expand to give an indeterminate or even macronodular pattern of cirrhosis. This is seen particularly following periods of abstention from alcohol.

6.2 Differential diagnosis

6.2.1 Alcoholic fatty liver

Numerous causes of fatty liver other than acute and chronic alcoholism are referred to in Chapter 9. In many of these the aetiology cannot be assessed from histopathology alone.

6.2.2 *Alcoholic hepatitis*

The clinical features and course of viral hepatitis are not dissimilar to alcoholic hepatitis in many respects; thus both may pursue a fulminating course or be characterized by persistent or progressive liver disease, sometimes with cholestasis and sometimes terminating in cirrhosis. In the early stages at least viral hepatitis differs structurally from alcoholic hepatitis by the absence of large droplet fatty change, while the inflammatory cell infiltrate is predominantly or entirely lymphocytic. Mallory bodies in hepatocytes are not a feature of viral infection but neither are they present in all cases of alcoholic liver damage. It is of course possible for the two conditions to coexist and it may be, in view of the difficulties in experimental production of alcoholic hepatitis, that liver-damaging factors such as hepatotrophic virus are necessary for the development of the human lesion, but this is unlikely.

The possibility that alcoholism is a cause of intrahepatic cholestasis has to be borne in mind. The presence of Mallory bodies may not be helpful in the diagnosis of such cases as they may be seen in some cases of primary biliary cirrhosis, although in the latter condition they are typically in a periportal location. They are rarely seen also in secondary biliary cirrhosis.

As long as the pathogenesis of hepatocyte damage in alcoholism remains obscure there can be no explanation for similar and even identical appearances in the liver in a number of apparently unrelated conditions discussed more fully in other chapters and which must also be considered in the differential diagnosis. These include:

(a) Chronic intestinal disease

Perhaps the best example is the liver damage which complicates certain cases of obesity treated by jejuno-ileal shunt and in which the histopathology is identical to severe alcoholic hepatitis (see Fig. 7.6). Other conditions include occasional cases of chronic inflammatory colonic disease and gastrocolic fistula.

(b) Pancreatitis

This is difficult to assess as many cases occur in alcoholics.

(c) Nutritional deficiency

Apart from kwashiorkor and the chronic digestive disorders referred to above, there is no evidence that malnutrition by itself is a cause of serious liver disorder in man. Moreover, it is recognized that alcoholic hepatitis can develop despite adequate food intake. Minor degrees of hepatitis and portal fibrosis may be found in the fatty livers of kwashiorkor (see Fig. 7.2) but may be a consequence of various infections to which these children are susceptible because of associated immunological deficiency. Cirrhosis is not a complication of kwashiorkor.

(d) Long-term administration of other substances which have a toxic effect on the liver, causing fatty change, chronic hepatitis, and eventually cirrhosis, as in experimental animals given repeated doses of carbon tetrachloride

This is most unlikely in human subjects but a similar condition may be found, for example, among patients with psoriasis treated for long periods with methotrexate. Compared with alcoholic hepatitis there may be a better defined periportal distribution of fatty change. Patients with angina pectoris treated with perhexilene may suffer liver damage similar to alcoholic hepatitis.

(e) Inherited metabolic disorders

Fatty liver is a feature of a number of these disorders and the following, which are more fully described in Chapter 9, may show also evidence of hepatitis and bear a close morphological resemblance to alcoholic liver disease:

(i) Wilson's disease. While this is unlikely to be confused with alcoholism on clinical grounds, it is interesting to note that, in some cases, fatty liver is present during the preclinical phase and is followed by hepatitis, hepatocyte degeneration, including the presence of Mallory bodies, and fibrosis leading to cirrhosis. Mallory bodies are found more frequently in periportal hepatocytes in Wilson's disease.

(ii) Porphyria cutanea tarda (p. 142, 200)

(iii) Chronic alpha-1-antitrypsin deficiency. Careful examination of the biopsy should reveal the characteristic inclusion bodies in periportal hepatocytes, and the possibility should be borne in mind in patients with coexisting chronic pulmonary disease.

(iv) Diabetes mellitus. Rare cases of maturity-onset diabetes and some cases of partial or complete lipodystrophy (lipo-atrophic diabetes).

(v) Abetalipoproteinaemia

6.2.3 *Alcoholic cirrhosis*

The majority of cases of micronodular cirrhosis can be attributed to alcohol, but it is possible that a few represent a late stage of the conditions mentioned under the differential diagnosis of alcoholic hepatitis. Steatosis is not invariably present and in these particular cases there is a histological resemblance to Indian childhood cirrhosis which is a non-fatty micronodular cirrhosis of unknown aetiology in which Mallory bodies are numerous. The cirrhosis of idiopathic haemochromatosis may be micronodular in type and the possible relationship of haemosiderin deposition to alcoholism has already been noted.

6.3 Prognosis

Fatty change by itself is probably unimportant in the development of progressive liver disease and is entirely reversible. Other forms of hepatocyte degeneration have more significance and a guarded prognosis should be given when they are prominent in liver biopsies, especially in acute cases. Compared with steatosis, hepatitis is much more resistant to abstention from alcohol, but recovery is possible and, in this respect, the prognosis is better than for other types of chronic hepatitis. It is very difficult to assess the risk of cirrhosis in individual cases. Prospective studies of human alcoholic liver disease are rare and one recent investigation (Sørensen *et al.*, 1984) found that only about 15% of 258 male alcoholics developed this complication over a 10–13 year period, a result unrelated to the severity or duration of alcoholism. In experimental alcoholic liver injury and in retrospective biopsy studies in man it has been suggested that pericentral venous fibrosis may be the lesion which determines whether liver disease in a particular alcoholic will progress to cirrhosis (van Waes and Lieber, 1977), but our experience of human cases indicates that it can regress (Fleming and McGee, 1984). Even in the absence of cirrhosis which, when established, is irreversible, there can be portal hypertension of prognostic significance. This may be a consequence of severe sinusoidal compression by hepatocytes swollen by fat and protein retention and AER hypertrophy (Vidins *et al.*, 1985). Portal hypertension can be severe in sclerosing hyaline necrosis which carries a serious prognosis because of this and because of parenchymal damage and a tendency to hepatic failure. Primary liver cell cancer may occur in chronic alcoholics, usually in those cases with macronodular cirrhosis or in association with haemochromatosis.

References

Barbatis, C., Morton, J., Woods, J. C. *et al.* (1986) Disorganisation of intermediate filament structure in alcoholic and other liver diseases. *Gut*, **27**, 765–70.

Bell, H. and Nordhagen, R. (1978), Association between HLA-BW40 and alcoholic liver disease with cirrhosis. *Br. Med. J.*, **1**, 822.

Berresford, P. A., Sunter, J. P., Harrison, V. *et al.* (1980), Histological demonstration and frequency of intrahepatocytic copper in patients suffering from alcoholic liver disease. *Histopathology*, **4**, 637–43.

Bruguera, M., Bordas, J. M. and Rodes, J. (1977), Asymptomatic liver disease in alcoholics. *Arch. Pathol. Lab. Med.*, **101**, 644–7.

Denk, H., Franke, W. W., Eckerstorfer, R. *et al.* (1979), Formation and involution of Mallory bodies ('alcoholic hyalin') in murine and human liver revealed by immunofluorescence microscopy with antibodies to prekeratin. *Proc. Nat. Acad. Sci.* USA, **76**, 4112–6.

Dutta, S. K., Mobrahan, S. and Iber, F. L. (1978), Associated liver disease in alcoholic pancreatitis. *Am. J. Digest. Dis.*, **23**, 618–22.

Fleming, K. A., and McGee, J O'D. (1984), Alcohol induced liver disease. *J. Clin. Pathol.*, **37**, 721–33.

French, S. W., Burbige, E. J., Tarder, G. *et al.* (1979), Lymphocyte sequestration by the liver in alcoholic hepatitis. *Arch. Pathol. Lab. Med.*, **103**, 146–52.

Kakumu, S. and Leevy, C. M. (1977), Lymphocyte cytotoxicity in alcoholic hepatitis. *Gastroenterology*, **72**, 594–7.

Morton, J. A., Fleming, K. A., Trowell, J. M. and McGee, J. O'D. (1980), Mallory bodies; immunohistochemical detection by antisera to unique non-prekeratin components. *Gut*, **21**, 727–33.

Randall, B. (1980), Fatty liver and sudden death: a review. *Hum. Pathol.*, **11**, 147–53.

Sørensen, T. I. A., Orholm, M., Bentsen, K. D. *et al.* (1984), Prospective evaluation of alcohol abuse and alcoholic liver injury in men as predictors of development of cirrhosis. *Lancet*, **ii**, 241–4.

Thomas, M., Halsall, S. and Peters, T. J. (1982), Role of hepatic acetaldehyde dehydrogenase in alcoholism: demonstration of persistent reduction of cytosolic activity in abstaining patients. *Lancet*, **ii**, 1057–8.

Uchida, T., Kronborg, I. and Peters, R. L. (1984), Giant mitochondria in the alcoholic liver diseases. *Liver*, **4**, 29–38.

van Waes, L. and Lieber, C. S. (1977), Early perivenular sclerosis in alcoholic fatty liver. *Gastroenterology*, **73**, 646–50.

Vidins, E. I., Britton, R. S., Medline, A. *et al.* (1985), Sinusoidal caliber in alcoholic and nonalcoholic liver disease: diagnostic and pathogenic implications. *Hepatology*, **5**, 408–14.

Zieve, L. (1958), Jaundice, hyperlipemia and hemolytic anemia: a heretofore unrecognised syndrome associated with alcoholic fatty liver and cirrhosis. *Ann. Intern. Med.*, **48**, 471–96.

Further reading

Hall, P. (ed.) (1985), *Alcoholic Liver Disease: Pathobiology, Epidemiology and Clinical Aspects*, Edward Arnold, London.

Lieber, C. S. (1984), Alcohol and the liver: 1984 update. *Hepatology*, **4**, 1243–60.

Review by an International Group (1981), Alcoholic liver disease: morphological manifestations. *Lancet*, **i**, 707–11.

7 Nutritional and digestive disorders

In view of the major role of the liver in the metabolism of protein, carbohydrate and fat, it would be reasonable to expect serious disorders of hepatic function, possibly with structural abnormalities, in patients suffering from malnutrition. This concept has been supported by the results of dietary experiments in laboratory animals; deficiency of certain amino acids in the food of young rodents leads to extensive liver necrosis which can be either fatal or heal with gross scarring of the organ; fatty change resulting in a finer type of cirrhosis develops from feeding diets deficient in certain lipotrophic factors, such as choline. The commonest cause of human chronic liver disease in the Western world is alcoholism, and some of these cases are clearly malnourished.

Despite these observations it seems that the role of malnutrition in human liver disease has been overemphasized. It is now apparent that the dietary experiments referred to have little relevance to human disease, and alcoholic liver damage can occur in the presence of adequate food intake. Nevertheless, there are a small number of conditions which have a nutritional basis for their development although the exact pathogenesis is unclear; for example, the liver damage associated with intestinal shunts and parenteral feeding. Similarly, nutritional factors cannot be excluded completely from consideration of the hepatic complications of diseases of other digestive organs, especially chronic inflammatory bowel disease, and these matters are considered in this Chapter.

7.1 Deficient food intake

7.1.1 Starvation

In marasmic patients with loss of body fat the liver is reduced in size and dark in colour. There is no histological evidence of hepatocyte abnormality apart from atrophy so that thin plates of liver cells are separated by

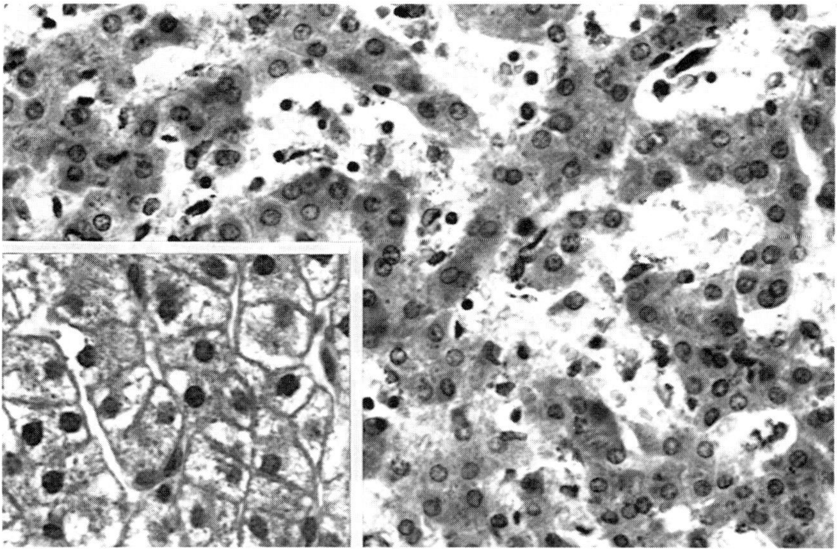

Fig. 7.1 Liver in marasmus. The cell plates are shrunken and separated by wide sinusoidal channels. (Insert) Normal liver at same magnification. (HE, × 400).

relatively wide sinusoidal channels (Fig. 7.1). Fatty change is usually minimal or absent and lipochrome pigment may be prominent, especially in centrilobular cells.

7.1.2 Protein-energy malnutrition (kwashiorkor)

This occurs in young children soon after weaning, who are fed a diet rich in starches or sugar and grossly deficient in protein. The liver is usually enlarged and pale due to fatty change which may be extreme. At first periportal hepatocytes show cytoplasmic fat as small vacuoles, these quickly coalesce and displace the nucleus to the periphery of the cell. The condition may spread to involve midzonal and centrilobular cells so that, in severe cases, the histological appearance of the liver resembles adipose tissue (Fig. 7.2). Hepatocyte necrosis is not a feature and hyaline degeneration is described rarely. Minor degrees of periportal and septal fibrosis are seen in some cases, but the condition is not regarded as pre-cirrhotic. Accordingly, liver dysfunction is not a prominent clinical feature, and the high morbidity and mortality are related much more to secondary infection associated with acquired immune deficiency. Histological examination of liver biopsies may show evidence of this, e.g. cytomegalic inclusion disease. The degree of hepatic steatosis seems to vary in different parts of the world and may reflect the predominant

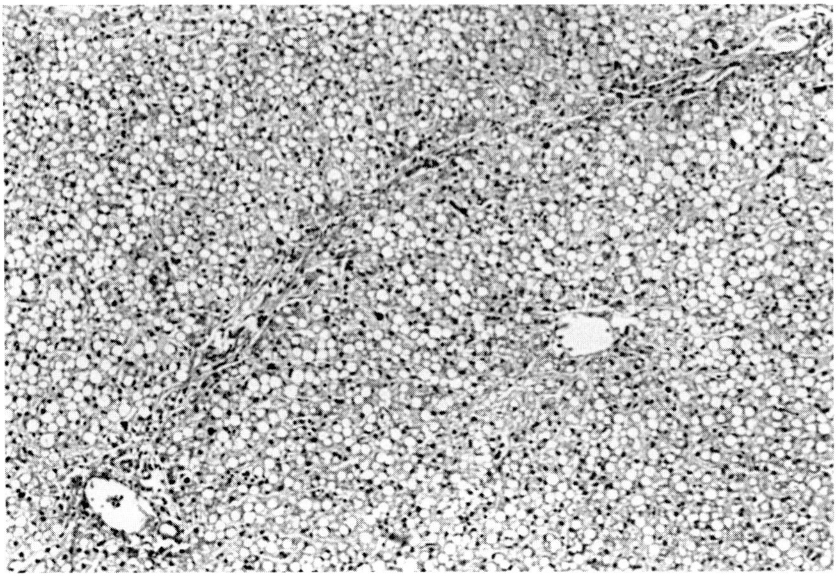

Fig. 7.2 Extensive fatty change in kwashiorkor. There is a tendency in this case to septal fibrosis. A few black granules of malaria pigment are also present (HE, × 102).

carbohydrate component of the diet, being high where this consists of sucrose rather than starch (Patrick *et al.*, 1973).

7.1.3 Illnesses characterized by severe vomiting and diarrhoea

This is well exemplified in young infants suffering from gastroenteritis often attributed to pathogenic strains of *E.coli*. As in kwashiorkor there is fatty change most marked in periportal hepatocytes (Fig. 7.3), the development of which could be aggravated by feeds of sucrose solution sometimes given to these babies. Probably it has little clinical significance and will resolve in a matter of weeks following recovery from infection and the resumption of normal feeding.

7.2 Obesity and related conditions

7.2.1 Obesity

Fatty liver may be found in persons who are overweight but is severe in only a minority of cases. Fatty change is often most prominent in centrilobular hepatocytes (Fig. 7.4). Some biopsies may show also

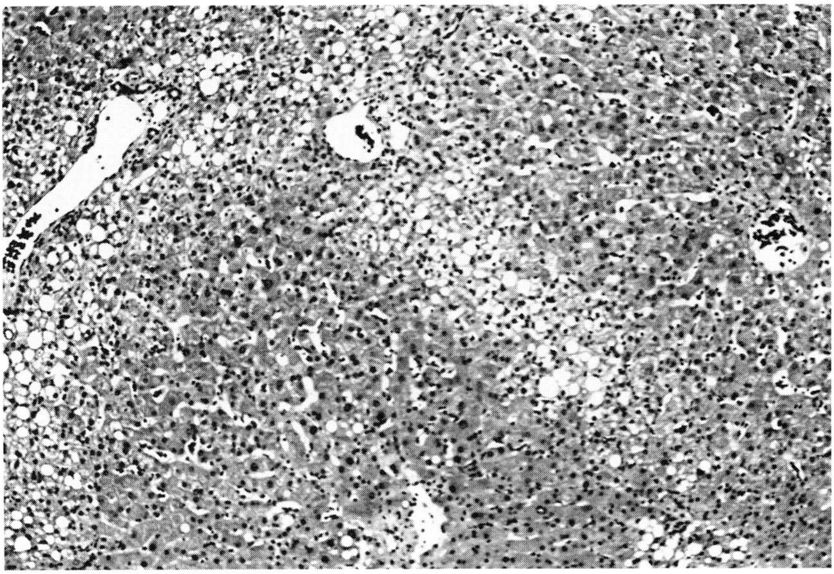

Fig. 7.3 Periportal fatty change in gastroenteritis (HE, × 100).

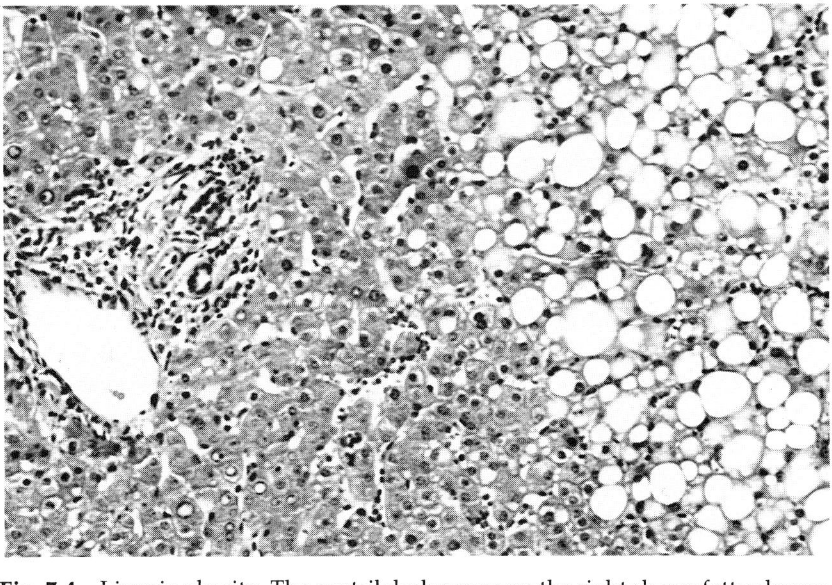

Fig. 7.4 Liver in obesity. The centrilobular zone on the right shows fatty change. A few periportal hepatocyte nuclei are vacuolated (HE, × 180).

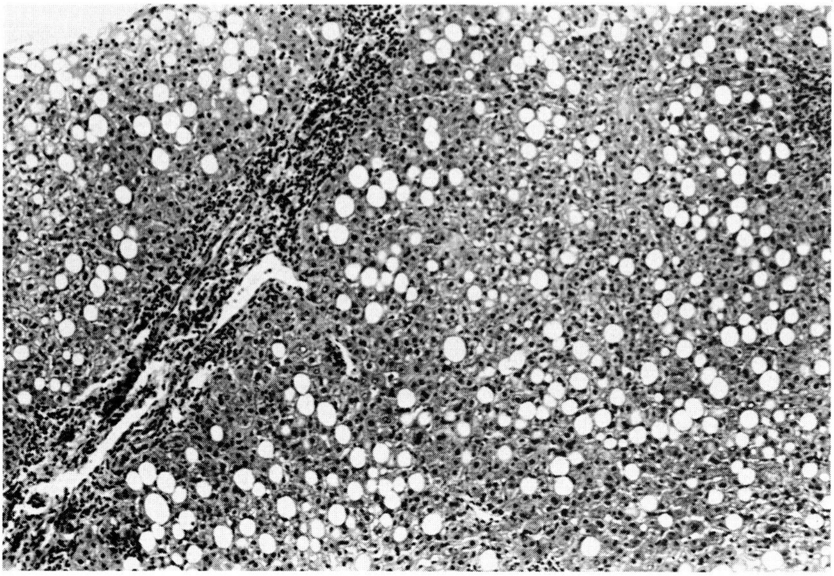

Fig. 7.5 Liver in obesity. In this case fatty change has a more diffuse distribution in the liver compared to Fig. 7.4, and there is mild portal inflammation (HE, × 125).

non-specific reactive hepatitis in the form of mild portal and focal inflammation (Fig. 7.5), the latter including fat granulomas (p. 130). Mallory bodies are reputed to occur in morbidly obese patients but this may be due to concomitant alcoholic abuse (Gluud *et al.*, 1984).

Similar changes may be found in association with *maturity-onset diabetes mellitus* and are related possibly to the obesity which frequently is a feature of this illness (Hornbøll and Olsen, 1982).

7.2.2 Jejuno-ileal shunt

This operation is performed as treatment for morbid obesity, mainly in young adults, and is often effective in procuring considerable weight loss. Unfortunately, in about a third of cases it is complicated by significant liver damage which can be fatal (Holzbach, 1977), and for this reason the practice has been discontinued in most medical centres.

Liver damage is most severe within a few months of operation when weight loss is maximal. Thereafter it may regress but cirrhosis has been recorded. The commonest change is the appearance or exaggeration of steatosis, but hepatitis may be present also, often accompanied by Mallory bodies and necrosis of hepatocytes and indistinguishable from

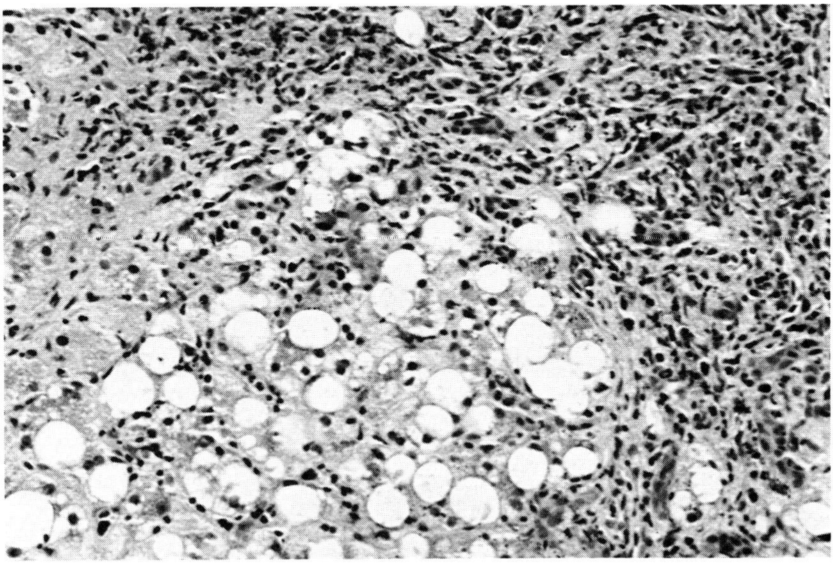

Fig. 7.6 Liver damage following jejuno-ileal shunt. Hepatocytes show fatty and hydropic change and a few indistinct Mallory bodies are present. There is extensive portal and periportal inflammation with fibrosis. The appearances are similar to those of severe alcoholic hepatitis, but there was no history of alcoholism in this case (HE, × 225).

severe alcoholic hepatitis (Fig. 7.6). In addition, there may be non-caseating granulomas not typical of fat granulomas.

Liver biopsy is important in assessing the severity of this liver damage, as its progress may be halted or reversed by taking down the shunt. Also biopsy is usually undertaken before operation as severe liver damage may develop more readily in patients with pre-existing mild hepatitis, but this is not invariably so.

Similar liver damage can complicate alimentary tract shunts from other causes such as *gastrocolic fistula* complicating peptic ulceration. It is described in patients with extensive *small bowel resection* (Puera *et al.*, 1980), an observation which throws doubt on the significance of absorption of toxins from non-functioning bowel as a cause of liver injury, but it would be inappropriate to discuss the complex pathogenesis here.

7.3 Parenteral feeding

Cholestatic jaundice has been observed from time to time in young, often premature, infants during parenteral feeding usually necessitated by abdominal surgery. A similar and probably identical condition occurs less frequently in older children and in adults undergoing this therapy (Salvian and Allardyce, 1980). Jaundice is uncommon during the first and second weeks of parenteral feeding and its occurrence thereafter bears some relationship to duration of therapy. Liver biopsy (Hughes *et al.*, 1983) shows centrilobular bile capillary cholestasis which may be the only abnormality. Portal inflammation and hydropic degeneration of hepatocytes may be present, the latter not confined to areas of cholestasis (Fig. 7.7). Kupffer cells are prominent, containing bile pigment and lipid material derived, possibly, from the infusion mixture. Bile ductular proliferation and cholestasis can occur but duct proliferation is rare (cf. extrahepatic biliary atresia). Other variable histological features are fatty change and giant cell transformation of hepatocytes.

The pathogenesis is unknown and may be multifactorial, which could explain the variable histological appearances (Bernstein *et al.*, 1977). The resolution of liver changes soon after discontinuation of parenteral feeding is suggestive of damage induced by some component or components of the infusion mixture. Secondary biliary cirrhosis can supervene if the feeding continues for long periods in the presence of cholestatic jaundice (Whitington, 1985).

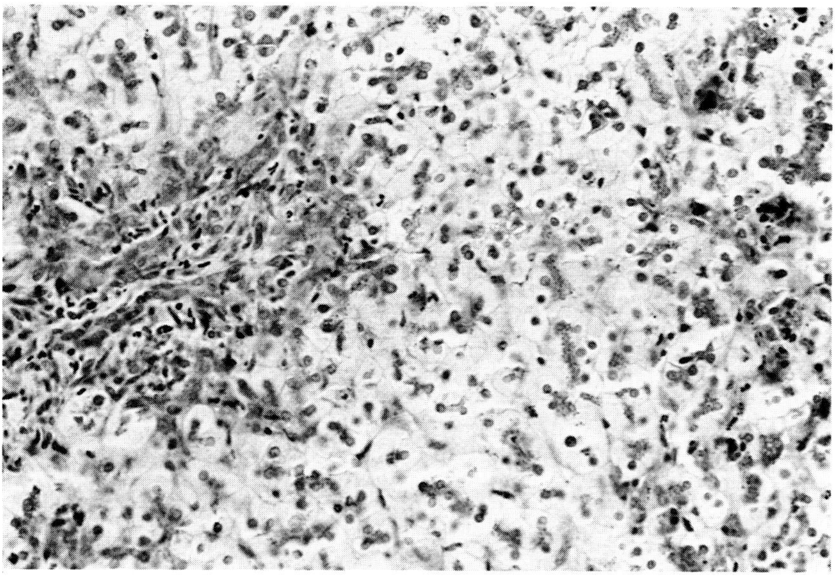

Fig. 7.7 Cholestatic hepatitis related to parenteral feeding. Centrilobular cholestasis is present (right). There is fairly acute portal inflammation (left) with some fibrosis and bile ductular proliferation. Hepatocytes are pale and swollen. From a 16-year-old girl who developed jaundice soon after commencing parenteral feeding following small intestinal resection (HE, × 247).

7.4 Other abdominal disease

7.4.1 Pancreatitis

Chronic pancreatitis may be accompanied by fatty liver and hepatitis, but in many cases the association could be explained by the effects of alcohol on both organs.

Acute haemorrhagic pancreatitis and extensive hepatic necrosis can coexist but the explanation is unclear. The former may also occur with cholangitis, no doubt related to common bile duct obstruction by gallstones.

7.4.2 Intestinal malabsorption

Associations with non-specific reactive hepatitis, CAH and cirrhosis have all been noted, but being of infrequent occurrence, their significance is uncertain.

The PAS-positive inclusion bodies characteristic of Whipple's disease have been noted in Kupffer cells.

7.4.3 Chronic inflammatory bowel disease

Apart from specific infections such as amoebiasis, there is a well-established association between hepatobiliary disease and both ulcerative colitis and Crohn's disease; about 10% of patients with either condition suffer from one or more of the following abnormalities of liver or bile ducts:

(a) Small duct primary sclerosing cholangitis (pericholangitis)

Chronic ulcerative colitis (UC) is associated with what was regarded as a benign condition, via. 'pericholangitis'. The latter term was used clinically to describe a poorly defined chronic hepatitis associated with ulcerative colitis. Because of the vagueness of defintion of pericholangitis it was omitted from the nomenclature recommended by the International Association for the Study of the Liver (Leevy, Popper and Sherlock, 1976). Wee and Ludwig (1985) have recommended, with good reason, that the term small duct primary sclerosing cholangitis (PSC) be used to replace what was referred to as pericholangitis.

Small duct primary sclerosing cholangitis occurs in 35% of UC with hepatobiliary disease. It is characterised by mild portal and periportal hepatitis with portal oedema. Changes in small interlobular bile ducts include: pleomorphic cholangitis in which there is degenerative change in

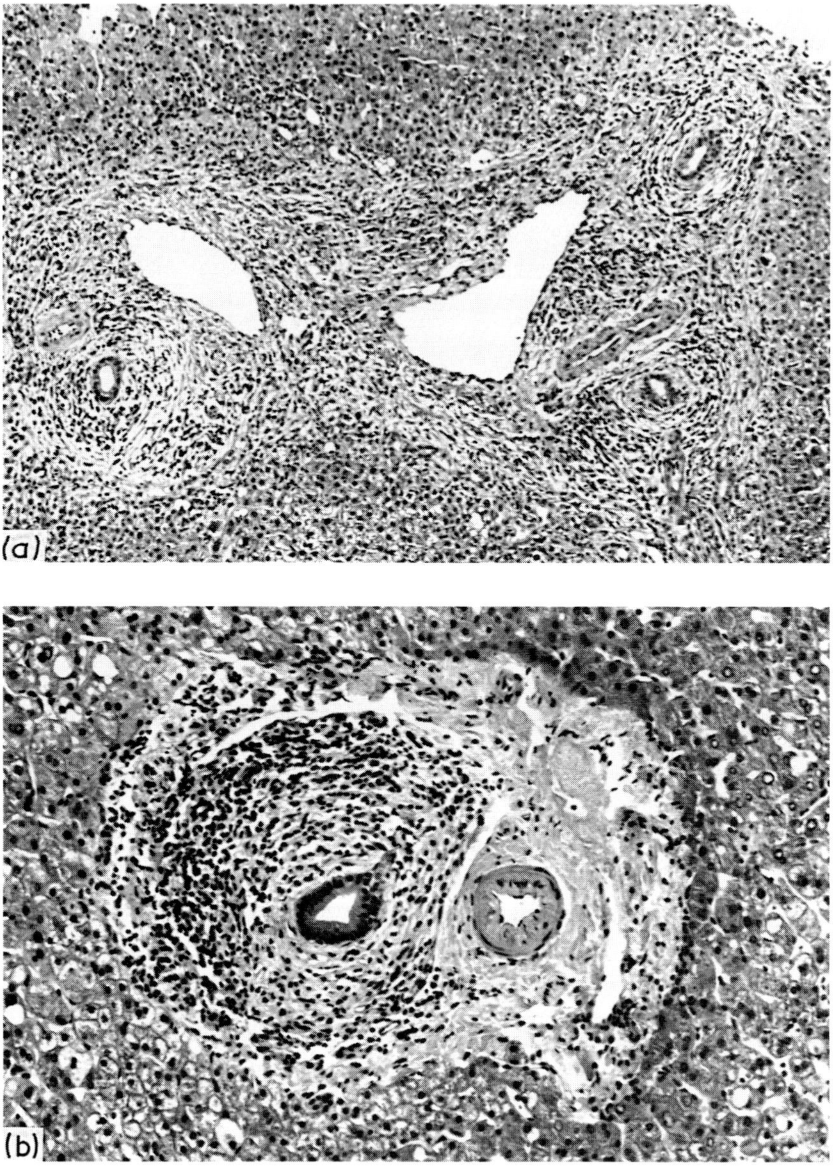

Fig. 7.8 (a) 'Pericholangitis' complicating ulcerative colitis. There is portal fibrosis with some infiltration by chronic inflammatory cells and a concentric arrangement of fibrous tissue around three small interlobular ducts. (b) 'Pericholangitis', from a case of ulcerative colitis. ((a) HE, × 90; (b) HE, × 180.)

bile duct epithelium and a surrounding chronic inflammatory infiltrate; fibrous cholangitis; and fibrous obliterative cholangitis with small nodular scars in portal tracts (Fig. 7.8). Loss of interlobular bile ducts and ductule proliferation is focal. Piecemeal necrosis is uncommon. Cholestasis is usually absent, althouth mild degrees of copper and CBP accumulation are relatively constant. Fatty change is mild when present. Lobular hepatitis is absent.

Small duct PSC is, therefore, qualitatively similar to large duct PSC (see below). The condition usually does not progress but occasional patients develop (or have concomitant undetected) large duct PSC. Biliary and gallbladder cancer has been documented. Only 8% develop cirrhosis.

Some of the more chronic cases may resemble primary biliary cirrhosis, (Wiesner et al., 1985) especially when there are dense collections of chronic inflammatory cells related to degenerate bile ducts, and when granulomas are present also. Small duct PSC tends to involve smaller ducts which, more often than not, show no epithelial damage but the distinction between the two conditions may not be possible on biopsy appearances alone; in small duct PSC the anti-mitochondrial antibody test is negative. Chronic active hepatitis and chronic persistent hepatitis may present similar difficulties in differential diagnosis; in these conditions periductal fibrosis is not a feature.

It is often stated that 'pericholangitis' arises from the action of bacteria or toxins on intra-hepatic bile ducts, but in some cases the brunt of the inflammatory process involves intra-hepatic blood vessels or lymphatics, and phlebosclerosis of portal veins may occur. Many opionions on the pathogenesis and clinical importance of 'pericholangitis' have been expressed and some pathologists have doubted the existence of such an entity. We believe that it is the explanation of liver dysfunction in some cases of UC. Chronic ulcerative cholitis, however, does not cause small duct PSC because the latter may occur in the absence of the former and UC may not be detected until after the bile duct disease is symptomatic. UC and PSC may, therefore, share a cause(s). Small duct PSC may persist after colectomy for UC.

At present, it is most reasonable to assume that small and large duct PSC are parts of the same disease spectrum.

(b) Large duct sclerosing cholangitis (Chapman *et al.*, 1980; Ludwig *et al.*, 1984; Barbatis *et al.*, 1985)

This implies gross fibrous thickening of the walls of bile ducts both within (Fig. 7.9) and without the liver. In the latter situation the affected ducts can have the gross appearance of a large artery; the lumen is seldom dilated but may be narrowed and, in the case of smaller intra-hepatic

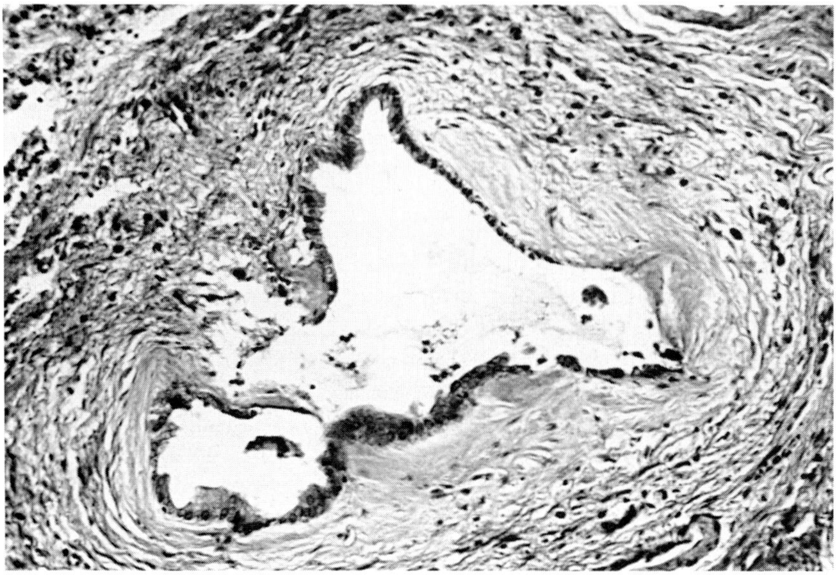

Fig. 7.9 Large duct sclerosing cholangitis. The wall of this septal bile duct shows fibrous thickening with hyaline degeneration. Much of the epithelial lining is atrophic. An idiopathic case which involved both extrahepatic and intrahepatic ducts (HE, × 180).

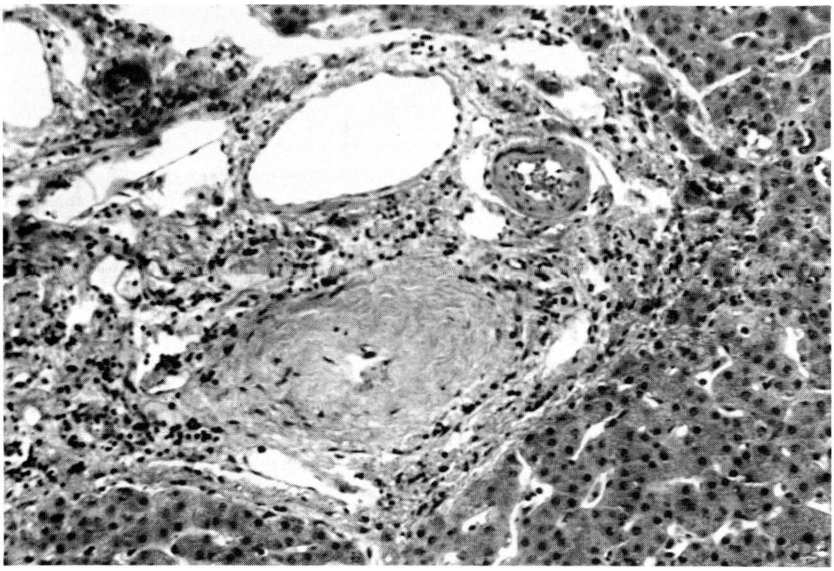

Fig. 7.10 Sclerosing cholangitis, from a case of ulcerative colitis. There is fibrous obliteration of an interlobular duct (HE, × 180).

Fig. 7.11 Sclerosing cholangitis. This common hepatic bile duct wall biopsy shows fibrous thickening, erosion of most of the epithelial surface (above) and epithelial-lined sinuses penetrating the muscle coat (as in chronic cholecystitis) (HE, × 45).

ducts, completely obliterated with fibrous tissue (Fig. 7.10). Fibrous constriction is segmental and can exist witin one part of the liver only, the intervening channels showing tubular or saccular dilatation (Ludwig *et al.*, 1986). Apart from some lymphocytic infiltration, active cholangitis is unusual except after surgical interference, but other parts of the liver may show 'pericholangitis' and ductular proliferation or cholestasis. Moderate copper and CBP is usually demonstrable in hepatocytes.

The morphological features in large duct PSC are qualitatively similar but quantitatively greater than in small duct PSC. Lobular hepatitis, similar to viral hepatitis, is greater in large duct diseases, as is piecemeal necrosis, and loss of bile ducts with nodular scars (Fig. 7.10). Biliary cirrhosis develops in about 60% and about 15% of patients develop bile/gall bladder cancer (Wee and Ludwig, 1985). The mean survival is about 3 years after diagnosis but this obviously depends on the stage at which diagnosis is made because some patients survive for more than a decade.

Because of the patchy distribution, the condition may be missed in aspiration biopsies of liver, and cholangiography is more valuable in diagnosis. Biopsies of affected extra-hepatic ducts may show an appearance similar to that of chronic cholecystitis with much fibrosis and glandular hypertrophy (Fig. 7.11); biopsy of areas of unusual segmental

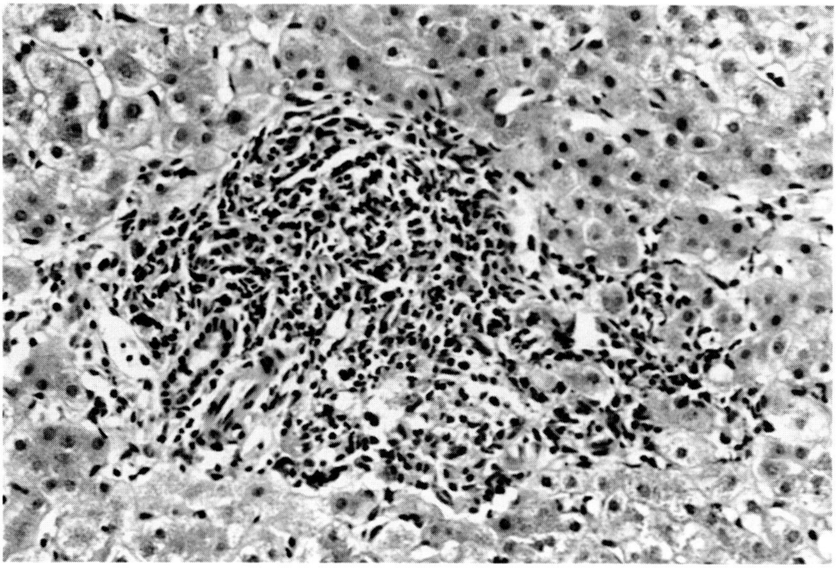

Fig. 7.12 Liver in Crohn's disease showing a poorly defined granuloma involving a portal tract (HE, × 260).

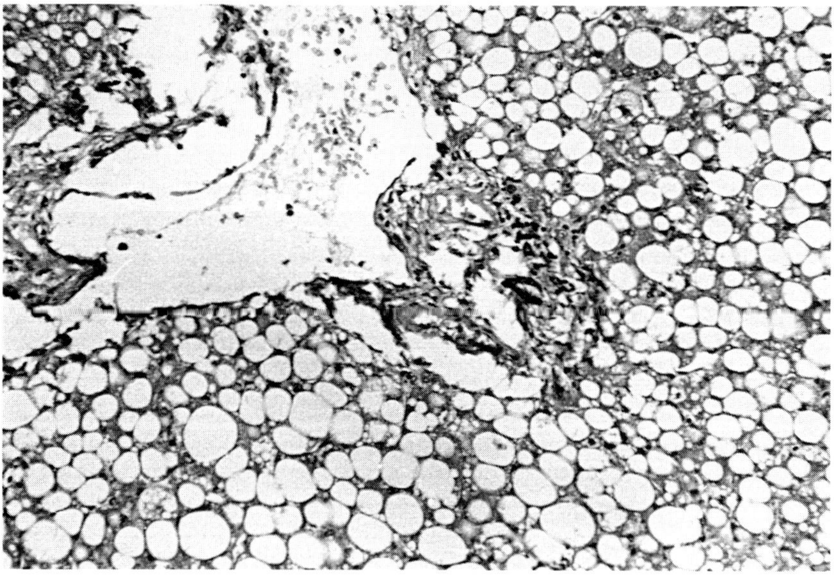

Fig. 7.13 Liver in ulcerative colitis. There is periportal fatty change in hepatocytes. The lipid is in the form of large droplets which have filled the cytoplasm and displaced the nucleus to the periphery of the cell (HE, × 360).

thickening of ducts should always be done at the time of ERCP or laparotomy to investigate the possible existence of a small scirrhous cholangiocarcinoma.

Sclerosing cholangitis may co-exist with other forms of pathological connective tissue overgrowth such as retroperitoneal and orbital fibrosis, Peyronie's disease and Riedel's thyroiditis. These conditions have an association with HLA-B8 but chronic inflammatory bowel disease is not present in every case. HLA-B8 prevelance is normal in UC uncomplicated by liver disease. Therefore, increased prevelance of HLA-B8 in UC may potentiate the risk of developing PSC.

Scarring of bile ducts secondary to surgical interference and cholelithiasis should be excluded.

(c) *Granulomatous hepatitis.*

The liver in a minority of cases of colitis, more commonly Crohn's disease, may contain non-caseating granulomas, situated either in portal tracts or within liver parenchyma. They may co-exist with sclerosing cholangitis. In some cases they are rather poorly defined and lack multinucleated giant cells (Fig. 7.12).

(d) Cholangiocarcinoma

This is a rare complication of ulcerative colitis, but the incidence is several times that occurring in the general population, especially in patients with associated sclerosing cholangitis. There is no association between this tumour and Crohn's disease.

(e) Other hepatic conditions

Fatty change is frequent, especially in chronic cases with malnutrition (Fig. 7.13). CPH and CAH are regarded as complications, but the significance of this is uncertain as in many reports no clear distinction is drawn between these entities and pericholangitis; furthermore CAH in PSC may be related to pertransfusion hepatitis in PSC. Other uncommon conditions include amyloid disease, portal thrombosis with or without infarction, pyogenic abscess and haemosiderosis.

References

Barbatis, C., Grases, P., Shepherd, H. H. *et al.* (1985), Histological features of sclerosing cholangitis in patients with chronic ulcerative colitis. *J. Clin. Pathol.*, **38,** 778–83.

Bernstein, J., Chang, C. H., Brough, A. J. *et al.* (1977), Conjugated hyperbilirubinemia in infancy associated with parenteral alimentation. *J. Pediatr.*, **90**, 361–7.

Chapman, R. W. G., Arborgh, B. A. M., Rhodes, J. M. *et al.* (1980), Primary sclerosing cholangitis: a review of its clinical features, cholangiography and hepatic histology. *Gut*, **21**, 870–7.

Gluud, G., Christoffersen, P., Anderson, T. *et al.* (1984), Occurrence and significance of Mallory bodies in morbidly obese patients. An immunohistochemical study. *Act. Path. Microbiol. Immunol. Scand.*, **92**, 39–43.

Holzbach, R. T. (1977), Hepatic effects of jejunoileal bypass for morbid obesity. *Am. J. Clin. Nutr.*, **30**, 43–52.

Hornboll, P. and Olsen, T. S. (1982), Fatty changes in the liver: the relation to age, overweight and diabetes mellitus. *Acta Pathol. Microbiol. Scand.* [A], **90**, 199–205.

Hughes, C. A., Talbot, I. C., Ducker, D. A. *et al.* (1983), Total parenteral nutrition in infancy: effect on the liver and suggested pathogenesis. *Gut*, **24**, 241–8.

Leevy, C. M., Popper, H. and Sherlock, S. (1976), Diseases of the liver and biliary tract: Standardisation of nomenclature, diagnostic criteria and diagnostic methodology. *Fogarty Int. Centre Proc.* **22**, 1–22.

Ludwig, J., Czaja, A. J., Dickson, E. R. *et al.* (1984), Manifestations of nonsuppurative cholangitis in chronic hepatobiliary diseases: morphological spectrum, clinical correlations and terminology. *Liver*, **4**, 105–16.

Ludwig, J., MacCarty, R. L., La Russo, N. F. *et al.* (1986), Intrahepatic cholangiectases and large duct obliteration in primary sclerosing cholangitis. *Hepatology*, **6**, 560–8.

Patrick, R. S., MacKay, A. M., Coward, D. G. and Whitehead, R. G. (1973), Experimental protein-energy malnutrition in baby baboons. *Br. J. Nutr.*, **30**, 171–9.

Puera, D. A., Stromeyer, F. W. and Johnson, L. F. (1980), Liver injury with alcoholic hyaline after intestinal resection. *Gastroenterology*, **79**, 128–30.

Salvian, A. J. and Allardyce, D. B. (1980), Impaired bilirubin secretion during total parenteral nutrition. *J. Surg. Res.*, **28**, 547–55.

Wee, A. and Ludwig, J. (1985), Pericholangitis in chronic ulcerative colitis: Primary sclerosing cholangitis of the small bile ducts. *Ann. Int. Med.*, **102**, 581–7.

Whitington, P. F. (1985), Cholestasis associated with total parenteral nutrition in infants. *Hepatology*, **5**, 693–6.

Wiesner, R. H., La Russo, N. F., Ludwig, J. *et al.* (1985), Comparison of the clinicopathologic features of primary sclerosing cholangitis and primary biliary cirrhosis. *Gastroenterology*, **88**, 108–14.

8 Liver injury from drugs and toxins

The liver has a valuable role in the detoxification of many substances intentionally or accidentally ingested and hepatocyte damage may arise from their accumulation in the liver. About 10% of all adult jaundice cases can be attributed to toxic liver injury. The subject, therefore, is of considerable importance to the diagnostic pathologist because of this high frequency and because of the serious consequences which may arise from failure in its detection; the continuous use of certain drugs or re-exposure to their toxic influence may cause severe and even fatal liver necrosis or, in some cases, the development of chronic hepatitis and cirrhosis.

A wide variety of histopathological changes may be seen depending on the type and dose of toxic agent responsible, the susceptibility of the patient, the modifying influence of other drugs and the coexistence of other disease in the liver or elsewhere. Thus, drug-induced injury may be enhanced by alcohol or by renal failure which allows the accumulation of the toxic substance to liver-damaging levels. Moreover, in some cases it may be impossible to attribute histopathological changes in the liver to a drug or to the primary disease for which the drug is given. Clearly, this is a difficult part of liver biopsy interpretation and in some cases the pathologist can only suggest to his clinical colleagues the possibility of toxic injury, a possibility that must be kept in mind in the differential diagnosis of many hepatic disorders.

Numerous hepatotoxic agents have been recognized but it is hardly possible to present an exhaustive list in a book of this kind as many reports are based on studies of single or small numbers of cases, often with inadequate pathological descriptions. It is customary to divide this group of liver diseases into those caused by substances which are deleterious to a susceptible minority of persons and those which have arisen in a more predictable manner. Liver biopsy can be of value in classifying cases in this way; thus, many examples of the predictable group have zonal necrosis of liver and many patients with unpredictable damage have lesions which mimic biliary disease from other causes or which mimic acute or chronic viral hepatitis. Apart from histopathology there may be other distinguishing features. The unpredictable group rarely includes children, and laboratory animals are not susceptible. The pathogenesis is often obscure, although in some the liver injury appears to be part of a generalized hypersensitivity reaction with skin rashes, pyrexia and eosinophilia. Chlorpromazine, a well-known unpredictable hepatotoxin, can damage the enzymic activity for active transport of cations across canalicular membranes with resulting cholestasis (p. 50) (Samuels and Carey, 1978), but this may occur only in a minority of persons who cannot convert it effectively to less toxic metabolites. Occasionally, the hepatic damage, however produced, will disappear despite the continuous or repeated intake of the offending agent. More commonly the opposite is true, and while relapse following re-exposure with a small challenge dose can be valuable in establishing the diagnosis, this course is not justified with drugs such as halothane which can cause fatal liver damage.

It is not surprising, therefore, that unpredictable liver injuries often present the pathologist with considerable diagnostic problems as the cause can be unsuspected clinically, while the severity is not dose related and there is usually a latent period of days or weeks between exposure and the appearance of symptoms. Moreover, as already stated, the histopathological features mimic those of various other liver diseases and the same drug can produce a variety of changes in different subjects. It is important always to bear in mind the possibility of drug-induced injury in almost every patient with liver disease in whom an alternative aetiology is not clearly apparent.

Unfortunately, there are a number of important examples of toxic liver injury of obscure pathogenesis which do not fall readily into either of these two categories despite their clinical and pathological features.

8.1 Predictable direct hepatotoxicity causing liver cell necrosis

This group presents the least diagnostic difficulty as the true nature of liver damage is often suspected clinically and the histopathological appearances are fairly characteristic. Many toxins are known to which all persons and most experimental animals are susceptible, although there can be species variation in severity. The extent of injury is also dose related and repeated exposure can lead to hepatic fibrosis and cirrhosis. Organs other than liver may be damaged also.

Most toxins in this group cause zonal necrosis of liver, usually centrilobular in distribution (Fig. 8.1). When severe the process may extend throughout entire lobules so that areas of submassive necrosis are seen (Fig. 8.2). Surviving parenchyma may show fatty change and in mild

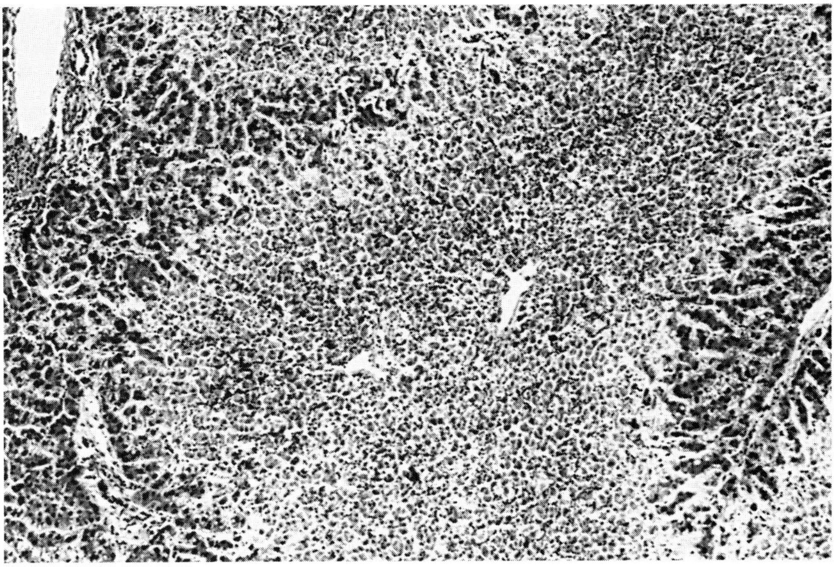

Fig. 8.1 Carbon tetrachloride liver damage in an alcoholic. There is severe centrilobular zonal necrosis extending into the midzonal areas of the liver (HE, × 100).

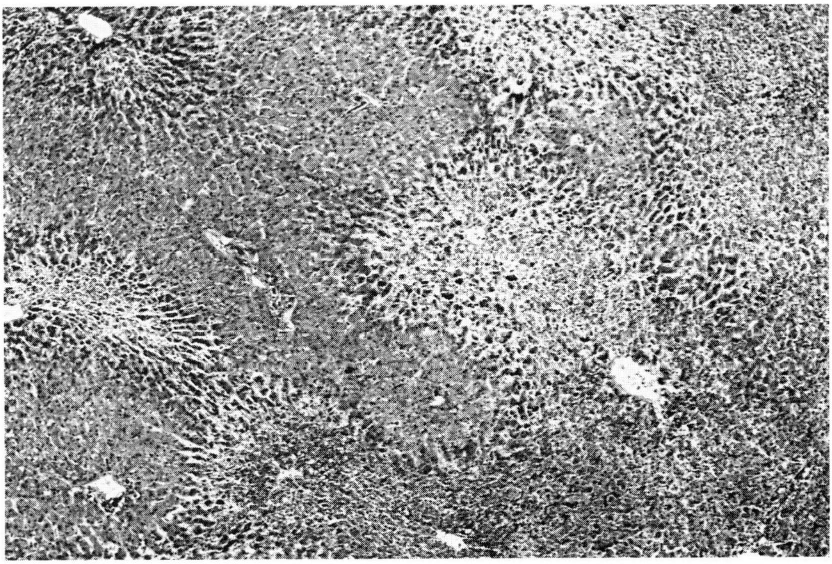

Fig. 8.2 Carbon tetrachloride liver damage. There is an area of confluent submassive necrosis (top right) which extended to the capsule of the liver (HE, × 45).

cases the picture is that of fatty liver alone. This fat may appear as small discrete droplets without displacement of the cell nucleus, especially in centrilobular zones (Fig. 8.6), and lipofuscin pigment may be prominent. Unlike acute liver disease from other causes there may be little or no hepatitis (Fig. 8.3).

Within hours of poisoning there is usually a febrile illness with upper abdominal pain, but jaundice and other evidence of liver damage is delayed until the second or third day. This correlates well with the maximum development of liver necrosis. Thereafter, in non-fatal cases, regeneration of surviving parenchyma, which begins at about the thirty-sixth hour, is well advanced and together with prominent macrophage activity is responsible for recovery which, in mild cases, can be complete within a week. Prominent Kupffer cells with ingested periodic acid-Schiff-positive debris, pigmented fat (ceroid) and haemosiderin may persist for several weeks longer.

A normal pattern of regeneration is assisted by survival of the reticulin framework of the liver. However, this may appear condensed or collapsed in the centrilobular zones during the early stages of recovery, but there is evidence for new collagen and mucopolysaccharide synthesis after a single incident of acute liver damage (McGee *et al.*, 1974). Certainly the continuous or intermittent exposure of experimental animals to such

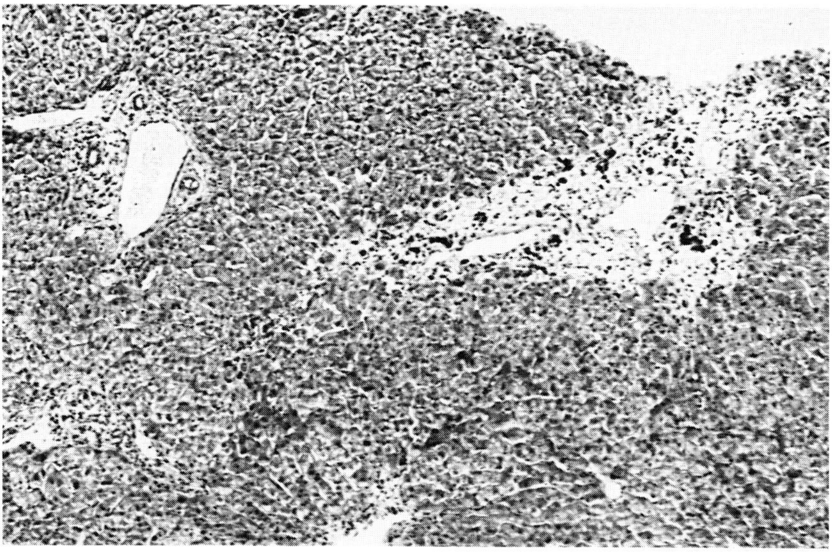

Fig. 8.3 Zonal necrosis (right) due to an overdose of paracetamol. There are no inflammatory cells in the portal tract or surviving parenchyma. Several deeply staining macrophages are present in the necrotic zone (PAS stain, × 90).

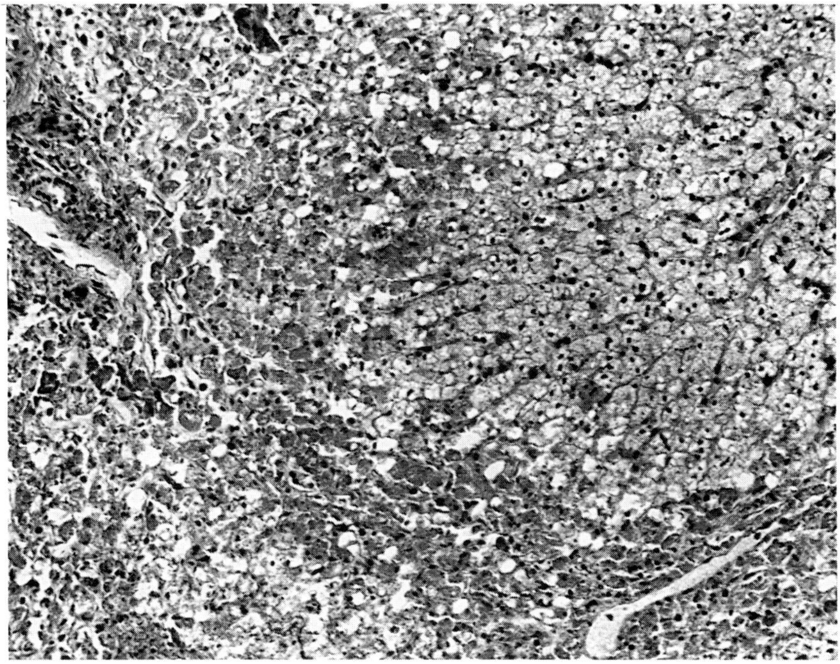

Fig. 8.4 Periportal zonal necrosis due to phosphorus poisoning. Necrotic hepatocytes are intensely eosinophilic while viable cells have a finely vacuolated cytoplasm due to small fat droplets (central vein – upper right) (HE, × 180).

poisons leads to hepatic fibrosis and cirrhosis, but this circumstance is unusual in man. Fine cirrhosis arising in this way should be differentiated from post-necrotic scarring, which in human cases can follow severe liver injury with areas of submassive cell destruction; intervening foci of surviving parenchyma undergo nodular regenerative hyperplasia so that a very coarse form of macronodular cirrhosis can result.

A few liver toxins are known, such as phosphorus and allyl formate, which cause most damage to periportal cells (Fig. 8.4). Midzonal necrosis is also described, and we have seen this induced in mice by bromobenzene, but have no experience of any human example.

Some predictable liver toxins are well known, e.g. chloroform and tannic acid, and their use as therapeutic agents has been discontinued for this reason. The uncommon cases now seen are often accidental or suicidal, for example, poisoning with toadstools, and following the ingestion or inhalation of carbon tetrachloride; man is relatively insusceptible to the latter hepatotoxin and severe liver damage is seen usually in combination with the intake of alcohol. Nowadays predictable liver injury is usually due to drugs like paracetamol; in normal doses it has little or no

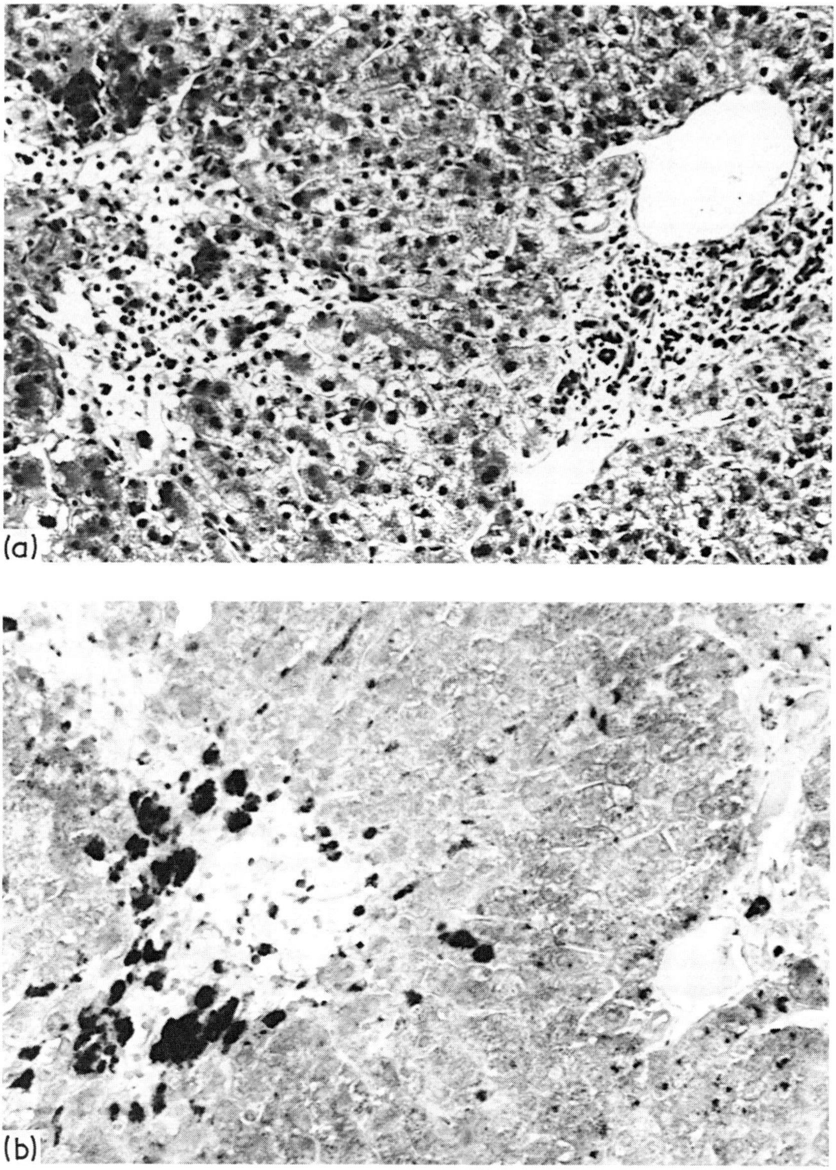

Fig. 8.5 (a) and (b), paracetamol liver damage. In the centrilobular zones (left) there are inflammatory cells including macrophages with ingested darkly staining pigmented lipid and loss of hepatocytes. ((a) HE, × 255; (b) Sudan black × 225).

toxic effect, but fatal liver damage is possible from the ingestion of as little as 10–20 regular dose tablets. Liver biopsy shows centrilobular loss of hepatocytes, prominence of pigmented macrophages and apparent condensation of centrilobular reticulin (Figs 8.3 and 8.5). Tromethamine (Tris) buffer added to fluids for intravenous use has caused haemorrhagic necrosis of liver in infants with respiratory distress syndrome.

8.2 Microvesicular fatty liver

The tetracyclines, and especially chlortetracycline, cause centrilobular fatty change with prominent lipochrome pigment rather than necrosis (Fig. 8.6); nevertheless, liver dysfunction can be severe or even fatal in patients who fail to excrete the drug administered intravenously because of chronic renal disease or pregnancy. A similar lesion is described occasionally in patients receiving sodium valproate as an anticonvulsant, and this can be fatal also, especially in children. Centrilobular zonal necrosis and cholestasis with cholangiolar proliferation are other features of some cases which might be the effect of an additional therapeutic agent (Powell-Jackson *et al.*, 1984). Microvesicular fatty change has been found in children dying of salicylate poisoning; the close clinical and pathological resemblance to Reye's syndrome (see Chapter 9) suggests a toxic cause for at least some examples of that disease (Starko and Mullick, 1983). The vomiting disease of Jamaica should be included here as the fatty liver damage in this condition is attributable to toxic metabolites of hypoglycon A present in Aiku apples.

Although of unknown aetiology it is appropriate also to include here idiopathic toxic fatty liver of pregnancy as microvesicular fatty change is usually the only evidence of hepatic disorder, but hepatocyte necrosis and cholestasis may coexist. Fortunately this is a rare complication of pregnancy as many cases are fatal, but minor examples may escape detection. The disease has to be differentiated from other forms of liver damage occurring in late pregnancy, especially those attributable to eclampsia and viral hepatitis (Sherlock, 1983). It is important to establish the diagnosis as early as possible as termination of pregnancy offers the best hope of recovery. Diagnosis depends greatly on liver biopsy but this carries the risk of serious bleeding because of associated disseminated intravascular coagulation; the problem can be overcome by recovery of hepatic tissue by the transjugular approach (p. 3). Some cases of the disease are attributed to tetracycline but in the majority the cause is unknown.

It should be remembered that microvesicular fatty change in association with lobular inflammation may indicate viral hepatitis, especially the nonA, nonB variety.

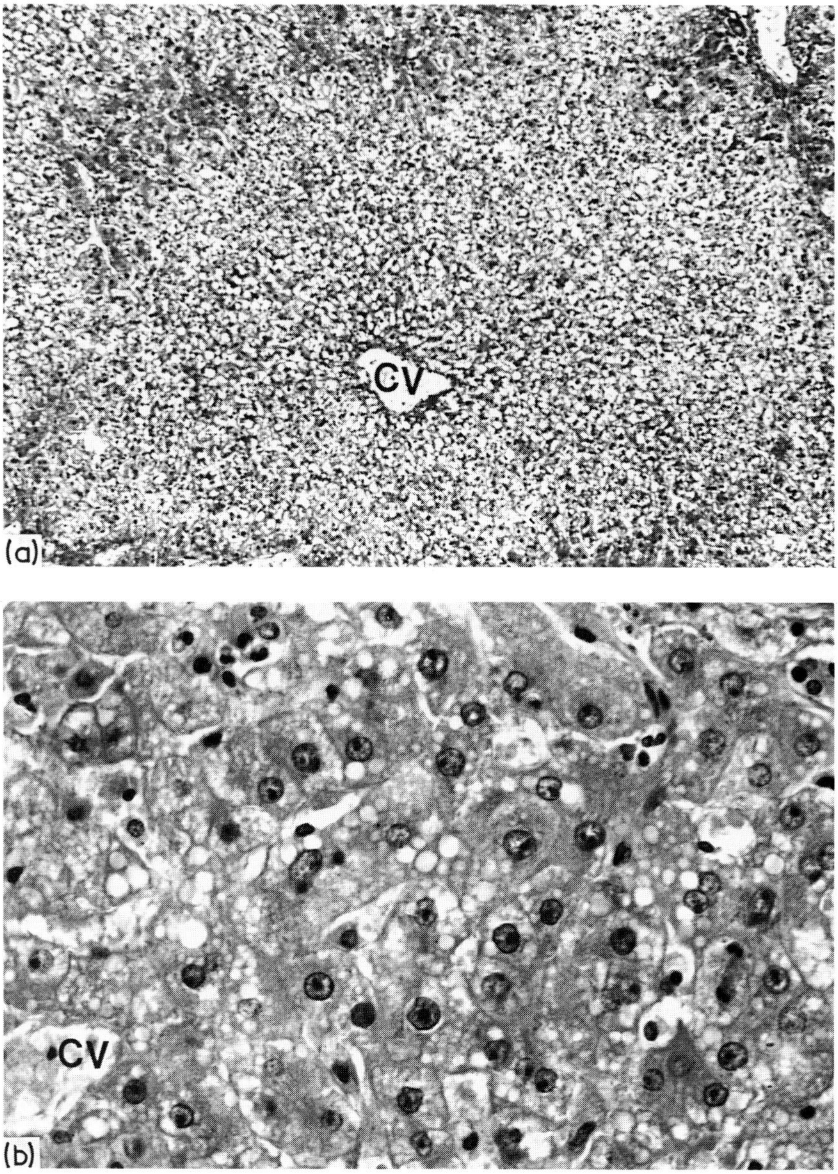

Fig. 8.6 (a) and (b) tetracycline liver damage. There is centrilobular and midzonal fatty change, the cytoplasmic fat consisting of small droplets without displacement of nuclei (CV, centrilobular vein). ((a) HE, × 72; (b) HE, × 450.)

8.3 Asymptomatic hepatomegaly

Clinical hepatic enlargement without dysfunction and without histological evidence of liver damage may follow the ingestion of certain substances by man and by experimental animals. Enlargement is due to an increase of hepatocyte cytoplasm usually without any corresponding increase in DNA and in mitotic activity, and so the process is one of hypertrophy rather than hyperplasia. The affected cytoplasm has a hyaline or finely granular appearance; compared with the ground-glass appearance due to hepatitis B infection, staining with eosin and with the acid-fuchsin component of Mallory stains is more intense and the liver is more uniformly affected (Fig. 8.7). The normal basophilic granular ergastoplasm of the hepatocyte appears to be displaced towards the centre of the cell. Electron microscopy reveals that these changes are due to considerable increase in the AER with marginal displacement of the granular component.

Many drugs and chemicals such as phenobarbitone and DDT together with some steroid and thyroid hormones have this effect on the liver. Moreover, the list includes certain other drugs such as chlorpromazine which also produce toxic liver injury in a susceptible minority of persons.

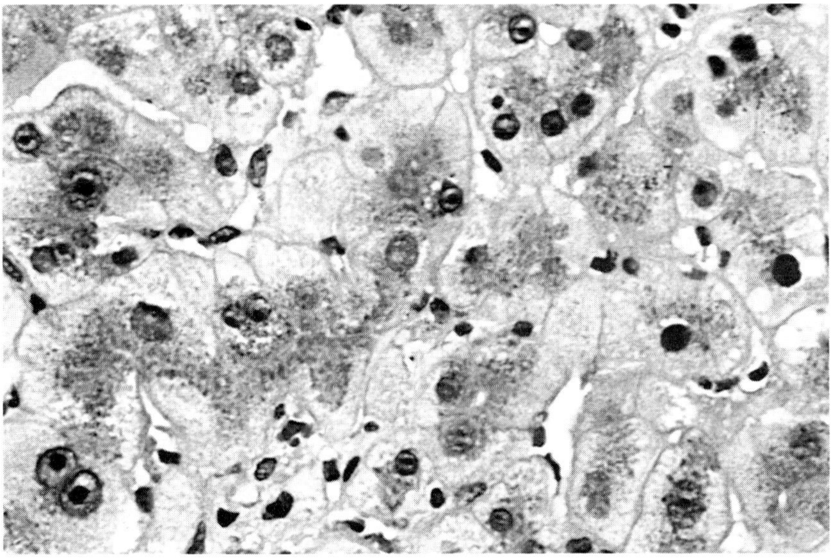

Fig. 8.7 Biopsy from a case of asymptomatic hepatomegaly due to phenobarbitone. Basopholic granules representing rough endoplasmic reticulum are confined to the pericanalicular hepatocyte cytoplasm, the remainder of which stains uniformly with eosin (HE, × 560).

Enzymes necessary for the metabolism and detoxification of many ingested chemicals are located in the agranular endoplasmic reticulum and their activity may increase with its hypertrophy. This can be beneficial in stimulating the elimination of the substance causing the change as well as other toxic drugs or chemicals taken concurrently. In certain cases, however, it is detrimental by stimulating the production of toxic metabolites from substances which are basically harmless to the liver; for example, carbon tetrachloride is not in itself hepatoxic but induces cellular damage through the action of its metabolic products. This can explain liver necrosis induced by certain monoamine oxidase inhibitors such as iproclozide. The damaging combination of isoniazid and rifampicin has been attributed to the stimulating effect of the latter drug on hepatic AER (Pessayre *et al.*, 1977).

8.4 Biliary disease

8.4.1 Disorders of bilirubin metabolism

Certain drugs may predictably give rise to jaundice in the absence of structural changes in the liver as seen by the light microscope. Apart from those causing a purely haemolytic jaundice, some induce functional upsets similar to those which characterize the two forms of Gilbert's syndrome. Thus there may be interference with transport of unconjugated bilirubin into or within hepatocytes, e.g. by flavispidic acid and cholecystographic media, or there may be interference with bilirubin conjugation, e.g. by novobiocin in infants. (It has been noted that contrast medium for cholangiography can also cause centrilobular zonal necrosis.) Salicylates and sulphonamides cause infantile jaundice by competing with unconjugated bilirubin for binding by serum albumin. Such cases are important in neonates because of the risk of kernicterus.

8.4.2 Intrahepatic cholestasis without hepatitis

These patients have an obstructive form of jaundice and so the effect of drugs must be taken into account in the differential diagnosis of this important clinical sign.

Cholestasis confined to centrilobular bile canaliculi and without any inflammatory reaction in the liver can be due to anabolic steroid therapy (Fig. 8.8). Rare cases are associated with the contraceptive pill and these same patients may experience jaundice in the late months of pregnancy. It is controversial whether this form of drug reaction should be regarded as unpredictable as all persons might suffer from cholestasis if they received large amounts of steroids. However, it is clear that only a small

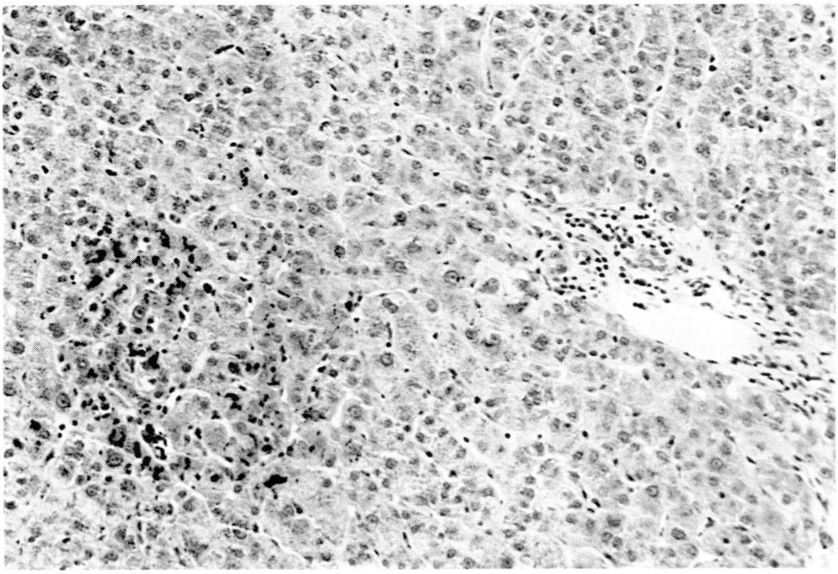

Fig. 8.8 Intrahepatic cholestasis without hepatitis associated with norethandrolone therapy. Stasis of dark bile pigment within centrilobular canaliculi is seen (left) (neutral red, × 180).

number of patients become jaundiced from normal therapeutic doses. Genetic factors appear to determine this susceptibility which may include a proneness to gallbladder disease.

8.4.3 Cholestatic hepatitis

This is the commonest form of unpredictable liver injury due to drugs. Clinically, there is obstructive-type jaundice with or without itching which may be preceded by a febrile illness with digestive upset. Many drugs may be responsible, the better known being chlorpromazine and other phenothiazine drugs, oral hypoglycaemic agents, *para*-amino salicylic acid, thiouracil and erythromycin esters.

Liver biopsy shows centrilobular cholestasis. In some cases mild hepatocyte damage and portal inflammation could be attributed to cholestasis as such, but they are often sufficiently severe to suggest an additional toxic influence of the drug on the liver (Fig. 8.9). Hydropic degeneration and pin-point necrosis of hepatocytes may be conspicuous but, unlike viral hepatitis, there is no cellular disarray of liver lobules or diffuse lymphocytic infiltration. On the other hand, cholestasis tends to be more prominent while the portal inflammatory infiltrate often includes neutrophil polymorphs and eosinophils.

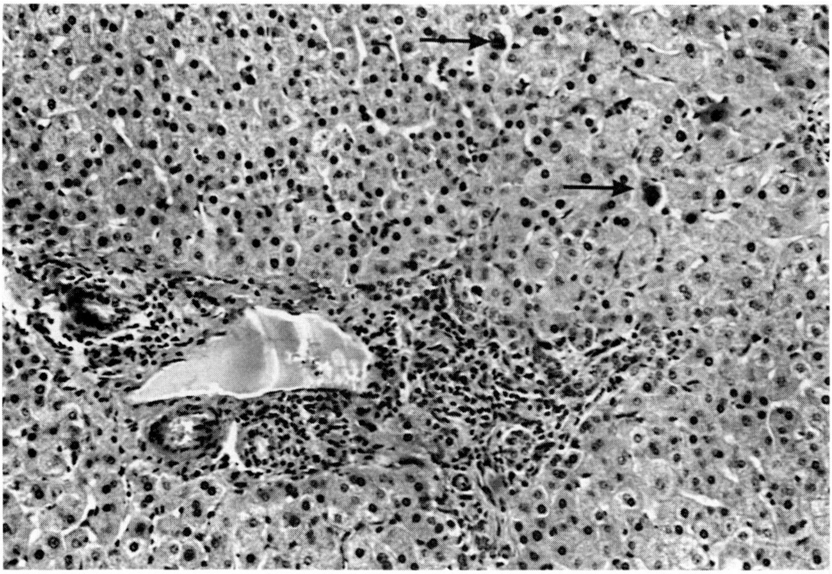

Fig. 8.9 Cholestatic hepatitis due to chlorpromazine. There is chronic portal inflammation and bile plugs in dilated canaliculi (arrows) (HE, × 180).

Even with continuous or repeated administration of these drugs the hepatitis may subside although cholestasis tends to persist. Mitotic activity in hepatocytes may be evident at any stage (Fig. 8.10). Portal fibrosis is not prominent in the majority of cases.

8.4.4 Biliary cirrhosis

Unusual cases are recorded in which an illness clinically like primary biliary cirrhosis can be dated from an attack of drug-induced cholestatic hepatitis (Read *et al.*, 1961). The condition may resolve gradually when the cause is recognized and removed, but in some cases it persists and progresses to established biliary cirrhosis.

Histologically the condition is not very similar to primary biliary cirrhosis as the centrilobular cholestasis persists, the lymphocytic and plasma cell infiltration of portal triads is less intense and granulomas related to segmental bile duct epithelial degeneration are not seen. Bile ductular proliferation is present, however, and adjacent portal areas can be bridged by fibrous tissue (Fig. 8.11). True cirrhosis with nodular regeneration of surviving parenchyma is described.

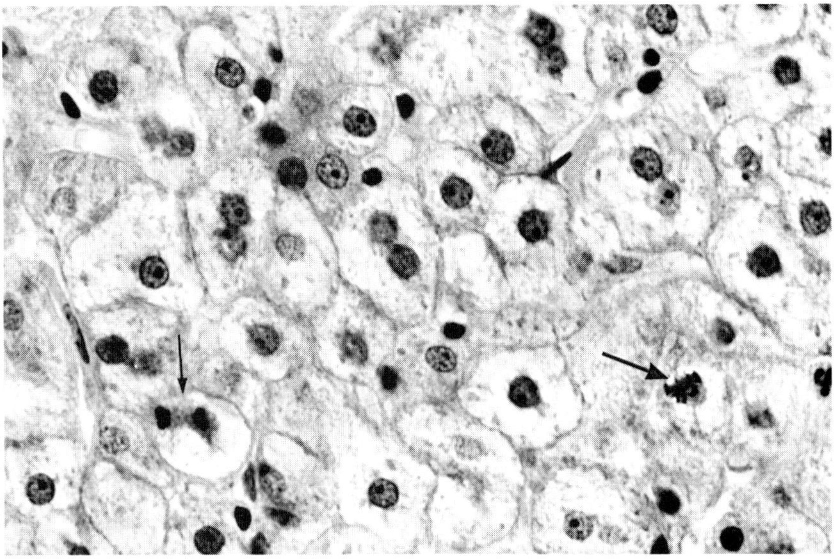

Fig. 8.10 Another field from the biopsy illustrated in Fig. 8.9 showing a mitotic figure (arrow) (HE, × 720).

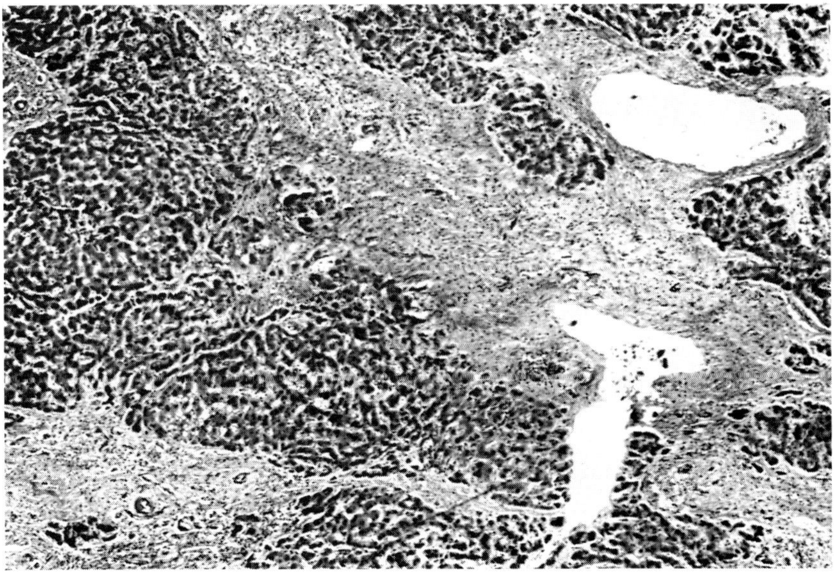

Fig. 8.11 Extensive hepatic fibrosis in a patient who had jaundice for 3 years associated with continuous intake of chlorpromazine. Bile capillary cholestasis was also prominent but is not evident at this magnification (HE, × 56).

8.4.5 *Cholangitis*

This is a very unusual manifestation of drug sensitivity and has been recorded in association with allopurinol-induced granulomatous hepatitis (Swank *et al.*, 1978). We have experience of similar cases responding to drug withdrawal.

8.5 Hepatitis

8.5.1 *Mild hepatitis without cholestasis*

These patients may have some clinical and biochemical evidence of hepatic disorder but it can be difficult to determine whether this is due to therapy or to the disease for which a particular drug is given, for example, cases of rheumatoid disease or disseminated lupus erythematosus given aspirin for prolonged periods (Seaman *et al.*, 1974). Improvement in liver function following drug withdrawal and relapse after a challenge dose are helpful in diagnosis.

Liver biopsy shows a spotty focal necrosis of single cells or small groups of hepatocytes distributed irregularly in the lobules and surrounded or

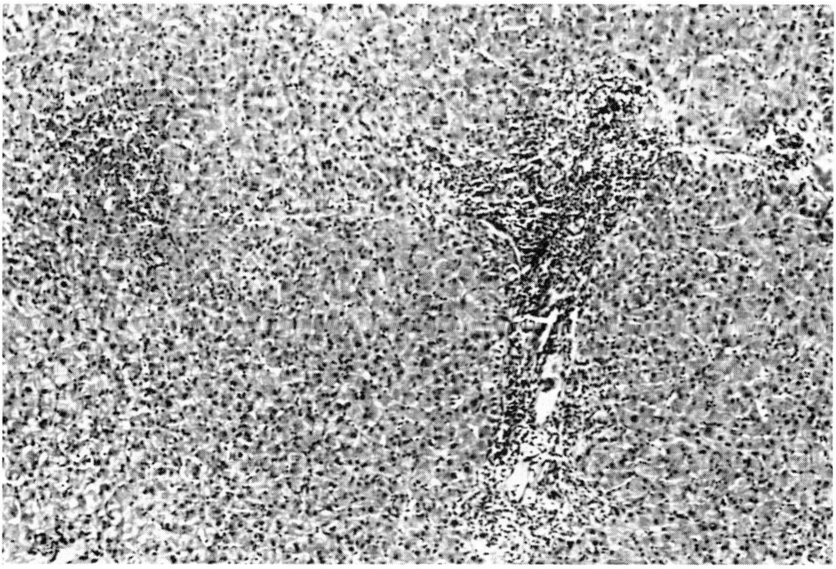

Fig. 8.12 Non-specific reactive hepatitis in a patient with liver dysfunction apparently related to penicillin sensitivity. Chronic inflammatory cells infiltrate the portal tract and a small area of focal necrosis. There was no cholestasis in this case (HE, × 90).

obscured by inflammatory cells, usually lymphocytes and histiocytes (Fig. 8.12). Similar cells may have infiltrated the portal tracts and eosinophils are sometimes prominent.

A variety of drugs, such as papaverine and penicillin, have this unpredictable effect. Salicylate liver injury is similar but is dose related. Round acidophil bodies measuring 1–12 μm in diameter have been noted within hepatocyte cytoplasm and extracellularly in aspirin-induced hepatitis (Seaman *et al.*, 1974). Comparable lesions in many heroin addicts are difficult to assess as they are not typical of viral hepatitis and evidence of virus B infection may be lacking although suspected clinically. NonA, nonB hepatitis has been implicated (Weller *et al.*, 1984).

8.5.2 Granulomatous hepatitis

This might be regarded as a variant of the last group in which some inflammatory cells including histiocytes form multiple granulomas within the liver parenchyma (Fig. 8.13). These lesions are usually small and may be indistinct. A variety of drugs are again implicated, including sulphonamides, phenylbutazone and allopurinol. Cases occurring in diabetic patients treated with sulphonylurea drugs may also have granulomatous lesions in the myocardium and other tissues.

Anti-epileptic drugs such as diphenylhydantoin occasionally induce large granulomatous lesions in the liver as well as in lymph nodes. Cellular pleomorphism may raise a suspicion of Hodgkin's disease, but the lesions regress when the drug is withdrawn.

8.5.3 Lesions resembling acute viral hepatitis

Considerable controversy has centred around this topic as it is claimed that certain drugs may potentiate pre-existing viral infection of liver or raise susceptibility to a new infection. However, the sporadic occurrence especially in patients after a second or later exposure to a suspected drug and the absence of laboratory evidence of hepatitis B virus infection favour the diagnosis of toxic liver damage. Some drugs in this group, such as halothane and cinchophen, can induce very severe hepatitis which may be fatal in a high proportion of cases. Microscopically the appearances are very similar or identical to viral hepatitis, with extensive ballooning and acidophilic degeneration and necrosis of hepatocytes, variable cholestasis, lymphocytic infiltration and an apparent increase in stroma (Fig. 8.14). In a minority of cases there are certain histological features which should raise the suspicion of drug-induced hepatitis: these are an irregular distribution of sharply demarcated liver damage, necrosis involving stroma as well as parenchyma, fatty change,

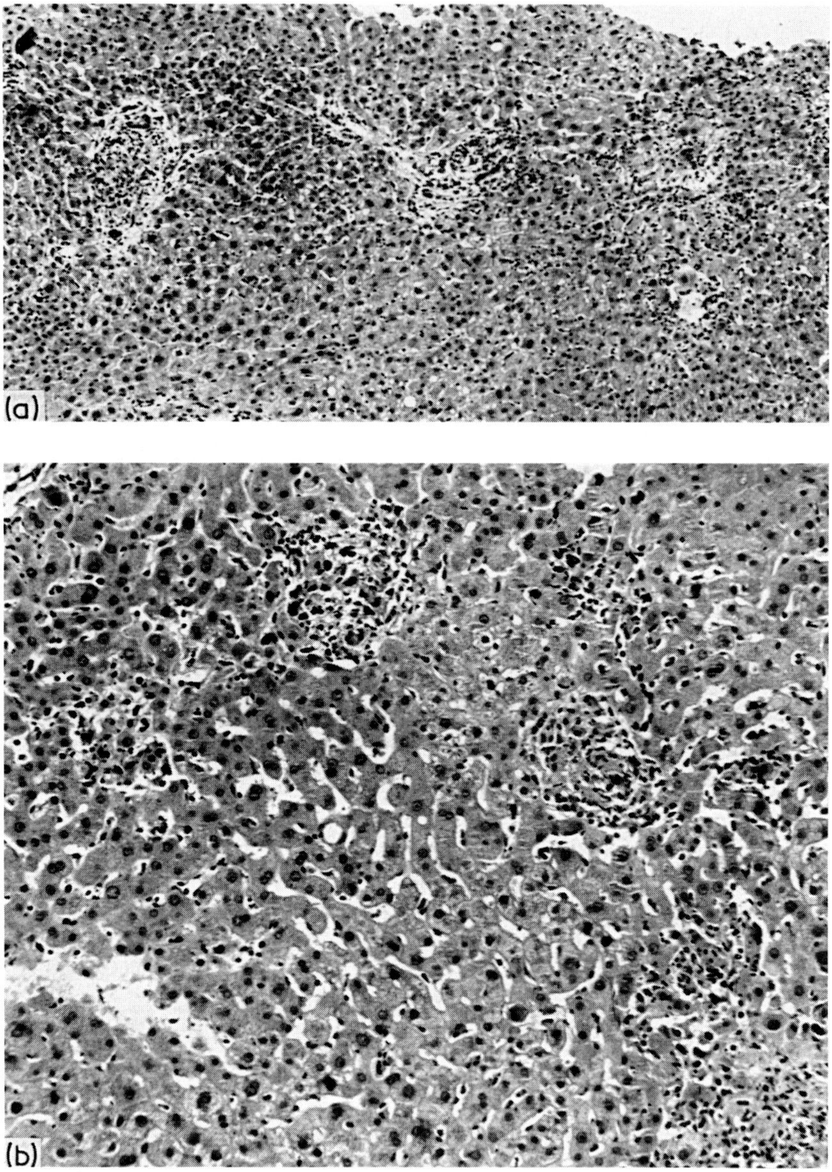

Fig. 8.13 (a) and (b) granulomatous hepatitis related to allopurinol therapy. ((a) HE, × 90; (b) HE, × 190.)

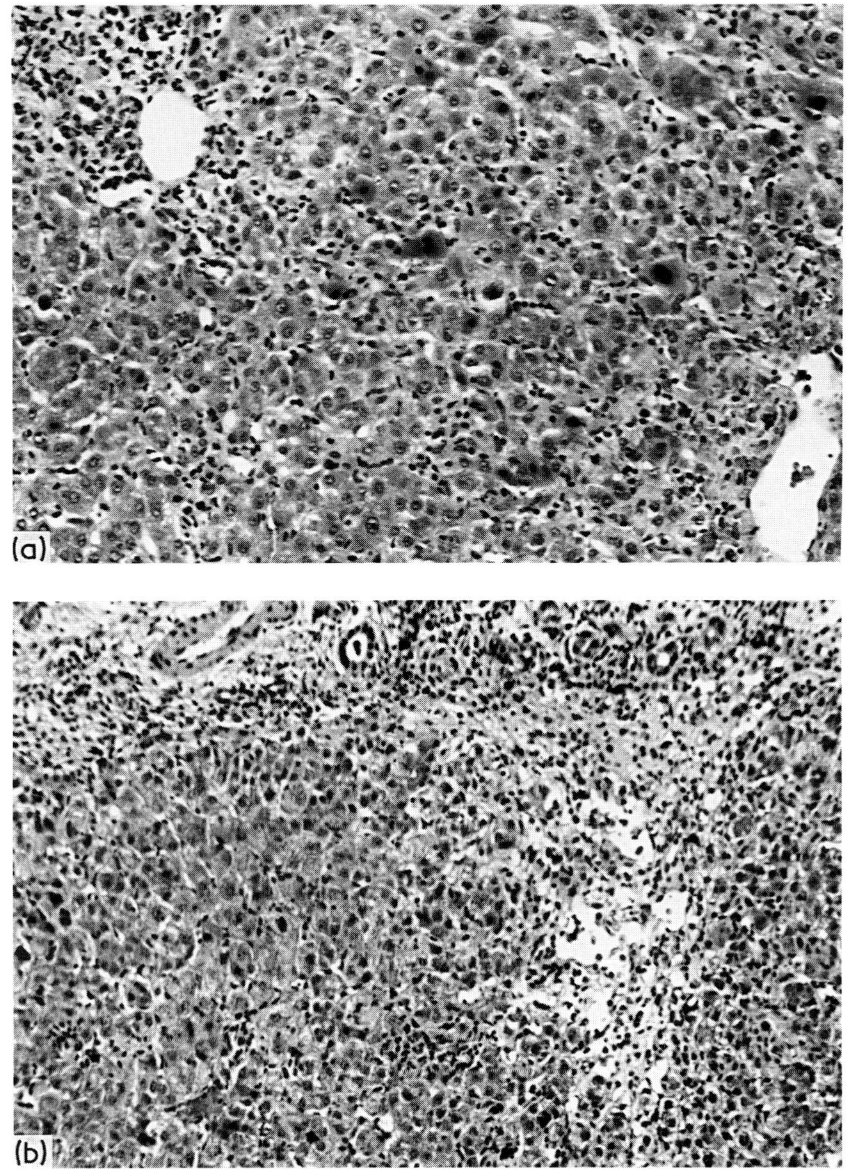

Fig. 8.14 Halothane hepatitis. Two fields from one biopsy showing portal and lobular hepatitis. Eosinophils were conspicuous: (a) acidophilic degeneration and necrosis of hepatocytes is prominent; (b) an area of bridging necrosis. ((a) and (b) HE, × 180.)

granulomas, prominent numbers of eosinophils and severity of liver damage or fibrosis out of keeping with a relatively mild clinical illness or duration of symptoms. The long-acting skeletal muscle relaxant, dantrolene, causes similar liver damage in 2% of recipients, in whom about one-third, usually female subjects, the outcome is fatal (Utili *et al.*, 1977). Phenylbutazone liver injury may take this form rather than granulomatous hepatitis. Fortunately, the majority of drugs in this list of unpredictable poisons, such as methyldopa and isoniazid, usually cause less severe changes from which recovery is the rule (Fig. 8.15). Whatever the initial severity, it is clearly very important to recognize these cases so that future exposure to the drug with liver injury of increasing intensity can be avoided.

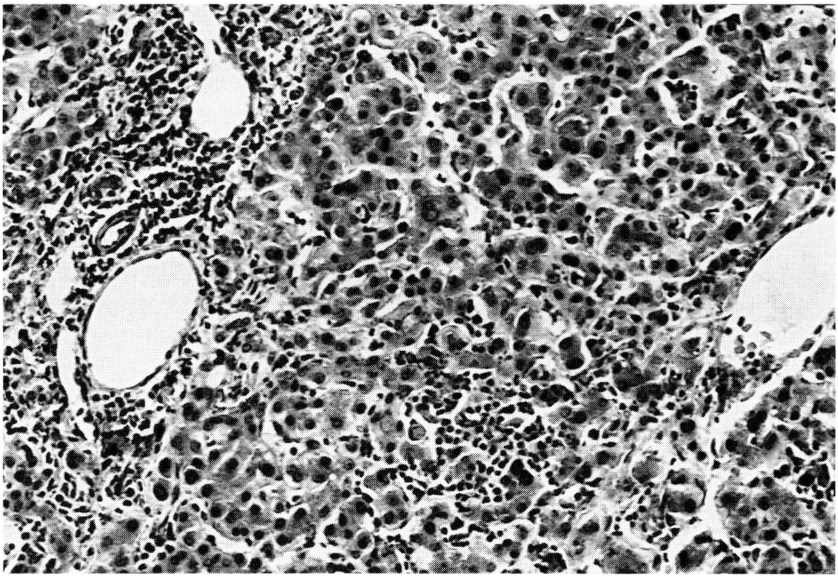

Fig. 8.15 Alpha-methyldopa liver injury. There is lymphocytic infiltration and hydropic degeneration of some hepatocytes as in viral hepatitis. Cholestasis was also present but is not readily seen in this photograph (HE, × 225).

8.5.4 Chronic active hepatitis and cirrhosis

The similarity between viral hepatitis and certain examples of drug-induced liver injury is further heightened by the possible development of persistent or aggressive CAH in both (Fig. 8.16). The risk in cases initiated by drugs is greatest where exposure is prolonged, as in cases of chronic urinary infection treated with nitrofurantoin and in the anaesthetist who developed cirrhosis after repeated exposure to halothane (Klatskin and

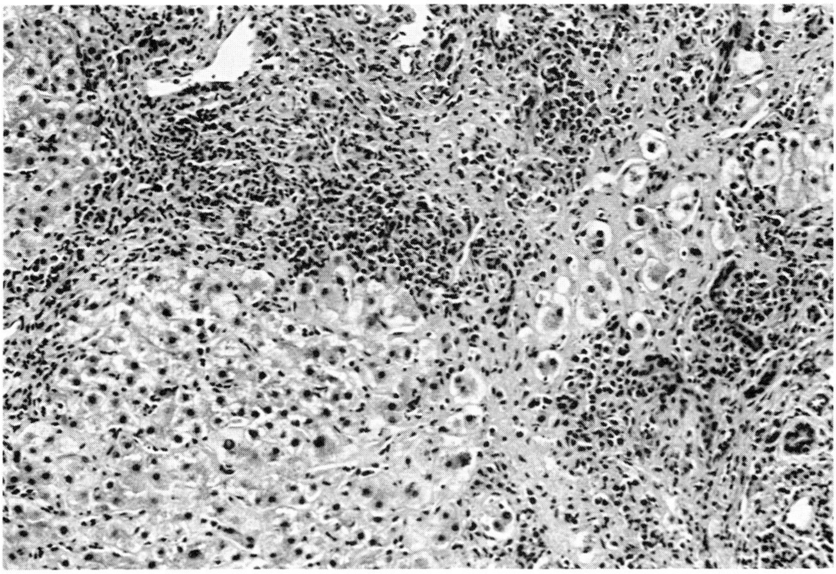

Fig. 8.16 Aggressive chronic active hepatitis associated with prolonged intake of alpha-methyldopa (HE, × 180).

Kimberg, 1969). The initial hepatitis may be relatively mild, as in those cases attributable to the laxative oxyphenisatin.

In spite of the close histological similarity to CAH, many of these drug-induced cases resolve when the causative agent is withdrawn. However, there is a risk of cirrhosis or post-necrotic scarring where initial damage has been severe with bridging or submassive necrosis.

8.5.5 *Lesions resembling alcoholic liver disease*

Certain hydrophobic drugs such as perhexilene, used in the treatment of angina pectoris, impair normal hepatic phospholipid metabolism causing accumulation of myelin bodies in lysosomes. While this is not typical of alcohol-induced liver disease, it is associated nevertheless with hepatitis, Mallory body production and micronodular cirrhosis (Pessayre *et al.*, 1979). There may be a close similarity here to the liver damage which can arise in hereditary lipodystrophy (see Chapter 9).

The continuous administration of methotrexate to patients with psoriasis and leukaemia is commonly associated with mild portal inflammation, fibrosis and fatty change, while in some cases there is evidence of CAH which can progress to cirrhosis. There can be a close histological resemblance to chronic alcoholic liver disease (Fig. 8.17) although Mallory

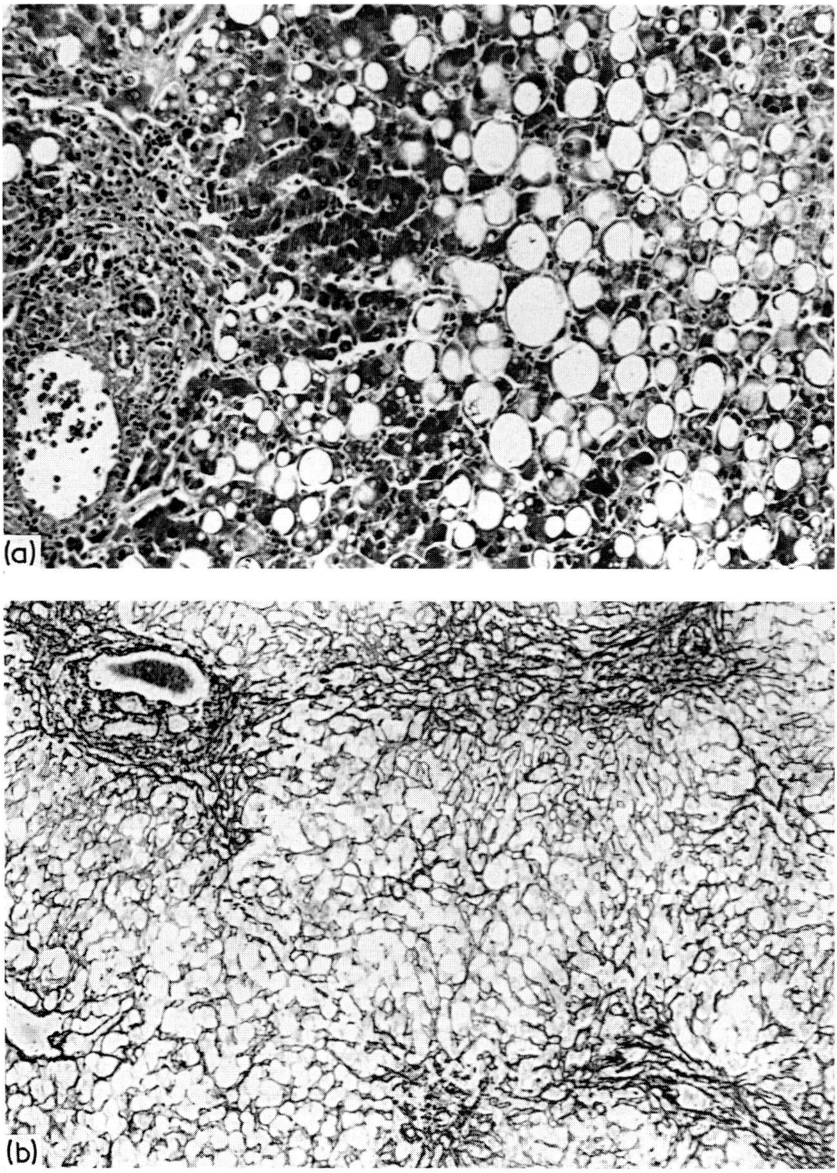

Fig. 8.17 (a) and (b) chronic methotrexate liver damage. There is fatty change together with chronic portal inflammation and fibrosis. ((a) HE, × 180; (b) reticulin stain × 72.)

bodies and neutrophil polymorph infiltration are not prominent. On the other hand, hepatocyte nuclear vacuolation is a feature of the methotrexate lesion. Susceptibility to liver damage seems to vary considerably, there being some direct correlation with age and mode of therapy; small continuous doses seem to be more hepatotoxic than intermittent large doses. Possibly previous alcohol consumption is also important, but the pathogenesis is uncertain. There is also a rough correlation with the quantity of drug administered and liver biopsy should be undertaken in patients who have received a total amount in excess of 1.5 g to exclude the presence of serious liver injury. The development of cirrhosis is slow and methotrexate therapy can be continued despite the presence of some liver damage if the dose is kept as low as possible and alcohol avoided (Zachariae *et al.*, 1980).

8.6 Physical liver injury

Traumatic injury may be followed by ischaemic necrosis of the liver associated with vascular damage usually adjacent to sites of laceration. Infiltration of viable tissue with neutrophil polymorphs can occur soon after injury (Fig. 8.18). This latter feature may be seen after mild trauma such as handling of the liver during abdominal operations and prior to wedge biopsy.

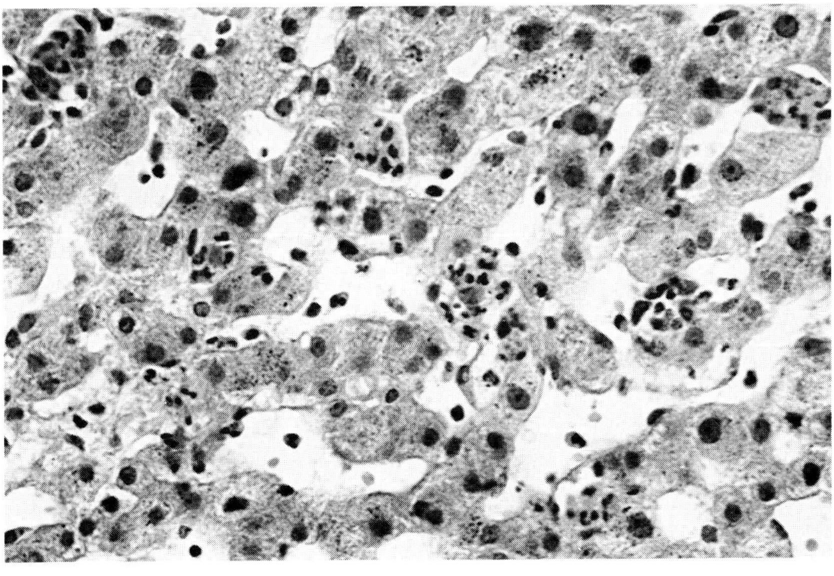

Fig. 8.18 Subcapsular neutrophil polymorph infiltration shown in wedge biopsy of liver and induced by relatively mild operative trauma (HE, × 450).

Surface necrosis of the liver may follow the application of high doses of radiation to the thorax and abdomen but this is now rarely if ever seen. Ionizing radiation may also induce vascular changes throughout the liver; these resemble veno-occlusive disease but can resolve spontaneously.

8.7 Miscellaneous changes

8.7.1 Foreign-body granulomas

These may be seen within the liver especially in the portal triads following intravenous injection of particulate matter such as ground-up tablets of narcotics taken by drug addicts (Fig. 8.19). Granulomas containing starch granules have been found following peritoneal dialysis. Mineral and vegetable oils may induce similar lesions in persons consuming large quantities of liquid paraffin or following the injection of drugs in an oily base as in the treatment of haemorrhoids (Fig. 8.20). Dincsoy *et al.* (1982) describe 44 cases, in many of whom the lesions were multiple and present in portal tracts and attached to the walls of hepatic venules. They were regarded as arising from the ingestion of mineral oils used as food additives or protective coatings and in polishing apples. More recently Cruickshank and Thomas (1984) have drawn attention to the high frequency of mineral oil granulomas found at autopsy in certain parts of

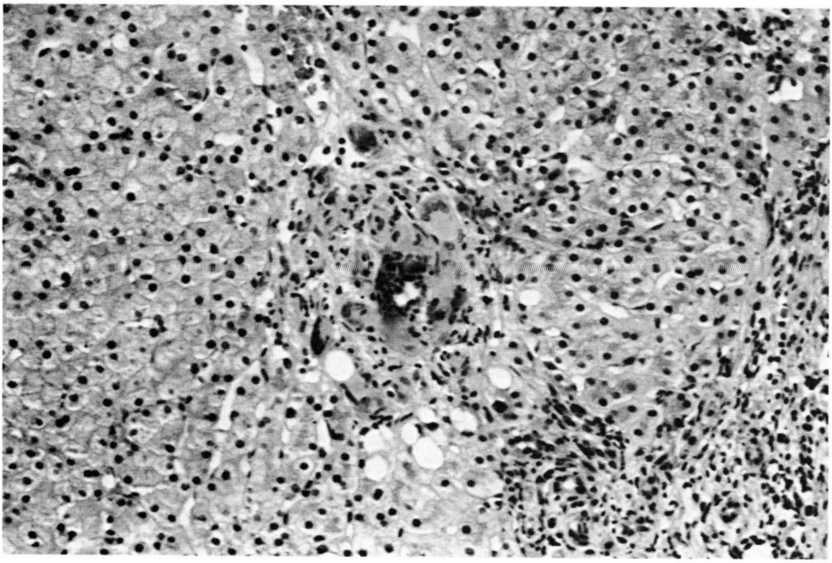

Fig. 8.19 Chronic hepatitis and foreign-body granuloma in the liver of a drug addict (HE, × 225).

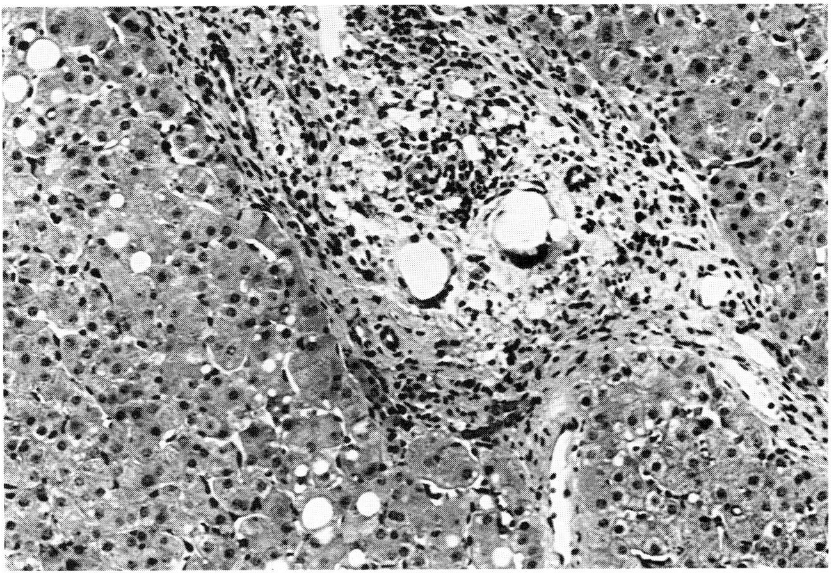

Fig. 8.20 Granulomatous portal hepatitis with foreign-body-type reaction to deposits of lipid, probably derived from ingested liquid paraffin (HE, × 200).

the world especially North America, Western Europe and Australia but not in Africa or in Central or South America. The spleen is most often involved but lesions have been found in liver, especially portal tracts, in about 50% of cases if care is taken to search for them. They are not fibrogenic and cause no serious disorder.

8.7.2 Lipochrome pigmentation

Granules of lipofuscin pigment may be conspicuous in hepatocyte cytoplasm in persons taking phenacetin. Pigment of a similar nature may be seen in Kupffer cells following infusions of dextran and other plasma expanders.

8.7.3 Haemosiderin pigmentation

Haemosiderin pigment in hepatocyte cytoplasm and Kupffer cells is seen following the ingestion of iron salts and after blood transfusions. Iron in Kupffer cells may also be conspicuous after the administration of chemicals which cause haemolytic anaemia, and in renal transplant patients on immunosuppression therapy.

8.7.4 *Anthracosis*

Granules of black carbon pigment ingested by macrophages may be seen in the portal tracts of some patients with severe pulmonary anthracosis (see Fig. 2.24). This is not associated with any other hepatic abnormality.

8.7.5 *Secondary phospholipidosis ('foam cell syndrome')*

There is hepatomegaly and abnormal liver function associated with fine vacuolation of hepatocyte cytoplasm, but all other liver cell types can be involved. This appearance is attributed to cytoplasmic autophagic vacuoles, multivesiculate bodies and enlarged secondary lysosomes containing myelin bodies. These myelin bodies may be seen as dark granules on light microscopy. They are present also in bone marrow and lymph nodes and stain blue with Giemsa stain. The condition is found in patients taking 4,4′-diethylaminoethoxyhexoestrol (coragil) as a coronary dilating agent. It seems to impair normal hepatic phospholipid synthesis and can affect experimental animals in the same way. Most reports come from Japan and we have no personal experience of the condition. The liver changes appear to be reversible when the drug is withdrawn (Itoh and Tsukada, 1973). Some other drugs, such as perhexilene and amiodarone, are now recognized which have a similar effect (Poucell *et al.*, 1984); the latter is a benzofuran derivative used to treat angina pectoris and cardiac arrythmia. They can also cause more serious hepatocyte damage with Mallory bodies and pericellular fibrosis which could be mistaken for alcoholic hepatitis (p. 181).

8.7.6 *Vascular lesions*

Budd–Chiari syndrome, veno-occlusive disease (Fig. 8.21) and peliosis hepatitis due to drugs are referred to in Chapter 10. Long-term methyl testosterone therapy may cause hepatocyte hyperplasia and accumulation of these cells under the endothelium of hepatic veins (Paradinas *et al.*, 1977).

8.7.7 *Perisinusoidal cell hyperplasia*

Carbon tetrachloride-induced acute liver injury stimulates the fibroblastic potential of these cells resulting in increased synthesis of collagen in Disse's space (McGee and Patrick, 1972). In hypervitaminosis A, and following the intake of certain drugs such as methotrexate, and chenodeoxycholic acid for gallstone dissolution, they also proliferate

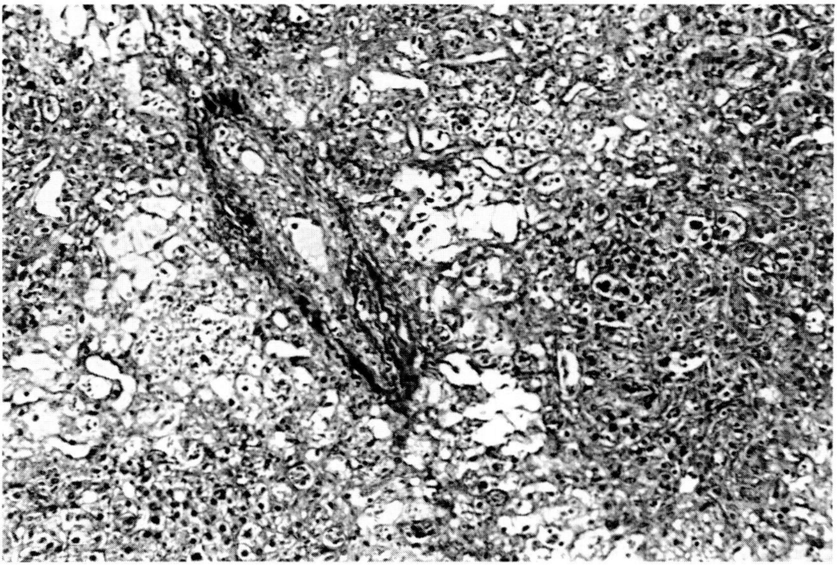

Fig. 8.21 Veno-occlusive disease induced by urethane therapy. There is considerable thickening of the centrilobular vein wall. Hydropic degeneration and necrosis of adjacent parenchyma are also apparent (MSB, × 132).

while retaining their fat-storage characteristic. In some of these cases there is hepatic fibrosis which may cause sinusoidal obstruction.

8.7.8 Primary liver tumours

Lesions associated with steroid therapy, thorotrast, vinyl chloride, etc., are referred to in Chapter 13.

8.7.9 Chemotherapy for malignant tumours

A minority of lymphoma cases so treated develop cholestasis and focal or irregular hepatic necrosis. Hepatic fibrosis may become prominent after successful elimination of the malignant condition, especially in children (Fig. 8.22) (Hutter *et al.*, 1960). It is impossible to state whether these histopathological appearances are related to the primary condition or to the chemotherapy. Centrilobular haemorrhagic necrosis or changes similar to veno-occlusive disease occur in a minority of cases of myeloma treated with urethane (Fig. 8.21). Hepatic arterial perfusion chemotherapy for secondary carcinoma may cause parenchymal damage and central venous sclerosis, especially following the use of mixtures containing 5-fluoro-2-deoxyuridine (Doria *et al.*, 1986).

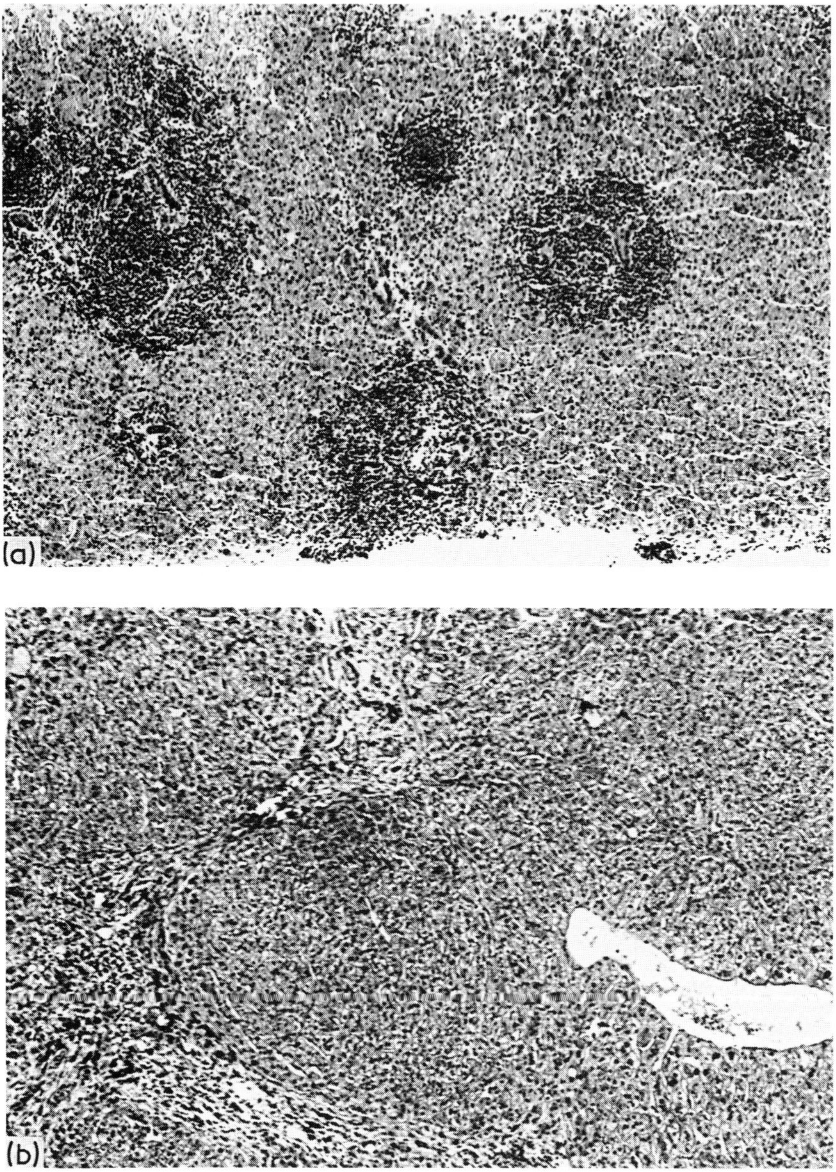

Fig. 8.22 (a) Follicular lymphoma involving liver; (b) liver 2 years later and after chemotherapy. The neoplastic condition is replaced by chronic portal inflammation and extensive fibrosis. ((a) and (b) HE, × 90.)

Fatal liver necrosis has been reported following withdrawal of chemotherapy in malignant disease. Possibly this is due to recovery of immune competence and rapid destruction of hepatocytes already infected with hepatitis virus (Galbraith *et al.*, 1975).

8.7.10 Drugs and pre-existing liver disease

Impaired liver function may enhance the toxic effects of drugs including those which may be given for the primary hepatic condition; thus liver damage in CAH is sometimes aggravated by mercaptopurine and azathioprine. Disulfiram and dipsam (cyanamide) used in the treatment of chronic alcoholism may occasionally cause severe liver damage which could be attributed wrongly to alcohol. However, disulfiram hepatitis is unlike alcohol hepatitis in being confined to portal and periportal zones and there may be large PAS-positive cytoplasmic inclusions especially in periportal cells which resemble Lafora bodies that occur in myoclonic epilepsy (Bruguera *et al.*, 1986). Toxic damage to tissues other than the liver may be enhanced by coexistent liver disease, as in ergot poisoning.

References

Bruguera, M., Lamar, C., Bernet, M. *et al.* (1986), Hepatic disease associated with ground-glass inclusions in hepatocytes after cyanamide therapy. *Arch. Pathol. Lab. Med.*, **110**, 906–10.

Cruickshank, B. and Thomas, M. J. (1984), Mineral oil (follicular) lipidosis: II histologic studies of spleen, liver, lymph nodes and bone marrow. *Hum. Pathol.*, **15**, 731–7.

Dincsoy, H. P., Weesner, R. E. and MacGee, J. (1982), Lipogranulomas in non-fatty human livers: a mineral oil induced environmental disease. *Am. J. Clin. Pathol.*, **78**, 35–41.

Doria, M. I., Shepard, K. V., Levin, B. *et al.* (1986), Liver pathology following hepatic arterial infusion chemotherapy. *Cancer*, **58**, 855–61.

Galbraith, R. M., Eddleston, A. L. W. F., Williams, R. *et al.* (1975), Fulminant hepatic failure in leukaemia and choriocarcinoma related to withdrawal of cytotoxic drug therapy. *Lancet*, **ii**, 528–30.

Hutter, R. V. P., Shipkey, F. H., Tan, C. T. C. *et al.* (1960), Hepatic fibrosis in children with acute leukaemia: a complication of therapy. *Cancer*, **13**, 288–307.

Itoh, S. and Tsukada, Y. (1973), Clinico-pathological and electron microscopical studies on a coronary dilating agent 4,4′-diethylamino ethoxyhexestrol-induced liver injuries. *Acta hepato-gastroent.*, **20**, 204–15.

Klatskin, G. and Kimberg, D. V. (1969), Recurrent hepatitis attributable to halothane sensitization in an anaesthetist. *N. Engl. J. Med.*, **280**, 515–22.

McGee, J. O'D. and Patrick, R. S. (1972), The role of perisinusoidal cells in hepatic fibrogenesis: an electron microscopic study of acute carbon tetrachloride liver injury. *Lab. Invest.*, **26**, 429–40.

McGee, J. O'D., Patrick, R. S., Rodger, M. C. and Luty, C. M. (1974), Collagen proline hydroxylase activity and ^{35}S sulphate uptake in human liver biopsies. *Gut*, **15**, 260–7.

Paradinas, F. J., Bull, T. B., Westaby, D. *et al.* (1977), Hyperplasia and prolapse of hepatocytes into hepatic veins during longterm methyltestosterone therapy: possible relationships of these changes to the development of peliosis hepatitis and liver tumours. *Histopathology,* **1,** 225–46.
Pessayre, D., Bentata, M., Degott, C. *et al.* (1977), Isoniazid-rifampicin fulminant hepatitis: a possible consequence of the enhancement of isoniazid hepatotoxicity by enzyme induction. *Gastroenterology,* **72,** 284–9.
Pessayre, D., Bichara, M., Feldmann, G. *et al.* (1979), Perhexiline maleate-induced cirrhosis. *Gastroenterology,* **76,** 170–7.
Poucell, S., Ireton, J., Valencia-Mayoral, P. *et al.* (1984), Amiodarone-associated phospholipidosis and fibrosis of the liver. *Gastroenterology,* **86,** 926–36.
Powell-Jackson, P. R., Tredger, J. M. and Williams, R. (1984), Hepatotoxicity to sodium valproate: a review. *Gut,* **25,** 673–81.
Read, A. E., Harrison, C. V. and Sherlock, S. (1961), Chronic chlorpromazine jaundice: with particular reference to its relationship to primary biliary cirrhosis. *Am. J. Med.,* **31,** 249–58.
Samuels, A. M. and Carey, M. C. (1978), Effect of chlorpromazine hydrochloride and its metabolites on Mg^{++} and Na^{+}, K^{+}-ATPase activities of canalicular-enriched rat liver plasma membranes. *Gastroenterology,* **74,** 1182–90.
Seaman, W. E., Ishak, K. G. and Plotz, P. H. (1974), Aspirin-induced hepatotoxicity in patients with systemic lupus erythematosus. *Ann. Intern. Med.,* **80,** 1–8.
Sherlock, S. (1983), Acute fatty liver of pregnancy and the microvesicular fat diseases. *Gut,* **24,** 265–9.
Starko, K. and Mullick, F. G. (1983), Hepatic and cerebral pathology findings in children with fatal salicylate intoxication: further evidence for a causal relation between salicylate and Reye's syndrome. *Lancet,* **i,** 326–9.
Swank, L. A., Chezfec, G. and Nemchausky, B. A. (1978), Allopurinol-induced granulomatous hepatitis with cholangitis and a sarcoid-like reaction. *Arch. Intern. Med.,* **138,** 997–8.
Utili, R., Boitnott, J. K. and Zimmerman, H. J. (1977), Dantrolene-associated hepatic injury: incidence and character. *Gastroenterology,* **72,** 610–6.
Weller, I. V. D., Cohn, D., Sierralta, A. *et al.* (1984), Clinical, biochemical, serological, histological and ultrastructural features of liver disease in drug abusers. *Gut,* **25,** 417–23.
Zachariae, H., Kragballe, K. and Sogaard, H. (1980), Methotrexate-induced liver cirrhosis. *Br. J. Dermatol.,* **102,** 407–12.

Further reading

Davis, M., Tredger, J. M. and Williams, R. (eds) (1981), *Drug Reactions and the Liver,* Pitman Medical, London.
Rubin, E. (1980), Iatrogenic hepatic injury. *Hum. Pathol.,* **11,** 312–31.
Sherlock, S. (1986), The spectrum of hepatotoxicity due to drugs. *Lancet,* **ii,** 440–4.
Strucker, B. H. C. and Spoelstra, P. (1985), in *Drug-Induced Disorders,* Vol. I (ed. M. N. Dukes), Elsevier Science, Amsterdam.

9 The liver in systemic and metabolic disorders

The liver becomes involved in many systemic and metabolic disorders. In some this involvement may dominate the clinical presentation, while in others it is slight or has little clinical significance. Our descriptions are confined to cases of particular importance because of their relative frequency or where liver biopsy has a role in reaching a correct diagnosis without which continuing morbidity or a fatal outcome may ensue through lack of appropriate specific therapy. Further information on the others, which are numerous but very uncommon, should be sought if required in appropriate paediatric or biochemistry texts.

9.1 Systemic disorders

9.1.1 Diabetes mellitus

Fatty change is frequently present in maturity-onset diabetes and can be severe; probably it is an accompaniment of obesity which is a common feature of such cases. It is less frequent and less severe in juvenile-onset (insulin-dependent) diabetics in whom hepatomegaly is due to excess glycogen in hepatocyte cytoplasm. Glycogenic vacuolation of nuclei (Fig. 9.1) is a common finding in diabetes, especially in association with fatty change, but it is found in other conditions such as Wilson's disease. Diabetic hyaline arteriolosclerosis may be evident in small branches of the hepatic artery.

The incidence of cirrhosis in diabetes is two to three times that of the rest of the community. The reasons for this are uncertain but it would appear that, in the majority, cirrhosis precedes the onset of diabetes (Batman and Scheuer, 1985).

In lipoatrophic diabetes (partial or complete lipodystrophy) there may be chronic liver disease similar to that due to alcohol abuse.

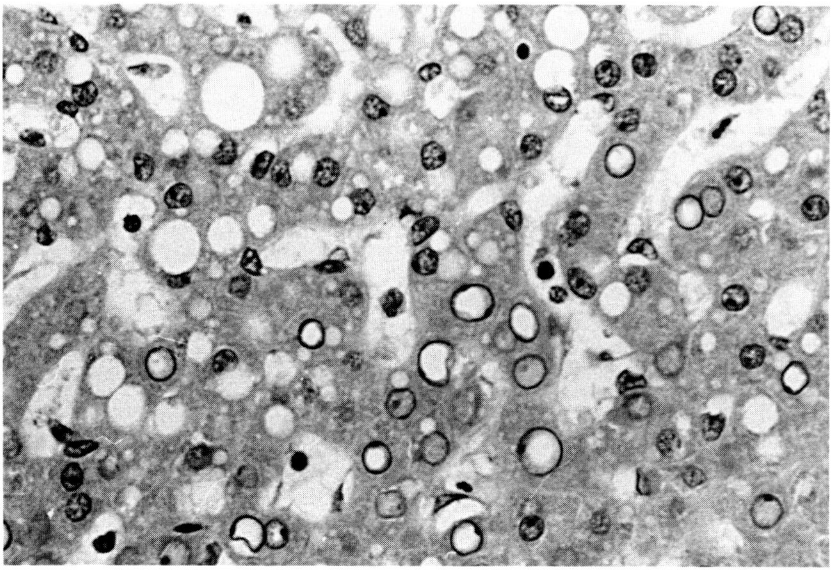

Fig. 9.1 Liver in diabetes mellitus. Most hepatocytes contain lipid vacuoles and many show glycogenic vacuolation of nuclei (HE, × 480).

9.1.2 Amyloidosis

The liver is involved in the majority of cases of both primary and secondary amyloidosis, but liver biopsy is necessary to confirm the diagnosis and assess its severity as the degree of liver dysfunction can be remarkably slight in advanced cases. Amyloid appears as a homogeneous pink extracellular material on routine HE sections. A number of stains have been used for its demonstration, including methyl violet and thioflavine T, but the most helpful are Sirius red and Congo red which stain amyloid pink. Under crossed prisms amyloid stained with the latter dyes shows a characteristic apple-green colour. It has a characteristic fibrillary ultrastructure, deposition beginning in Disse's space and leading to atrophy and loss of adjacent hepatocytes (Fig. 9.2), but in some milder cases it may have a periportal distribution. It is responsible also for severe hyaline thickening of the walls of portal tract blood vessels, especially arterioles, and in some cases this is the only hepatic manifestation (Fig. 9.3). It has been held that the distribution is predominantly vascular in primary and myeloma amyloidosis but it is probably not possible to differentiate primary and secondary amyloidosis from histopathology of the liver alone (Chopra *et al.*, 1984).

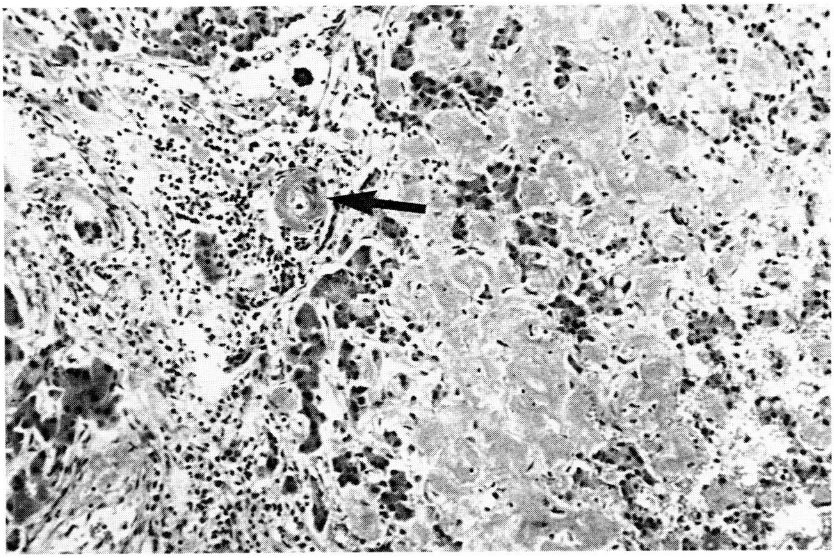

Fig. 9.2 Secondary amyloidosis of liver. There is amyloid deposition in a hepatic arterial branch (arrow) which appears hyalinized. In this case, amyloid has been extensively deposited along sinusoids out to the periphery of the lobule. There is marked hepatocyte atrophy in relation to the sinusoidal amyloid (HE, × 200).

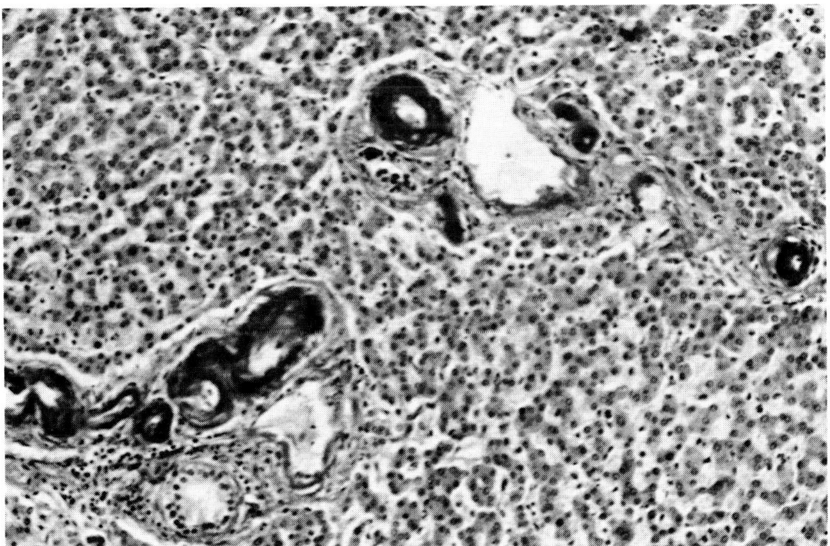

Fig. 9.3 Amyloid disease confined to intrahepatic branches of the hepatic artery (Sirius red, × 149).

Intrahepatic cholestasis has an uncommon association with primary amyloidosis and carries a very poor prognosis. Otherwise extensive amyloid disease can persist for many years without the development of fibrosis or cirrhosis.

Rarely, hyaline globules, 5–40 μm in diameter with histochemical and ultrastructural features of amyloid, may be present in Disse's space and as aggregates within portal tracts. They may coexist with non-globular amyloid elsewhere but their significance is unknown (Kanel *et al.*, 1981).

9.1.3 *Pregnancy*

The commonest causes of liver dysfunction in pregnancy are conditions such as viral hepatitis which are not exclusive to the pregnant state. The following disorders are unique to pregnant females:

(a) Intrahepatic cholestasis of pregnancy

This is rare and is characterized by the onset of pruritis and jaundice between the sixth and thirty-ninth week of pregnancy. On biopsy there is mild focal cholestasis most prominent in the centre of the lobule. Liver dysfunction disappears after delivery and has no long-term sequelae for the mother or child. Cholestasis may recur with subsequent pregnancies or if the mother is on oral oestrogenic contraceptives.

(b) Toxic fatty liver of pregnancy

This is extremely rare and carries a high mortality. It presents between the thirty-sixth week and term with persistent vomiting and abdominal pain followed by jaundice and renal failure. The liver is smaller than normal and yellow due to marked centrilobular fatty change and cholestasis. The lipid is present in numerous small vacuoles around a central nucleus; the appearances resemble tetracycline hepatotoxicity (see Fig. 8.6). In those patients who survive the liver returns to normal.

(c) Toxaemia of pregnancy

This is often accompanied by mild liver dysfunction but when jaundice occurs this signifies a poor prognosis. In contrast to autopsy findings, liver biopsy may be unimpressive showing only a few fibrin thrombi in peripheral sinusoids unassociated with inflammation. In fatal cases of eclampsia, fibrin thrombi occlude many sinusoids leading to periportal or submassive or massive necrosis with haemorrhage (see Fig. 10.22).

(d) Hyperemesis gravidarum

If severe, this condition may give rise to mild jaundice; in those cases which have been biopsied liver morphology is entirely normal.

9.1.4 *Reye's syndrome*

This is an uncommon febrile illness with encephalopathy and a high mortality rate which afflicts infants and, less commonly, older children and young adults, either sporadically or in small groups. It has a worldwide distribution but there are few reports from Western Europe. Patients are not usually jaundiced but do have abnormal liver function with elevated transaminase levels. Grossly, the liver is almost white or pale yellow-orange due to severe microvesicular fatty change, and so liver biopsy, which is necessary for the diagnosis although carrying some risk of haemorrhage, reveals a foamy appearance of all hepatocytes (Fig. 9.4). In some cases the fat is 'concealed', being seen as tiny droplets only in frozen sections stained for fat. In some others the periportal hepatocytes have larger fat globules or there may be periportal zonal necrosis, usually without much evidence of hepatitis. Autopsy will also reveal steatosis at other sites, particularly renal tubular epithelium. Recovery can take place within a few days with no residual hepatic damage.

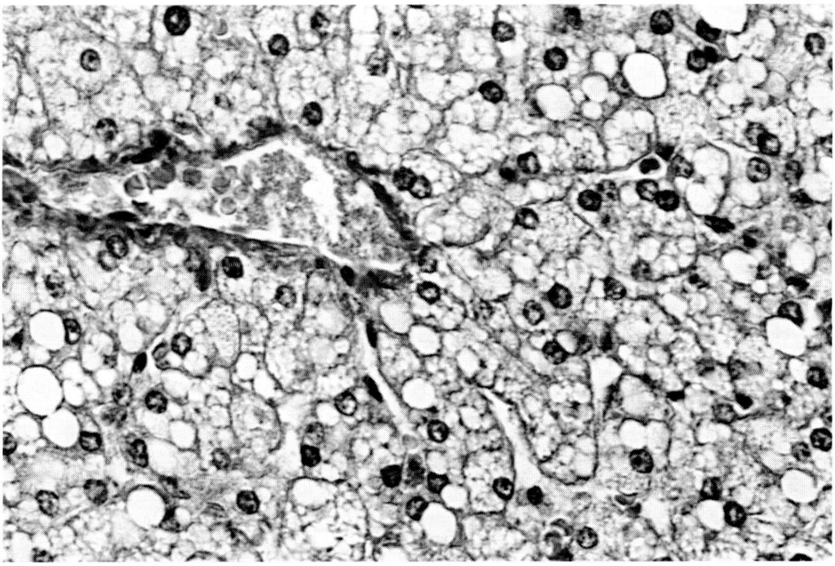

Fig. 9.4 Reye's syndrome. Some hepatocytes have prominent fat globules but the majority have a fine vacuolated appearance due to microvesicular fat which does not displace cell nuclei (HE, × 560).

As in all such conditions of unknown aetiology, there are several conflicting hypotheses relating to the cause. Many cases follow an acute viral infection, especially influenza B or varicella, but infective agents have been recovered rarely from the liver. Microvesicular fatty change is suggestive of a toxic cause and salicylates in particular have been incriminated in some cases (Partin *et al.*, 1982). Hypoglycaemia, hyperammonaemia and fatty acidaemia are other features suggestive of a primary metabolic disorder and certain cases have demonstrable abnormalities in urea cycle enzyme activities, especially ornithine transcarbamylase deficiency. Ultrastructural studies of liver biopsies invariably show mitochondrial abnormalities which could arise from various causes and be responsible for the same liver structural and metabolic defects.

9.2 Metabolic disorders

9.2.1 Iron storage disorders

(a) Idiopathic haemochromatosis

This is an inherited disease associated with excessive absorption of normal amounts of dietary iron from the bowel and its deposition from saturated transferrin as haemosiderin in the parenchymal cells of liver and many other organs. The mode of inheritance is probably autosomal recessive and there is an association with histocompatibility antigens A3 and B14 (not demonstrable in secondary haemochromatosis). It is not common and usually presents clinically between 40 and 60 years of age, especially in males, either by liver dysfunction or by cardiac failure or diabetes mellitus. As shown by liver biopsy, younger asymptomatic members of such families may have stainable haemosiderin in periportal hepatocytes (Fig. 9.5).

In well-established cases, biopsy shows profuse haemosiderin pigment in hepatocyte cytoplasm; it may be most prominent in periportal cells or have a diffuse lobular distribution. It is generally absent from Kupffer cells except where focal phagocytosis of degenerate hepatocytes has occurred, when it will be seen also in portal tract macrophages. Portal fibrosis develops with short projections of connective tissue from widened portal tracts giving these a holly-leaf appearance (Fig. 9.6). These projections expand to form bridging lesions and eventually there is cirrhosis of micronodular type (Fig. 9.7). Haemosiderin is often present in the epithelial lining of proliferating biliary channels in the fibrous septa (Fig. 9.8) and may be encrusted on collagen and elastic fibres (Fig. 9.9). Some advanced cases have a mixed or macronodular pattern of cirrhosis

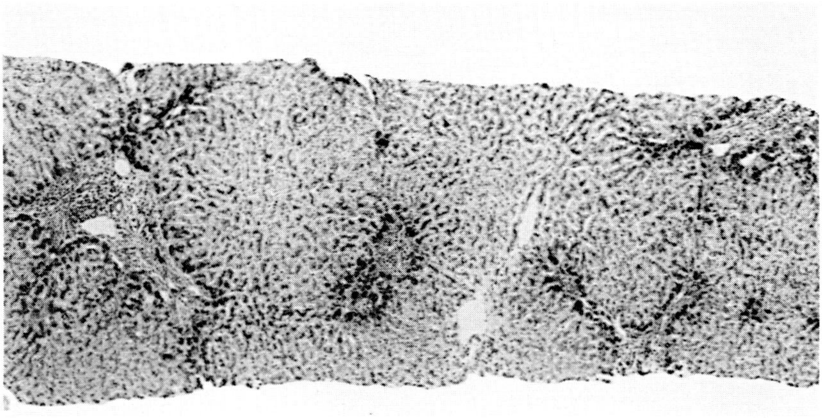

Fig. 9.5 Liver biopsy from asymptomatic 22-year-old brother of established case of idiopathic haemochromatosis. Dark-staining haemosiderin pigment is present in periportal hepatocytes and there is minimal portal inflammation (Perls' stain, × 40).

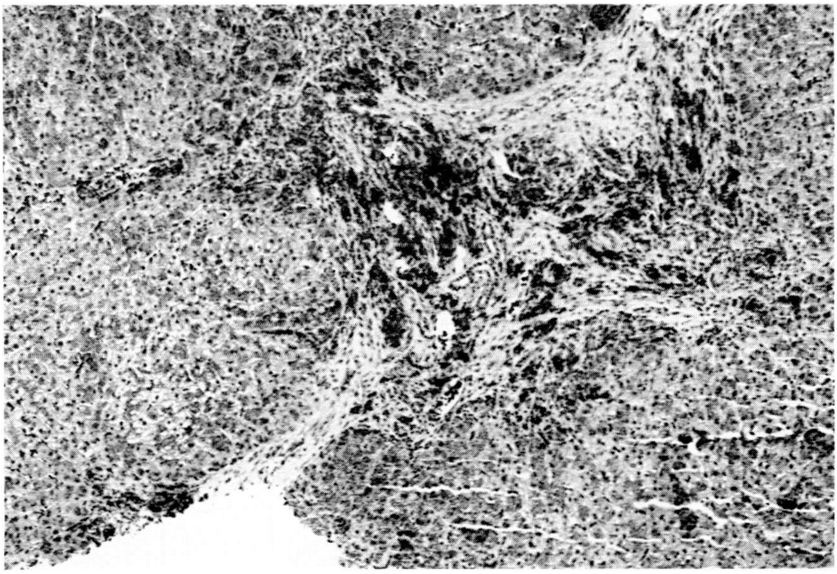

Fig. 9.6 Portal fibrosis in haemochromatosis with short blunt projections of fibrous tissue extending into liver parenchyma (Perls' stain, × 90).

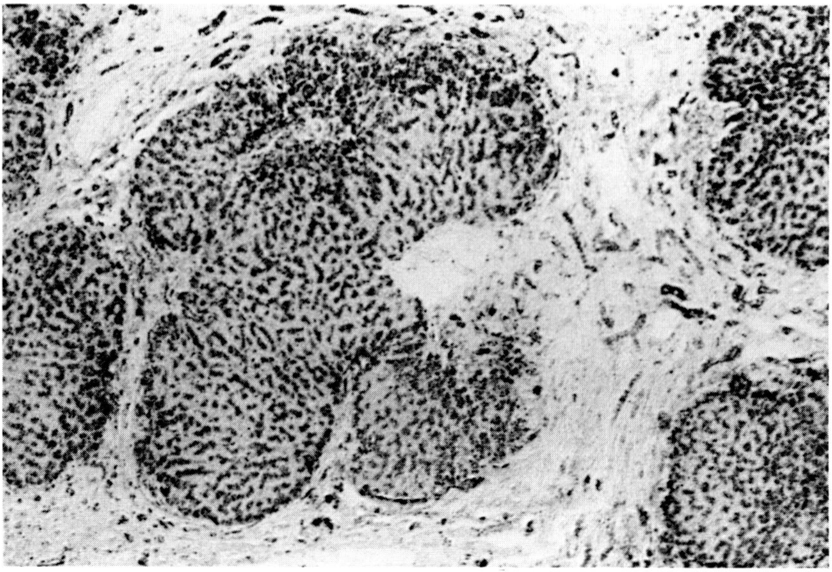

Fig. 9.7 Idiopathic haemochromatosis with cirrhosis (Perls' stain, × 52).

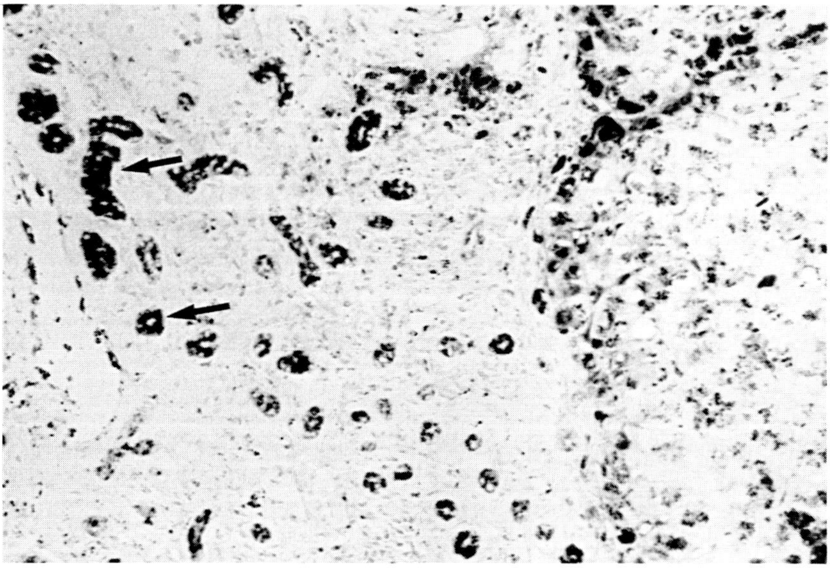

Fig. 9.8 Idiopathic haemochromatosis. In this fibrotic liver haemosiderin is present in highest concentration in bile ductular epithelium (arrows) (Perls' stain, × 200).

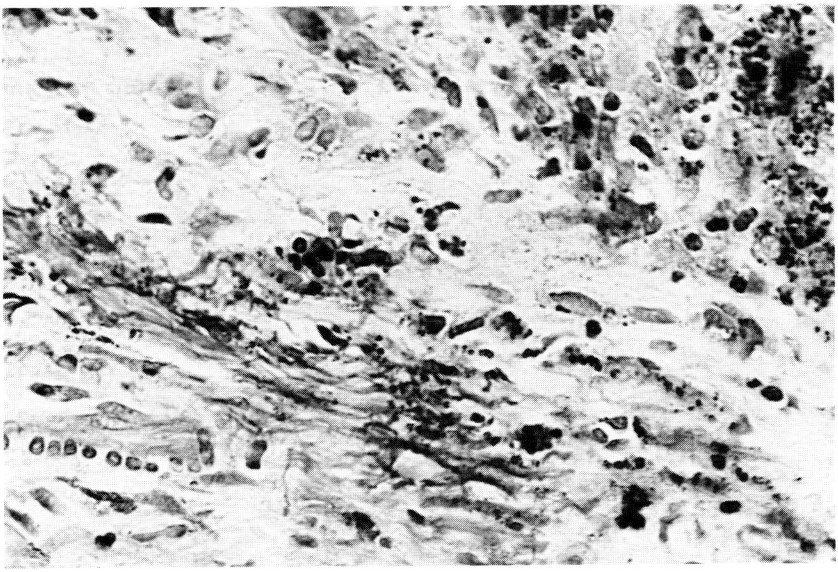

Fig. 9.9 Idiopathic haemochromatosis. Connective tissue fibres have haemosiderin encrusted on their surfaces (Perls' stain, × 560).

and there is an association with primary liver cell cancer; haemosiderin is absent from the cancer cells (see Fig. 13.9).

(b) Secondary haemochromatosis

(i) Cases associated with anaemia. This occurs especially where there is defective erythropoiesis accompanied by exaggerated absorption and mobilization of iron and its deposition, particularly in periportal liver cells. Iron deposition is exaggerated also in cases with severe haemolysis and in those receiving multiple blood transfusions when haemosiderin is seen especially in Kupffer cells. Again there is widening of portal tracts due to fibrosis, which expands outwards with some destruction of limiting cell plates and proceeds to cirrhosis. Cirrhosis is seen most frequently in primary haemolytic anaemia due to thalassaemia and following multiple transfusions. The histopathological features can be very similar to idiopathic haemochromatosis except that iron deposition in Kupffer cells is more prominent and the cirrhosis more active (Fig. 9.10), but the distinction can be difficult in the absence of other clinical information.

(ii) Cases secondary to excessive intake of iron. This can be iatrogenic and is seen in the Bantu from contamination of beverages with iron

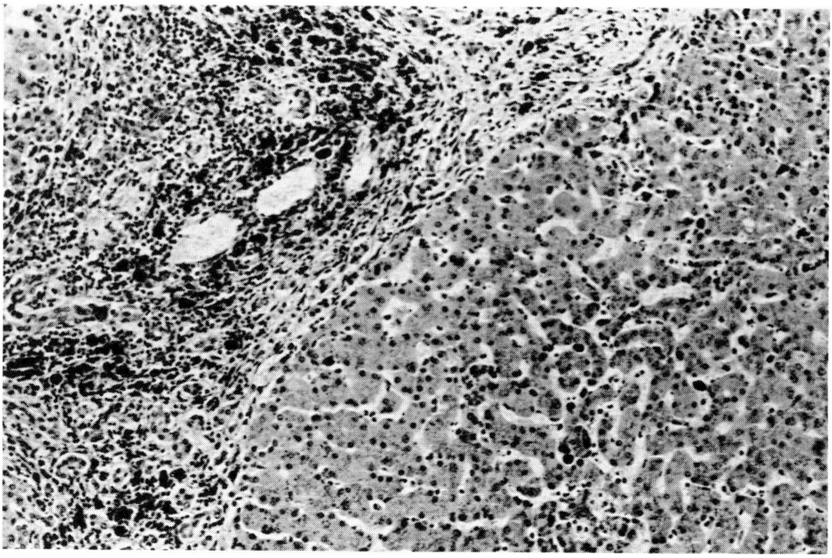

Fig. 9.10 Haemochromatosis secondary to multiple blood transfusions. There is severe portal fibrosis and inflammation including numerous macrophages with ingested dark haemosiderin pigment, present also in Kupffer cells. Most hepatocytes had fine granules of haemosiderin not evident at this magnification (HE, × 134).

derived from cooking or storage utensils. There may be a simple hepatocyte siderosis but severe fibrosis or cirrhosis occurs in a minority of cases.

(iii) Cases secondary to cirrhosis from other causes. This is seen in some cases of alcoholic cirrhosis with folate deficiency and in association with natural or acquired porto-caval shunts. Compared with idiopathic haemochromatosis with cirrhosis, the amount of haemosiderin is generally less, it is often confined to the periphery of cirrhotic nodules (Fig. 9.11) and does not usually involve bile duct epithelium or connective tissue fibres.

(iv) Porphyria cutanea tarda. Hepatic deposition of iron is prominent in many of these cases but the relationship to the commonly associated alcoholism is uncertain; there may also be a more direct association with the porphyrin metabolic abnormality as removal of hepatic iron by phlebotomy can be followed by clinical improvement and relapse is induced by subsequent enteral administration of iron.

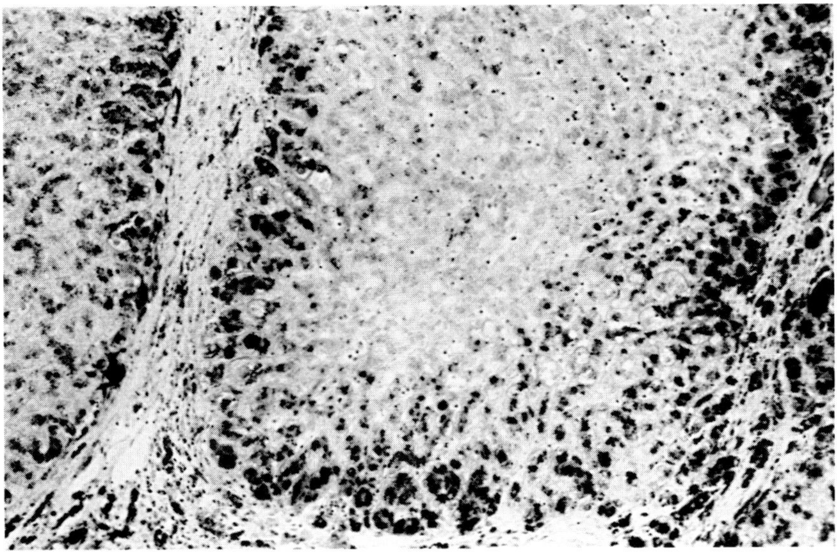

Fig. 9.11 Alcoholic cirrhosis with prominent haemosiderin deposition at periphery of nodules (Perls' stain, × 110).

(c) Haemosiderosis

This term indicates the presence of iron deposits in the liver but without any evidence of toxic damage.

(i) Liver parenchymal siderosis. This may be seen in early mild idiopathic haemochromatosis and during a relapse of pernicious anaemia (Fig. 9.12).

(ii) Kupffer cell siderosis. Haemosiderin pigment in Kupffer cells may be the only biopsy finding of note in some cases of severe haemolytic anaemia and following blood transfusion (Fig. 9.13). It occurs in some cases of anaemia associated with chronic infection and malignancy when there seems to be a resistance to the release of iron from reticulo-endothelial cells. It may occur also in association with extramedullary haematopoiesis and after parenteral iron administration. It may be a consequence of the destruction and phagocytosis of parenchymal cells, for example during recovery from viral hepatitis and obstructive biliary disease. A little haemosiderin pigment may be seen also in hepatocytes in some of these cases.

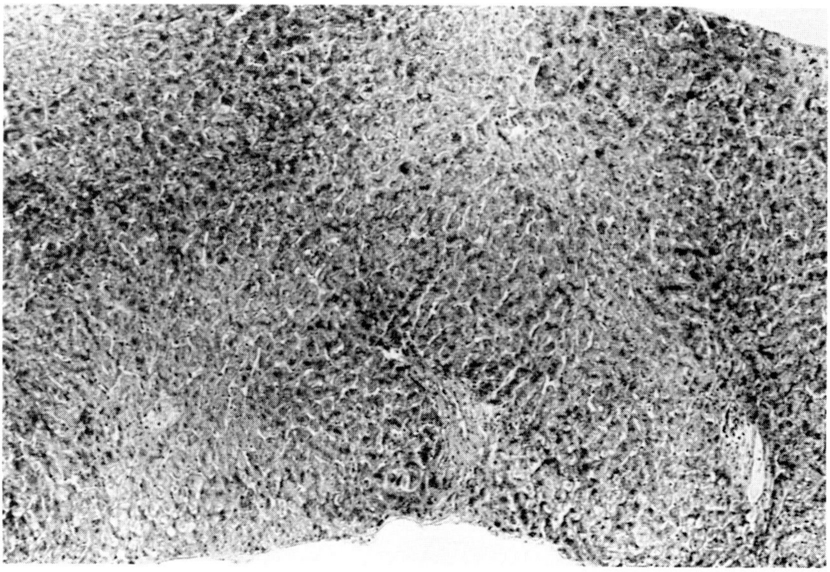

Fig. 9.12 Hepatic haemosiderosis during pernicious aenamia relapse (Perls' stain, × 80).

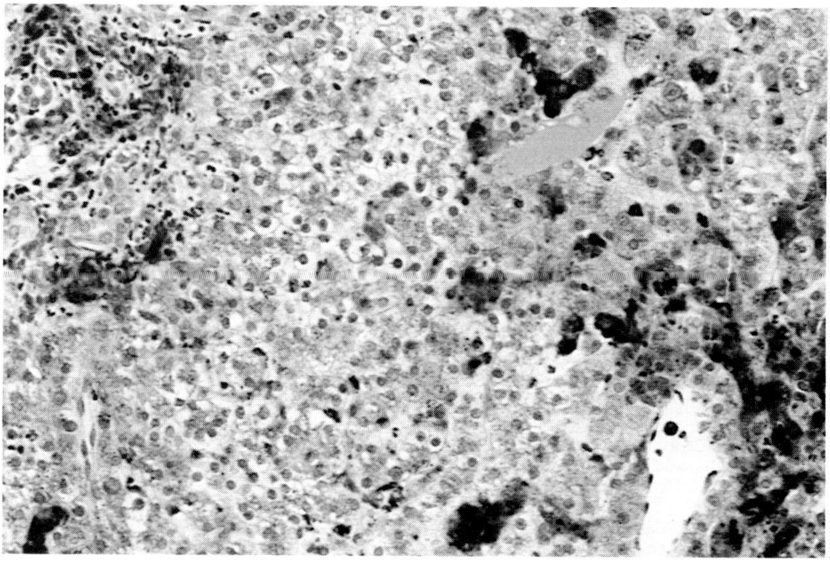

Fig. 9.13 Kupffer cell haemosiderosis following blood transfusion (Perls' stain, × 225).

9.2.2 Copper storage disorders (Sternlieb, 1980)

(a) Wilson's disease (hepatolenticular degeneration)

Wilson's disease is an inborn error of copper metabolism in which excessive amounts of the metal are deposited in liver, brain and iris; in the latter it is responsible for Kayser-Fleischer rings. The disease is inherited as a recessive condition and presents clinically from 6 to 40 years. It is not generally realized that even in children about 50% of patients seek advice because of liver rather than neurological disease. Liver biopsy of young siblings of established cases may show steatosis and glycogen vacuolation of hepatocyte nuclei (Fig. 9.14) but the excess copper is distributed rather diffusely and demonstrable only by chemical analysis. As the disease progresses there is evidence of chronic hepatitis and fibrosis (Fig. 9.15), leading to micronodular or macronodular cirrhosis. Diffusely distributed copper-rich lipofuscin granules in hepatocytes may now be demonstrable by rubeanic acid or rhodanine staining (Fig. 9.16) and CBP by orcein staining, but the distribution can be irregular and the results of these procedures unreliable. Electron microscopy shows complex hepatocyte inclusions (Fig. 9.17) which are probably lysosomal in type; this is a helpful finding in adults presenting with an apparent cryptogenic

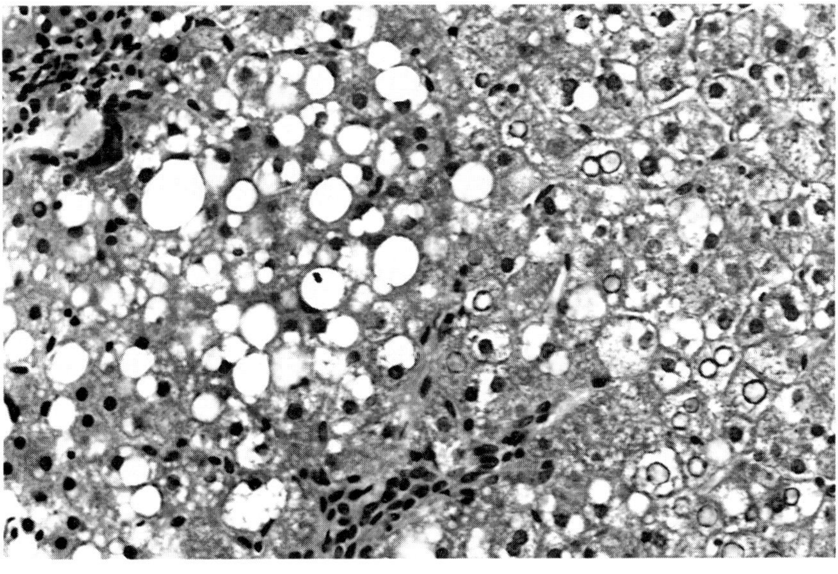

Fig. 9.14 Biopsy from young sibling of a fatal case of Wilson's disease. Steatosis and glycogen inclusions in hepatocyte nuclei are prominent. Portal inflammation is slight and there was minimal liver dysfunction (HE, × 286).

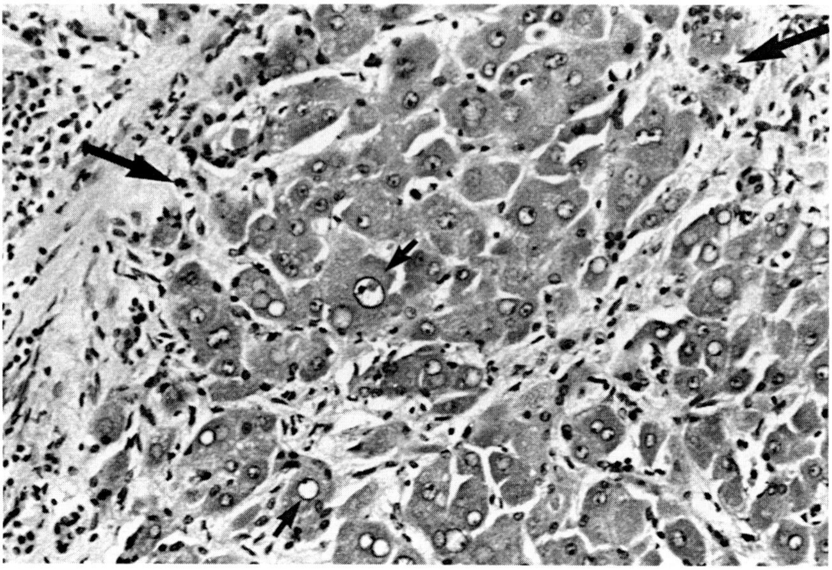

Fig. 9.15 Liver in Wilson's disease. There is chronic active hepatitis with piecemeal necrosis (large arrows) and glycogenic vacuolation of nuclei (small arrow). Fatty change and Mallory bodies which are also features of Wilson's disease are not present in this biopsy (HE, × 480).

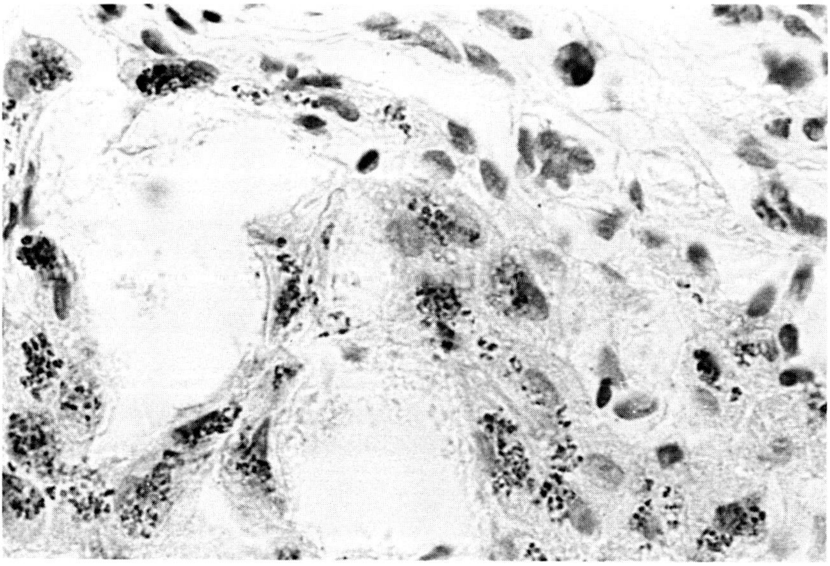

Fig. 9.16 Liver in Wilson's disease. Copper-containing granules are evident in many hepatocytes. This may occur only in isolated foci as in this case (rubeanic acid, × 640).

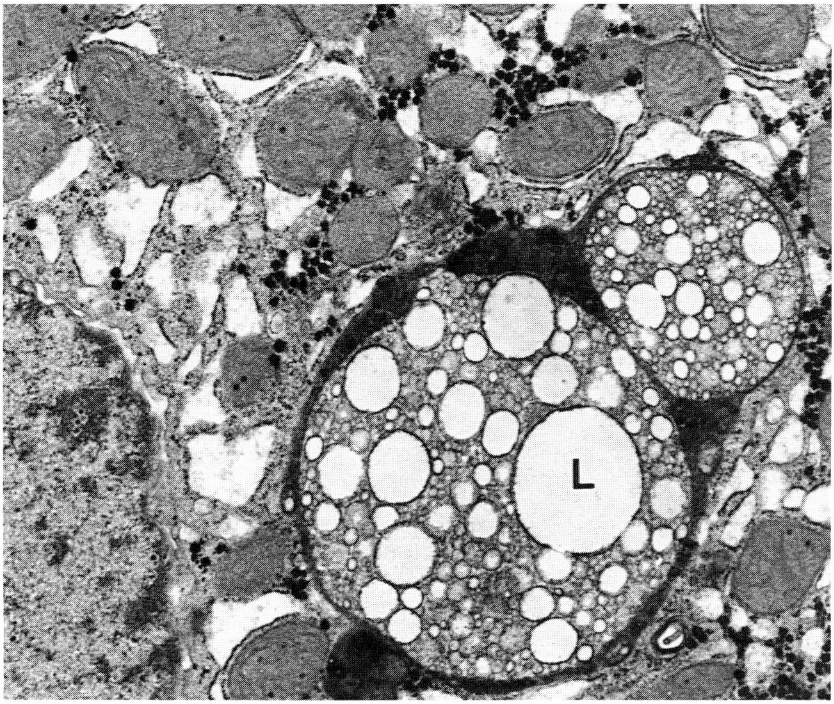

Fig. 9.17 Electron micrograph of the liver in Wilson's disease. Complex lysosomal structures (L) in hepatocytes are typical of this disease (uranyl acetate and lead citrate, × 19 800).

cirrhosis. Sometimes liver damage is severe and Wilson's disease should always be considered in young persons with severe aggressive CAH. In some others, the histopathology mimics alcoholic hepatitis because of fatty change, fibrosis and Mallory bodies.

(b) Copper retention in other liver diseases

As bile is the main vehicle for copper excretion, hepatic copper and CBP can be detected readily in many cases of chronic biliary diseases, especially primary biliary cirrhosis. It is uncertain what role the retained copper may have in contributing to hepatocyte damage in such cases. Compared to Wilson's disease there is no cerebral damage, the serum ceruloplasmin level is normal or elevated and the copper and CBP have a periportal distribution in the liver; the incidence of positive staining for these substances is also higher in biliary disease. Copper retention is also demonstrable in certain other chronic liver diseases such as alcoholic cirrhosis but is rarely, if ever, found in acute liver injury or in CPH or CAH

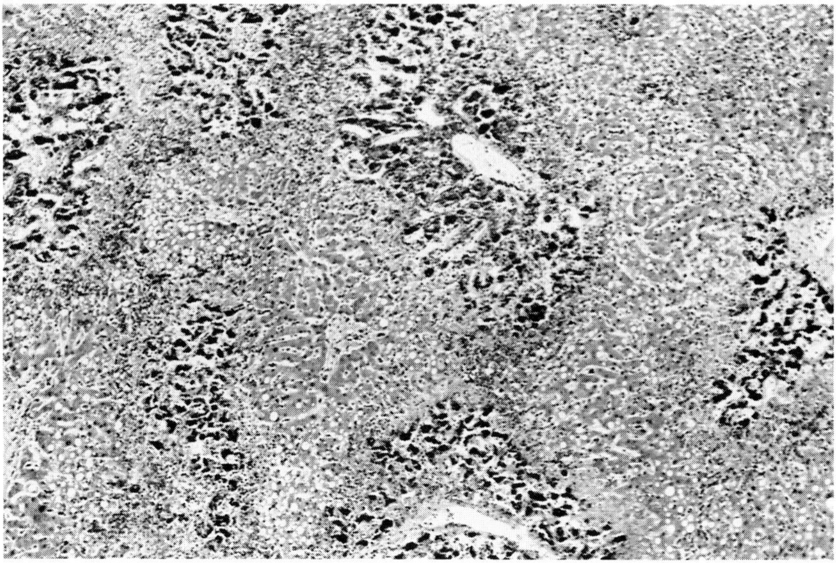

Fig. 9.18 Prominent dark granules of calcium salts have a centrilobular distribution in the liver of a fatal case of renal failure associated with secondary hyperparathyroidism and hypervitaminosis D (HE, × 56).

in adults. It is a prominent feature in Indian childhood cirrhosis where again there is no cerebral lenticular degeneration or depression of serum ceruloplasmin.

9.2.3 *Calcinosis of liver*

Rarely in calcinosis, HE sections of the liver show bluish-black granules of calcium salts, especially in the centrilobular zones (Fig. 9.18).

9.2.4 *Disorders of bilirubin metabolism*

(a) Conditions associated with unconjugated hyperbilirubinaemia

Gilbert's syndrome is the diagnosis for fairly common mild attacks of jaundice in young adults with no haemolytic disorder or other detectable reason for their unconjugated hyperbilirubinaemia. Liver biopsy may show prominent lipofuscin granules in hepatocytes but usually there is no structural abnormality to be seen on light microscopy. AER hypertrophy is a striking feature of some cases (Fig. 9.19) (McGee *et al.*, 1975), but its complete absence in others has suggested the possibility of more than one mechanism in the syndrome (Dawson *et al.*, 1979); some cases

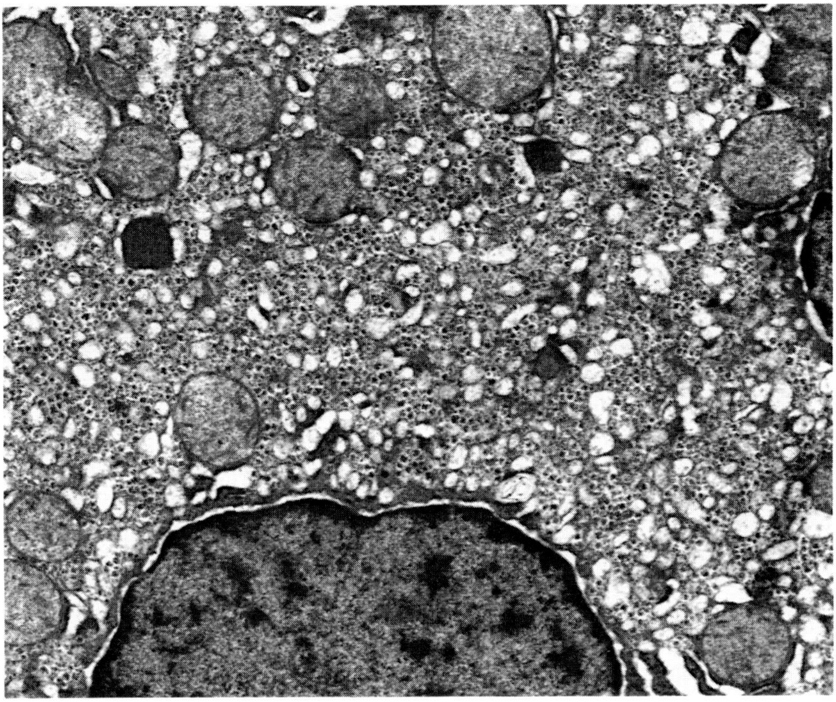

Fig. 9.19 Ultrastructure of hepatocytes in Gilbert's syndrome. The hepatocytes contain abundant smooth endoplasmic reticulum (uranyl acetate and lead citrate, × 20 000).

could arise from defective uptake of serum bilirubin by hepatocytes and others by faults in its transportation or conjugation within the cell.

Defective bilirubin conjugation has more serious significance in infants, especially in those suffering from Crigler–Najjar syndrome in whom there is often severe and fatal hepatocyte damage and kernicterus. Less severe and transient defective bilirubin conjugation in neonates may be an effect of maternal circulating steroid-like substances (Lucey–Driscoll syndrome) or from ingestion of steroids in maternal milk. Jaundice in infants may follow administration of certain drugs such as aspirin which compete with bilirubin for the conjugating activity of hepatocytes.

(b) Conditions associated with conjugated hyperbilirubinaemia without biliary obstruction or intrahepatic cholestasis

Such cases have persistent hyperbilirubinaemia with intermittent attacks of jaundice. Unlike Gilbert's syndrome the circulating bilirubin is

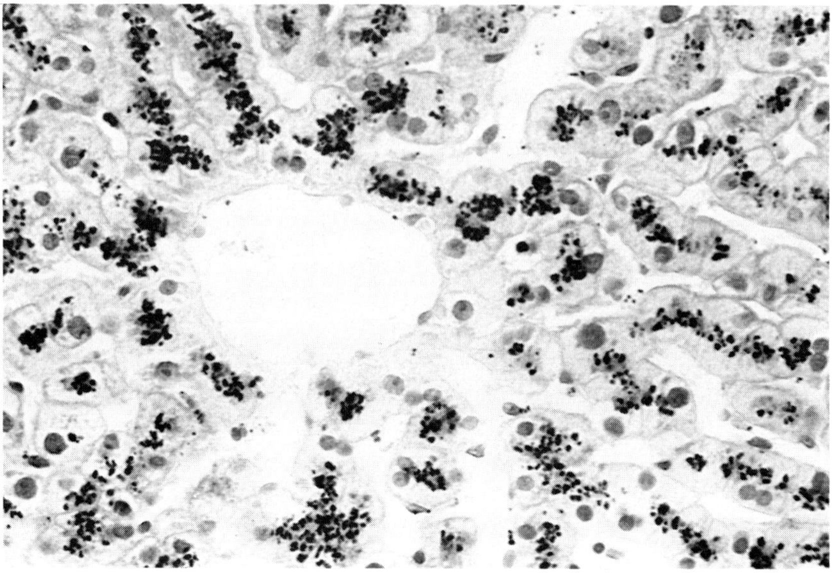

Fig. 9.20 Dubin–Johnson syndrome. The centrilobular hepatocytes contain abundant dark-brown granules which are Fontana positive (Fontana, × 400).

conjugated in type and liver biopsy may show coarse granules of dark brown or black pigment, not bile, in hepatocyte cytoplasm (Dubin–Johnson syndrome) (Fig. 9.20). This pigment gives a dark brown or grey colour to the liver on naked-eye examination. It is well seen in HE preparations and stains readily by the Fontana method. The condition is associated with defective secretion of bile pigment and bromsulphthalein but there is no impairment of bile salt secretion and no cholestasis. Some cases have all the features of this disease but lack the pigmentation (Rotor syndrome).

(c) Inherited cholestatic syndromes

These are described in Chapter 3.

9.2.5 *Disorders of porphyrin metabolism*

(a) Hepatic porphyrias (Cortés *et al.*, 1980; Lefkowitch and Grossman, 1983)

Of the various porphyrias in which the major accumulation of porphyrin occurs in the liver, significant hepatic damage may be found in acute intermittent porphyria and in porphyria cutanea tarda, especially the latter. Frequent biopsy findings are fatty change, deposition of haemosiderin in hepatocytes and Kupffer cells, glycogen inclusions in hepatocyte nuclei and acidophilic necrosis of single hepatocytes. Not infrequently, there are granuloma-like aggregates of Kupffer cells and chronic inflammatory cells in the liver lobules, and these contain haemosiderin and ceroid pigment. There may also be centrilobular fibrosis and portal inflammation but rarely of the severity of CAH. In some cases, the appearances are very similar to alcoholic hepatitis with Mallory bodies and it may be significant that porphyria cutanea tarda can occur in alcoholics. Over the years there is gradual deterioration in liver structure and function, often terminating in cirrhosis.

A constant feature of every case is red fluorescence of the biopsy tissue exposed to ultraviolet light. This is evident in the gross specimen and is well seen in unfixed frozen or alcohol-fixed sections. Fluorescence is related to water-soluble birefringent needle-shaped cytoplasmic inclusions of porphyrin in both hepatocytes and Kupffer cells. These needles are about the length of hepatocyte nuclei.

(b) Erythropoietic porphyrias

In erythropoietic protoporphyria there is often severe liver damage including hepatitis or cirrhosis. The main histological feature is deposition of dark brown protoporphyrin pigment in bile canaliculi, interlobular bile ducts and Kupffer cells. In unfixed frozen sections this gives intense red fluorescence on exposure to ultraviolet light; it is birefringent and under polarized light consists of granules and small Maltese cross structures. Sheaves of electron-dense crystals may be seen on electron microscopy of hepatocytes and Kupffer cells.

9.2.6 Disorders of carbohydrate metabolism

(a) Glycogen storage disease

Depending on the specific enzyme defect there are ten recognized forms of the disease with some liver involvement in all except type V and type VII. All arise in childhood causing hepatomegaly and stunted growth, and in the majority there is some degree of hypoglycaemia, hyperlipidaemia and mild liver dysfunction. Liver biopsy can be helpful in the diagnosis but this depends ultimately on the demonstration of a specific enzyme defect in the biopsied tissue. These enzymes are very labile and it is necessary to use only fresh material which is quickly frozen and transported rapidly to an appropriate centre for biochemical analysis.

In type I (von Gierke's disease) deficiency of glucose 6-phosphatase leads to gross accumulation of glycogen in hepatocytes which are swollen with clear cytoplasm, small centrally placed nuclei and distinct margins, and this imparts a mosaic-like appearance to the liver parenchyma (see Fig. 2.7). There may be numerous nuclear glycogen inclusions especially in periportal cells. In patients who survive for many years single or multiple liver cell adenomas may appear and, occasionally, liver cell carcinoma. Chronic hepatitis and cirrhosis are not sequelae but many die in childhood from infection, keto-acidosis or gout. Type III (Forbes' disease) and type VI (Hers' disease) are similar but less severe so that survival into adult life is usual but with the risk of hepatic fibrosis or even cirrhosis. In type II (Pompe's disease) there is no excess hepatic glycogen but marked hepatocyte vacuolation from lysosomal enlargement. In type IV (brancher enzyme deficiency) the defective glycogen molecules are relatively insoluble and accumulate in periportal hepatocytes as hyaline inclusions causing nuclear displacement; these resemble Lafora bodies (p. 223) and are strongly PAS positive. Extensive fibrosis and cirrhosis may ensue.

(b) Galactosaemia

Due to deficiency of galactose-1-phosphate uridyl transferase, galactose accumulates in liver parenchyma and other tissues unless this sugar or its usual source, lactose, be excluded from the diet. The enzyme defect can be demonstrated in hepatic tissue, but much more readily in circulating erythrocytes. Jaundice appears within a week of birth and there may be early death from infection, liver failure or haemorrhage. Some cases survive for many months when they develop chronic liver disease including cirrhosis, and also cataract from glycogen accumulation in the lens.

Liver biopsy in the early stages shows fatty change and cholangitis with prominent bile ductular proliferation and cholestasis in these channels. Within a few weeks there is a characteristic acinar arrangement of many hepatocytes, some containing bile plugs, while steatosis persists (Fig. 9.21). Portal fibrosis progresses to cirrhosis in 3–6 months. Early diagnosis is very important as removal of the cause through exclusion of natural milk from the diet can be followed by reversal of these liver changes and clinical improvement. Resistance to galactose seems to develop in survivors at a later date.

(c) Hereditary fructose intolerance

This has similar clinical and pathological features to galactosaemia but is due to deficiency of fructose-1-phosphate aldolase. Fructose from the

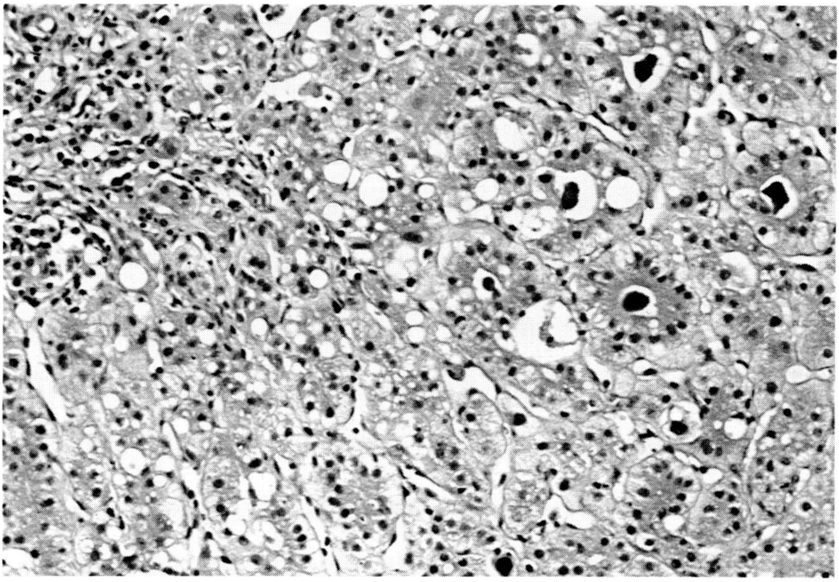

Fig. 9.21 Galactosaemia. There are several acinar structures lined by hepatocytes and containing dark bile pigment. Fatty change and mild portal inflammation are also present (Van Gieson's stain, × 225).

diet, usually from its precursor sucrose, accumulates in the tissues and there is fructosaemia and fructosuria. The illness may become less severe in children surviving more than 6 months. Liver biopsy shows fatty change, fibrosis or cirrhosis and sometimes giant cell transformation.

9.2.7 *Disorders of glycoprotein metabolism*

(a) Alpha-1-antitrypsin deficiency

Alpha-1-antitrypsin is a serum glycoprotein with protease-inhibiting properties. It is synthesized in the liver and accumulates in hepatocytes in alpha-1-antitrypsin deficiency. The protease inhibitor (Pi) system has 30 alleles identifiable by electrophoresis and in normal persons the MM phenotype is the most common. Alpha-1-antitrypsin deficiency is associated with the Z allele and in those having the homozygous ZZ phenotype the glycoprotein in serum is only about 20% normal; in heterozygotes with one Z gene the level is about 60% normal.

Apart from liver damage there is an association between alpha-1-antitrypsin deficiency and panlobular pulmonary emphysema involving the lower lobes and, occasionally, nephritis. Both hepatic and pulmonary

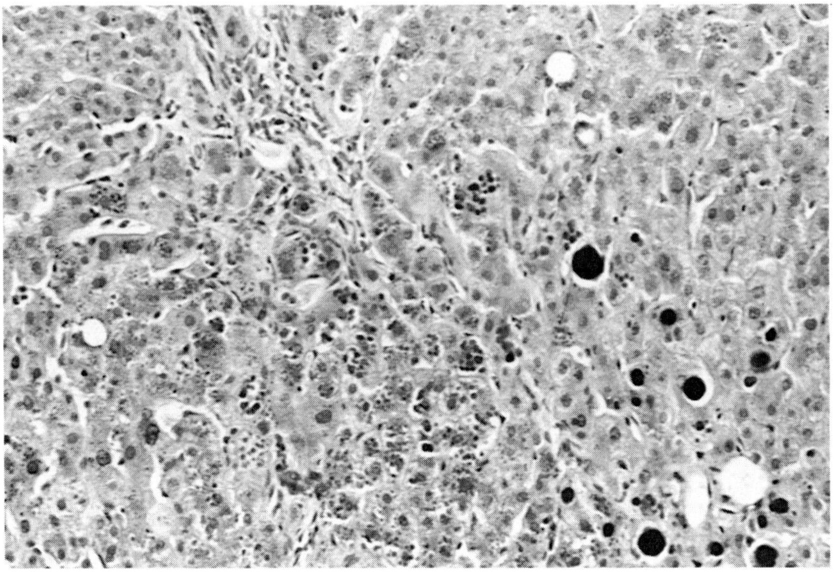

Fig. 9.22 Alpha-1-antitrypsin deficiency. There are several prominent spherical inclusion bodies and a large number of tiny globules within hepatocytes (PAS-diastase, × 225).

disease can coexist in one patient or be present separately. Liver involvement is recognized by PAS-positive diastase-resistant round bodies in hepatocyte cytoplasm, each measuring 1–40 μm in diameter (Fig. 9.22). Multiple inclusions may be seen in a single cell and the affected cells tend to have a periportal distribution or, in cirrhotic cases, are at the periphery of nodules. However, the distribution throughout the liver is often irregular and may be missed in a needle biopsy specimen. These hyaline bodies represent alpha-1-antitrypsin retained within dilated rough endoplasmic reticulum cisternae (Fig. 9.23); this glycoprotein may be abnormally deficient in sialic acid but, nevertheless, can be positively identified by immunohistochemical means (Fig. 9.24). In a minority of adult cases, probably not more than 20%, there is cirrhosis, usually of macronodular type, and this can progress to hepatocellular carcinoma. The inclusions just described should not be confused with morphologically similar but

Fig. 9.23 Electron micrograph of a hepatocyte from a case of alpha-1-antitrypsin deficiency. The antiproteinase (arrow) is present in large dilated sacs of rough ▶ endoplasmic reticulum (uranyl acetate and lead citrate, × 12 000).

Fig. 9.24 Liver in alpha-1-antitrypsin deficiency. The alpha-1-antitrypsin is demonstrated by the immunoperoxidase procedure in peripheral hepatocytes ▶ (arrows) (× 480).

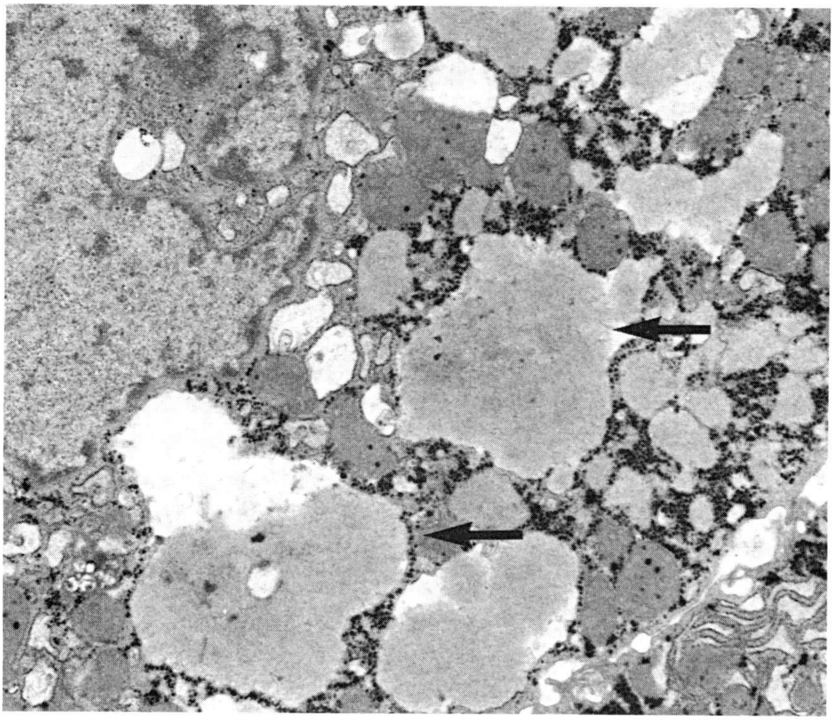

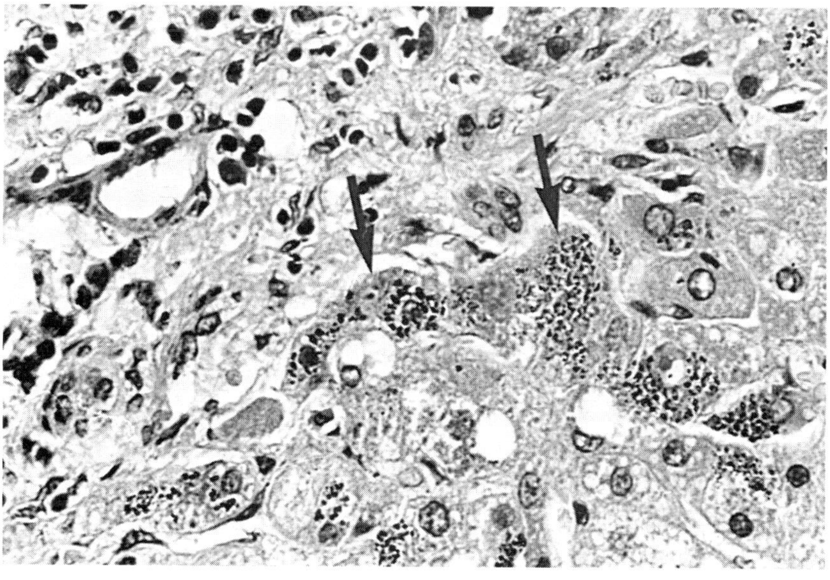

immunohistochemically negative bodies sometimes seen in centrilobular hepatocytes in venous congestion and in hepatocellular carcinoma (Qizilbash and Young-Pong, 1983).

Alpha-1-antitrypsin deficiency may also affect neonates and young infants causing jaundice and other evidence of liver dysfunction. Liver biopsy may show cholestasis or giant cell transformation but the typical hyaline inclusions are not seen before the age of 3 months. Some of these infants develop cirrhosis with severe or fatal liver failure, a histological feature being the presence of very thick bands of fibrous tissue between nodules. The majority recover normal liver function and it is not possible to assess prognosis from liver biopsies taken during the acute phase of the illness.

(b) Mucopolysaccharidoses

Of the seven recognized varieties there is liver involvement with similar histological features in all except types V and VI. Liver biopsy is rarely undertaken as a correct diagnosis is usually made from the clinical features, e.g. in Hurler's and Hunter's syndromes, and demonstration of the specific enzyme defect in leucocytes or cultured fibroblasts. When carried out it shows gross hepatocyte swelling with clear cytoplasm due to retained acidic mucopolysaccharide which stains with Alcian blue or colloidal iron but only faintly with PAS. Much of this material is water soluble and it is necessary to use an alcoholic fixative such as Carnoy's fluid. The mucopolysaccharide is present also in vacuolated Kupffer cells and portal tract macrophages. There is a tendency to fibrosis, and cirrhosis may occur in some older children or adults.

Similar histological features may be present in other related but very uncommon inherited metabolic illnesses such as the mucolipidoses, fucosidosis and aspartylglucosaminuria.

(c) Mucoviscidosis (cystic fibrosis) (Psacharopoulos *et al.*, 1981)

Liver involvement is not infrequent but sometimes difficult to assess because of its patchy distribution, often with no clinical hepatic dysfunction, and lack of correlation with the severity of pancreatic and pulmonary lesions. The diagnostic biopsy finding is amorphous inspissated mucus within intrahepatic bile ducts (Fig. 9.25) that is strongly PAS positive and may contain some acidic mucopolysaccharide and bile pigment. The affected ducts are dilated, sometimes to a marked degree, with atrophic epithelial lining. Their presence is associated with portal and periportal fibrosis and ductular proliferation. Fibrosis expands in an irregular fashion (Fig. 9.26) and septa come to encircle nodules of

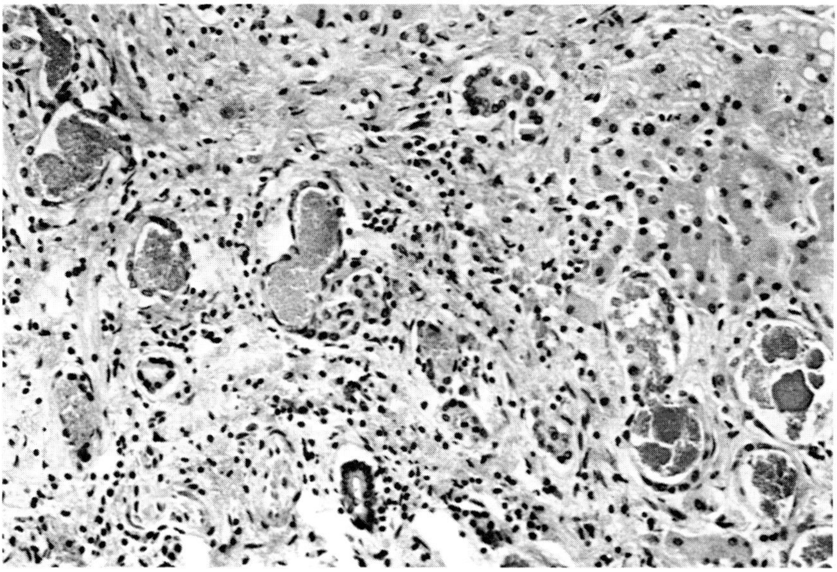

Fig. 9.25 Mucoviscidosis. Inspissated mucus is contained within proliferated biliary channels which are dilated and lined by flat and degenerate epithelium. There is portal fibrosis (HE, × 225).

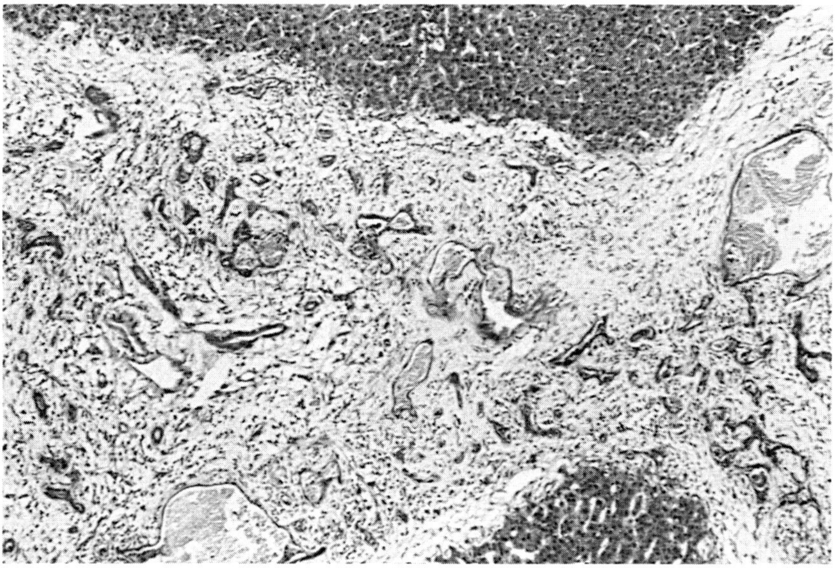

Fig. 9.26 Mucoviscidosis. An area of focal biliary fibrosis (PAS stain, × 70).

parenchyma giving rise to 'focal biliary fibrosis' or 'focal nodular cirrhosis' often separated by wide areas of normal liver. These latter features are usually seen in older children or adults but there can be severe portal hypertension in younger patients.

9.2.8 *Disorders of amino acid metabolism*

(a) Tyrosinosis

This is a condition in which the specific enzyme defect has not yet been defined. It presents either acutely in early infancy with hepatic failure and jaundice, or later in childhood with cirrhosis and renal rickets. In the acute form the hepatocytes show fatty change, portal tract fibrosis and the hepatocytes at the fibrous parenchymal junction form pseudoacini similar to those seen in galactosaemia (Fig. 9.21). In older children a cirrhosis may be evident with prominent bile duct proliferation; large regenerating nodules resembling hepatoma may also be present and frank hepatocellular cancer appears in older individuals.

(b) Cystinosis

Here again the basic biochemical abnormality is uncertain. L-cystine crystals are deposited in many organs including kidneys which can lead to renal failure, and in conjunctiva, cornea and bone marrow. In the liver small groups of coarse cystine crystals (Fig. 9.27) may appear to lie free or be within Kupffer cells and portal tract macrophages. They tend to be more numerous in centrilobular zones. Being refractile they are well seen by polarizing microscopy (Fig. 9.28). The biopsy tissue should be fixed in an alcoholic solution as the crystals are water soluble. They do not invoke any inflammatory reaction or parenchymal degenerative change in the liver.

(c) Primary hyperoxaluria

Polarizing microscopy shows oxalate crystals in the walls of hepatic arterial branches but this is of little clinical significance compared with the renal involvement.

(d) Methionine activating enzyme deficiency

This disease presents clinically like tyrosinosis with hepatosplenomegaly, malabsorption, ascites and rickets with high urinary levels of methionine but normal urinary tyrosine. The liver shows a micronodular

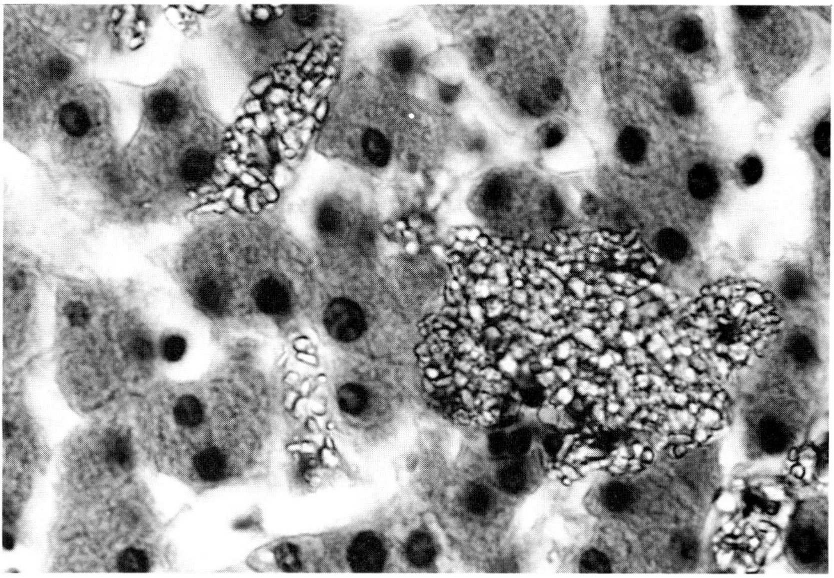

Fig. 9.27 Cystinosis. Several groups of cystine crystals are seen (HE, × 900).

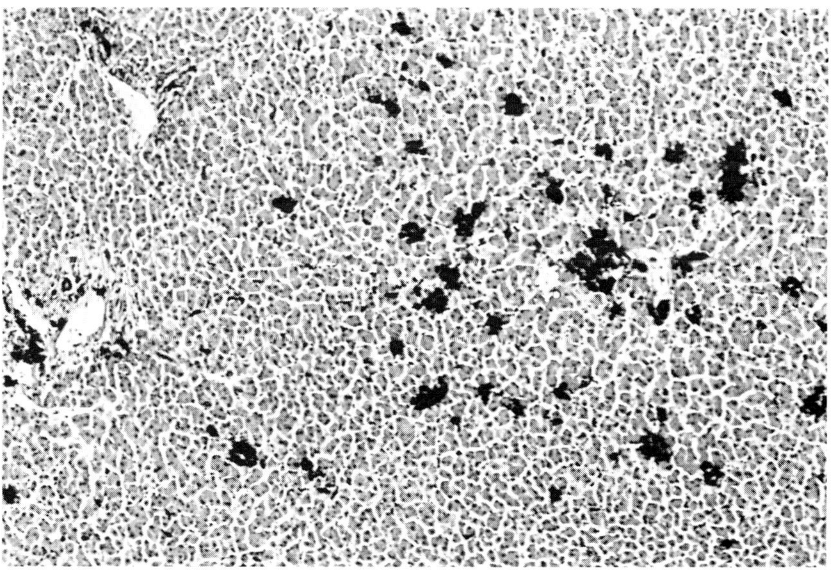

Fig. 9.28 Cystinosis. Cystine crystals appear black by polarizing microscopy. They are numerous in the centrilobular zone (right) and in portal tract macrophages (× 90).

cirrhosis and this is associated with a deficiency of hepatic methionine activating enzyme.

(e) Congenital hyperammonaemic syndromes

There are a large number of these uncommon conditions causing serious illness in infancy. Some are associated with deficiency of urea cycle enzymes, especially ornithine transcarbamylase and arginosuccinase. In most cases the liver shows some degree of fatty change and possibly there is a relationship with Reye's syndrome in some of these.

9.2.9 Disorders of lipid metabolism

(a) Abetalipoproteinaemia

This is a familial condition in which the basic defect is impaired synthesis of a particular apolipoprotein which is an integral part of chylomicrons and low-density lipoprotein. This causes an inability to form chylomicrons and to transport triglyceride from hepatocytes, and so hepatocytes accumulate triglyceride in the form of large fat vacuoles which have no special morphological features. The liver lesion may progress to a micronodular cirrhosis in which the hepatocytes contain a hyaline material similar to that found in alcoholic liver disease.

Although liver involvement is common in this disease, the clinical presentation is usually dominated by malabsorption, retinitis and ataxia. These features and the fatty liver may respond to medium-chain triglyceride therapy.

(b) Tangier disease

This is an inherited disease of lipid storage first described in a child from Tangier Island, Virginia. It is associated with almost complete absence of high-density plasma lipoproteins, and low plasma cholesterol and phospholipids. The tonsils are characteristically flecked with yellow-orange spots. The liver, which is frequently enlarged, contains foamy macrophages filled with cholesterol crystals. The disorder is benign apart from possible predisposition to atherosclerosis.

(c) Wolman's disease

This is a rare condition due to deficiency of acid lipase demonstrable in leucocytes, cultured fibroblasts and hepatic tissue. Liver biopsy shows prominent vacuolation of hepatocytes and Kupffer cells which may be

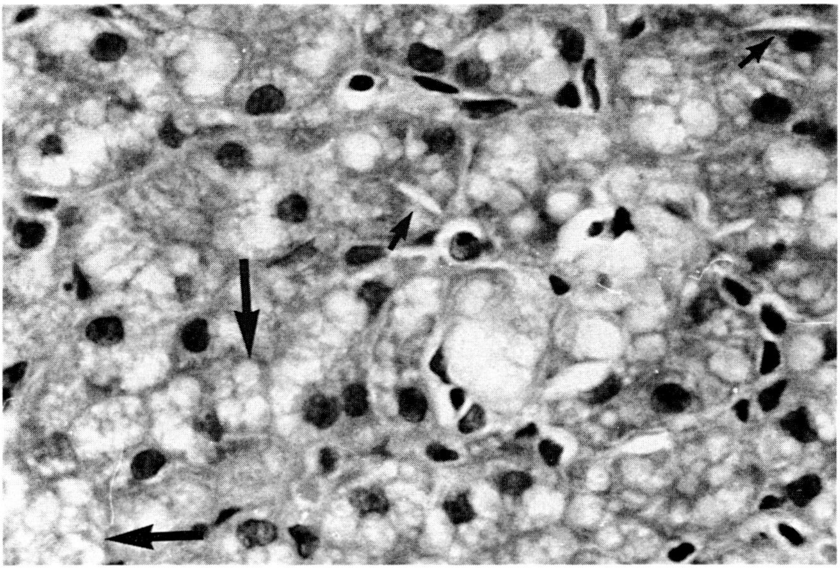

Fig. 9.29 Liver in Wolman's disease. The liver cells contain fine lipid vacuoles (large arrows) which are best demonstrated in frozen sections with oil red 0. Cholesterol crystals (small arrows) are also present in hepatocytes; these crystals are better seen in frozen sections examined in polarized light (HE, × 520).

difficult to distinguish from each other (Fig. 9.29). This is due to accumulation of triglyceride and cholesterol, the latter mainly in histiocytes. Cholesterol crystals may be seen readily in frozen sections by polarized light and in electron micrographs (Fig. 9.30). Portal fibrosis can develop but many cases die in early infancy. Foamy cells are seen also in bone marrow sections but there are no skin xanthomas. Adrenal calcification is a feature.

(d) Cholesterol ester storage disease

Acid lipase deficiency is present but the condition is less severe than Wolman's disease and compatible with a normal life-span. Triglyceride accumulation in hepatocytes is less marked but cholesterol crystals are readily seen in hepatocytes and Kupffer cells. Hepatic fibrosis is not a feature and adrenal calcification is absent.

(e) Xanthomatous neuropathy

Xanthomas have been found rarely involving unmyelinated nerves within portal tracts. They consist of accumulation of cholesterol and other

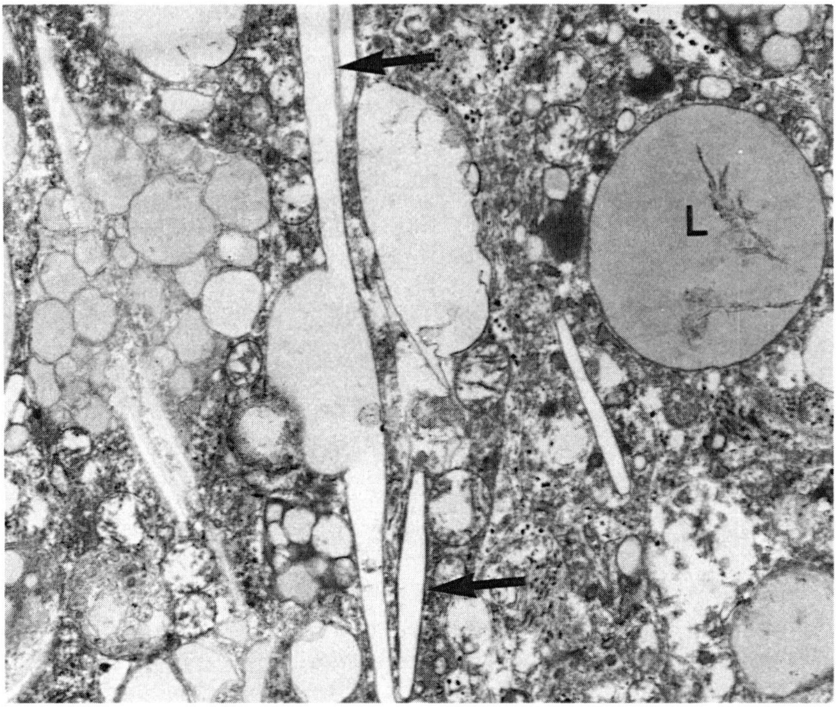

Fig. 9.30 Electron micrograph of the liver in Wolman's disease. The hepatocytes contain numerous cholesterol crystals (arrows) and membrane-bound lipid (L) (uranyl acetate and lead citrate, × 12 000).

lipids at these sites. They may be associated with diabetes and other conditions with hyperlipidaemia (Ludwig *et al.*, 1982).

(f) Cerebrotendinous xanthomatosis (CTX) (Salen *et al.*, 1978)

In this condition a block in bile acid synthesis causes overproduction of cholesterol which is deposited in almost all tissues and the excretion of large amounts of bile acid precursors. Liver biopsy shows little of note except profuse, very fine granules of golden-brown pigment in hepatocyte cytoplasm which are best seen in 1 μm sections stained with toluidine blue. The pigment may be a product of non-metabolized bile acids. Diffuse fibrosis may be present.

(g) Niemann–Pick disease

This disease may present as asymptomatic hepatomegaly or less commonly as neonatal hepatitis with obstructive jaundice. Neurological

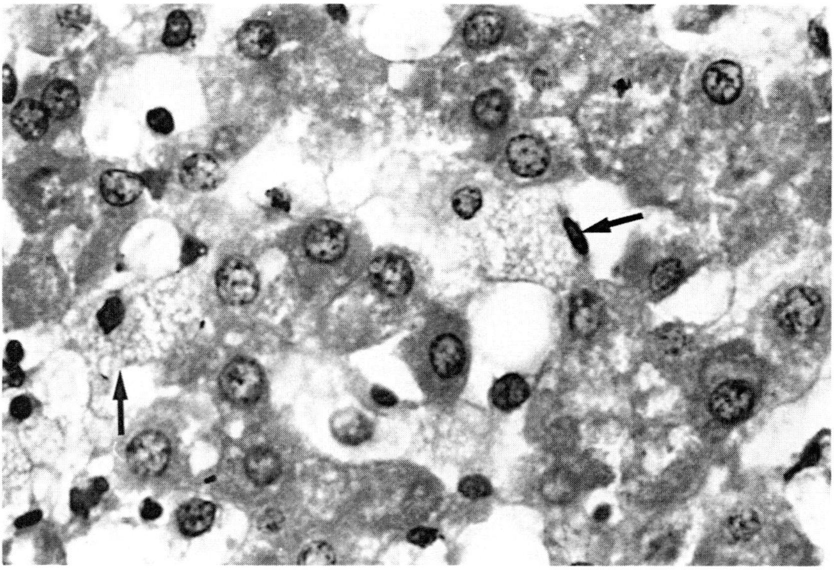

Fig. 9.31 Liver in Niemann–Pick disease. There are enlarged sinusoidal cells (arrows) containing numerous small, clear vacuoles (HE, × 640).

deterioration becomes evident in the first year of life and progresses to death usually before 4 years. There are five subtypes of the disease but the commonest, type A, is characterized by a deficiency of lysosomal sphingomyelinase and therefore sphingomyelin accumulates. On liver biopsy hepatocytes are vacuolated and large foamy reticuloendothelial cells are scattered among them (Fig. 9.31) containing refractile droplets which are sudanophilic in frozen sections and sometimes PAS positive. On electron microscopy the foam cells are packed with whorls of concentric membranes; hepatocytes contain similar membrane-bound inclusions, and may become almost indistinguishable from the foamy cells on light microscopy.

(h) Gaucher's disease

This, too, is an inborn error of metabolism in which cerebrosides accumulate in the reticuloendothelial system due to a deficiency of glucocerebrosidase. There are three forms (I–III), but liver biopsy shows the same features in each group. Enlarged Kupffer cells are found throughout the liver but particularly in nests around the central veins (Fig. 9.32); these Gaucher cells may be 100 μm in diameter and have a fibrillar cytoplasm, in contrast to the foam cells found in Niemann–Pick disease. They are

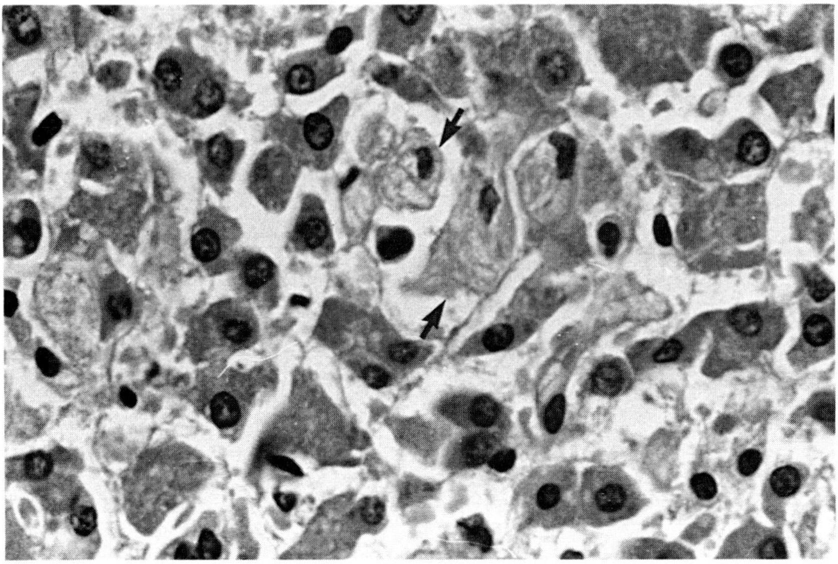

Fig. 9.32 Liver in Gaucher's disease. There are six Gaucher cells in a single sinusoid (arrows). These cells (unlike Niemann–Pick cells) are not vacuolated but rather have a 'folded' cytoplasm (HE, × 640).

PAS positive and rich in acid phosphatase. Hepatic fibrosis may be present and this, together with sinusoidal compression by Gaucher cells, may explain the portal hypertension which is sometimes present.

(i) Fabry's disease

Deficiency of ceramide trihexosidase causes accumulation of ceramide hexose (a sphingomyelin compound) especially in the kidneys and the cardiovascular system with fatal degenerative changes in these organs. Liver biopsy shows prominence of Kupffer cells because of ingested cholesterol and glycolipid which is PAS positive. Similar hepatic changes may be seen in Farber's disease due to lysosomal acid ceramidase deficiency.

(j) Metachromatic leukodystrophy

Deficiency of aryl sulphatase causes accumulation of cerebroside-3-sulphate in the nervous system, kidneys and other organs. Large foamy macrophages containing this material are seen in the walls of the gall bladder and intrahepatic bile ducts and less commonly in other liver cells.

(k) Gangliosidoses

Various gangliosidoses are described which are neurodegenerative disorders. In type II GM_2 (Sandhoff's disease) due to hexosamidase deficiency, triglyceride droplets in hepatocytes are membrane-bound because of their location within lysosomes.

(l) Familial hepatosteatosis

In this disease, male infants suffer from hepatomegaly, jaundice and kernicterus. Severe large droplet fatty change affects the liver and other organs may be fatty also. It is rapidly fatal and the cause unknown.

9.2.10 Miscellaneous

Myoclonus epilepsy (Lafora's disease)

The presence of Lafora bodies is the main histological feature, not only in brain but in many other organs including liver. These are well-defined intracytoplasmic inclusions of uncertain nature, although an abnormal polyglucosan has been suggested. In affected hepatocytes they are strongly PAS positive and stain also with Lugol's iodine, colloidal iron and methenamine silver nitrate; the nucleus is compressed to the margin of the cell. The distribution is mainly periportal and there is no associated hepatitis or fibrosis.

Morphologically similar lesions may be seen in the liver in type IV glycogen storage disease and in patients taking disulfiram or dipsam in the treatment of alcohol addiction.

9.3 Fatty liver

Fatty change is seen frequently in liver biopsies, either in association with other histopathological features or as the only structural abnormality. In almost every case the fat is triglyceride and sudanophilic, but special stains are hardly necessary for its demonstration. There are numerous causes and it would be inappropriate to present the complex pathogenesis in this book. Minor degrees usually have no significance and even extensive fatty change can exist without biochemical evidence of liver functional abnormality other than changes in the rather sensitive bromsulphalein excretion test. Jaundice may accompany severe fatty change which develops rapidly as in acute alcoholism, and in these cases liver biopsy may also show intrahepatic cholestasis. Liver dysfunction is found also when the cytoplasm contains tiny vacuoles of fat and the nucleus is placed centrally in the hepatocyte (Fig. 9.4). In the much more common but less important type of fatty change there are large cytoplasmic droplets with displacement of the nucleus to the margin of the cell (see Fig. 6.1).

The lobular distribution of fatty change should be noted, although its value in clinical assessment may be limited as it seems to vary among individual cases having the same aetiology, and the pattern is lost in severe steatosis from whatever cause when the liver becomes diffusely involved. None the less, the following structural classification of fatty change gives a rough guide to aetiology.

(a) Random lobular distribution of large droplet fat

This is typical of alcoholic liver disease.

(b) Large droplet fat in periportal cells

This is seen frequently and in association with a variety of illnesses such as gastrointestinal and pancreatic disorders especially with vomiting and diarrhoea (e.g. infantile gastroenteritis), anaemia and various wasting diseases as in advanced carcinomatosis, certain types of toxic liver injury such as methotrexate poisoning, occasional cases of obesity in whom it can be exaggerated by jejuno-ileal shunts, and some types of inherited metabolic disorders such as abetalipoproteinaemia. A periportal distribution is also typical of the fatty liver in kwashiorkor, although in many cases it is diffuse.

(c) Large droplet fat in centrilobular hepatocytes

This is found in most patients with obesity, again exaggerated by jejuno-ileal bypass. It may represent minor damage in some patients with toxic liver injury or in ischaemia associated with shock or venous congestion.

(d) Small droplet fat

This has more serious implications and is seen typically in tetracycline liver damage, in idiopathic fatty liver of pregnancy and in Reye's syndrome. The distribution is usually diffuse but it may be more prominent in centrilobular cells. In some cases of Reye's syndrome fatty change or even necrosis may be more prominent in periportal cells.

(e) Fatty liver in inherited metabolic disorders

Diffuse or focal fatty change in hepatocytes and liver mesenchymal cells is a feature of many of these diseases described in this chapter. In some, triglyceride droplets are present in hepatocyte lysosomes and so are membrane bound. More complex lipids may accumulate in liver cells in certain disorders such as cholesterol ester storage disease, Gaucher's disease, etc.

References

Batman, P. A. and Scheuer, P. J. (1985), Diabetic hepatitis preceding the onset of glucose intolerance. *Histopathology*, **9**, 237–43.

Chopra, S., Rubinow, A., Koff, R. S. *et al.* (1984), Hepatic amyloidosis: a histopathologic analysis of primary (AL) and secondary (AA) forms. *Am. J. Pathol.*, **115**, 186–93.

Cortés, J. M., Oliva, H., Paradinas, F. J. *et al.* (1980), The pathology of the liver in porphyria cutanea tarda. *Histopathology*, **4**, 471–85.

Dawson, J., Seymour, C. A. and Peters, T. J. (1979), Gilbert's syndrome: analytical subcellular fractionation of liver biopsy specimens. *Clin. Sci.*, **57**, 491–7.

Kanel, G. C., Uchida, T. and Peters, R. L. (1981), Globular hepatic amyloid: an unusual morphologic presentation. *Hepatology*, **1**, 647–52.

Lefkowitch, J. H. and Grossman, M. E. (1983), Hepatic pathology in porphyria cutanea tarda. *Liver*, **3**, 19–29.

Ludwig, J., Dyck, P. J. and La Russo, N. F. (1982), Xanthomatous neuropathy of liver. *Hum. Pathol.*, **13**, 1049–51.

McGee, J. O'D., Allan, J. G., Russell, R. I. and Patrick, R. S. (1975), Liver ultrastructure in Gilbert's syndrome. *Gut*, **16**, 220–4.

Partin, J. S., Partin, J. C., Schubert, W. K. *et al.* (1982), Serum salicylate concentrations in Reye's disease. *Lancet*, **i**, 191–4.

Psacharopoulos, H. T., Howard, E. R., Portman, B. *et al.* (1981), Hepatic complications of cystic fibrosis. *Lancet*, **ii**, 78–80.

Qizilbash, A. and Young-Pong, O. (1983), Alpha-1-antitrypsin liver disease: differential diagnosis of PAS positive diastase-resistant globules in liver cells. *Am. J. Clin. Pathol.*, **79**, 697–702.

Salen, G., Zaki, F. G., Sabesin, S. *et al.* (1978), Intrahepatic pigment and crystal forms in patients with cerebrotendinous xanthomatosis (CTX). *Gastroenterology*, **74**, 82–9.

Sternlieb, I. (1980), Copper and the liver. *Gastroenterology*, **78**, 1615–28.

Further reading

Bianchi, L. *et al.* (eds) (1983), *Liver in Metabolic Diseases* (*Falk Symposium* 35), MTP Press, Lancaster.

Stanbury, J. B. *et al.* (eds) (1983), *The Metabolic Basis of Inherited Disease*, 5th edn, McGraw Hill, New York.

10 Vascular disorders of the liver

Hepatocytes, like all specialized parenchymal cells, suffer readily from anoxia, and a normal blood flow is essential for the maintenance of adequate liver function. Impaired circulation may be a contributory factor in producing or sustaining hepatocyte damage in liver diseases not normally regarded as primarily vascular in nature, such as cirrhosis from any cause. This Chapter deals mainly with those hepatic conditions which can be attributed principally to failure of the circulation or to diseases of hepatic vessels. It includes causes of portal hypertension but excludes its effect on vessels outwith the liver, as an account of this and of cardiovascular changes resulting from liver failure are outwith the scope of this book.

10.1 Lesions of the hepatic artery and its branches

10.1.1 Structural changes in vessel walls

Small branches in portal triads may show hyaline arteriolosclerosis, especially in diabetic subjects, and amyloid change can be confined to these vessels (Fig. 10.1). Polyarteritis may involve hepatic arterial branches (Fig. 10.2) but is often missed in needle biopsies. A less acute granulomatous arteritis responding well to steroid therapy has been described. Both saccular and dissecting aneurysms also occur, the latter being very uncommon.

10.1.2 Arterial obstruction

This may complicate abdominal trauma or follow surgical ligation which can occur accidentally or be undertaken as treatment for hepatic aneurysm, primary liver carcinoma, etc. Other causes include septic emboli and thrombosis, as in polycythaemia and some cases of polyarteritis and aneurysm. The extent of liver damage is very variable: it

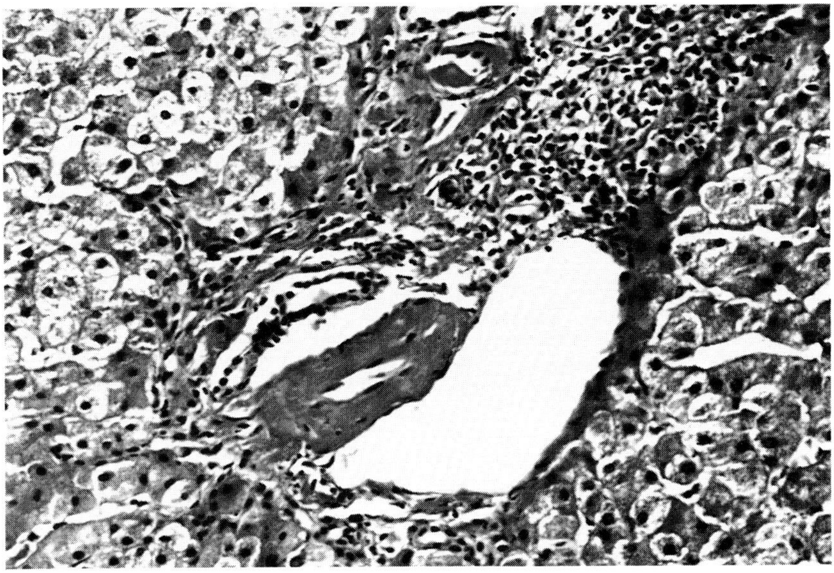

Fig. 10.1 Amyloid degeneration of a small hepatic arterial branch next to a dilated portal vein. The wall of the affected vessel is grossly thickened (HE, × 267).

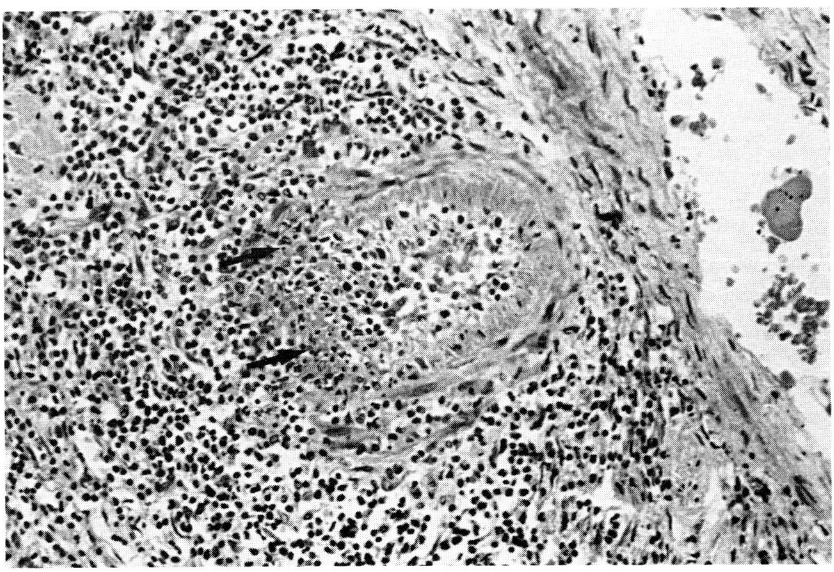

Fig. 10.2 Polyarteritis nodosa. The wall of a large branch of the hepatic artery has been destroyed; in this area there is fibrinoid necrosis and polymorph infiltration of the vessel wall (arrows) (HE, × 360).

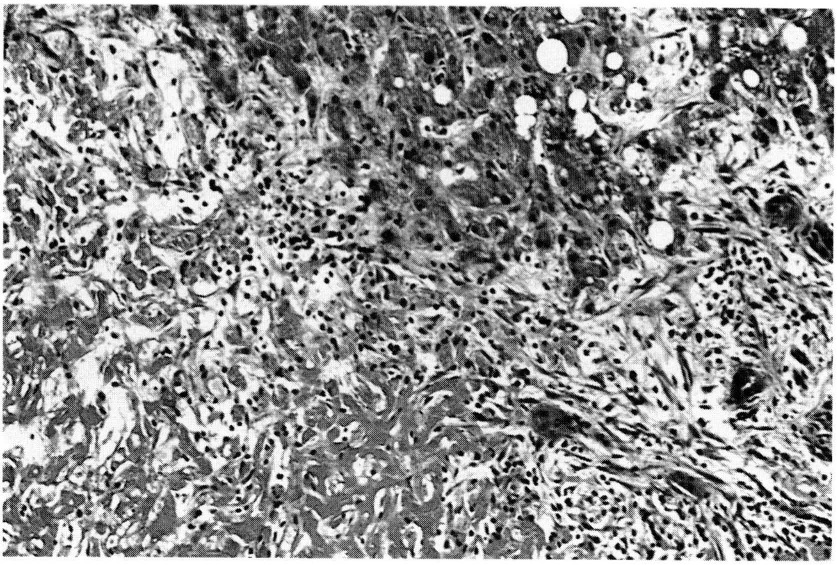

Fig. 10.3 Infarction of liver caused by traumatic injury to branches of hepatic artery. There is coagulation necrosis of hepatic parenchyma seen in the lower part of the photograph (HE, × 180).

may be minimal or there may be extensive and fatal necrosis. In the majority of planned surgical cases it is well tolerated (Kim *et al.*, 1973). Typically, there is a pale coagulative necrotic infarct (Fig. 10.3) which may be red in colour initially because of temporary sinusoidal congestion. An outer margin of centrilobular or midzonal necrosis is usually seen, and after 48 hours there is infiltration with acute inflammatory cells forming a cuff around the necrotic tissue. In the early stages portal tracts and the walls of hepatic veins within the lesion resist the necrotic process. Infiltration with lymphocytes, macrophages and fibroblasts is evident at the outermost margin after 3 or 4 days. The obstructive lesion may be seen in adjacent vessels, but in some cases of hepatic infarction the cause is not discovered.

10.2 Lesions of the portal vein and its branches

10.2.1 Extrahepatic lesions

Obstruction of the vessel or its main branches outwith the liver from extrinsic or intrinsic causes need not be associated with any structural changes in the liver parenchyma despite severe portal hypertension usually without ascites, but portal tract fibrosis is said to develop in some cases. In portal hypertension with increased blood flow in the portal system and persistence of the umbilical vein in the falciform ligament (Cruveilhier–Baumgarten syndrome), the liver structure is unaffected. Where there is increased flow related to splenomegaly, the basic disorder, e.g. Hodgkin's disease, may also be evident in liver biopsies undertaken to investigate portal hypertension.

10.2.2 *Intrahepatic obstructive lesions*

Cirrhosis with distortion and compression of portal venous radicles plus arteriovenous shunting is, of course, the main intrahepatic cause of portal hypertension, the level of which can rise further in cases complicated by extrahepatic portal and splenic venous thrombosis. In many non-cirrhotic cases it is impossible to state whether intrahepatic portal thrombosis has caused, or is the result of, high portal venous pressure. The same difficulty applies in understanding the pathogenesis of *non-cirrhotic portal fibrosis* ('idiopathic portal hypertension'; 'tropical splenomegaly'), which is a recognized accompaniment of portal hypertension in India but is probably of world-wide incidence (Boyer *et al.*, 1967; Aikat *et al.*, 1979). In this condition there is subintimal cellular proliferation in medium-sized veins (Fig. 10.4), which may lead to their complete obliteration (Fig. 10.5).

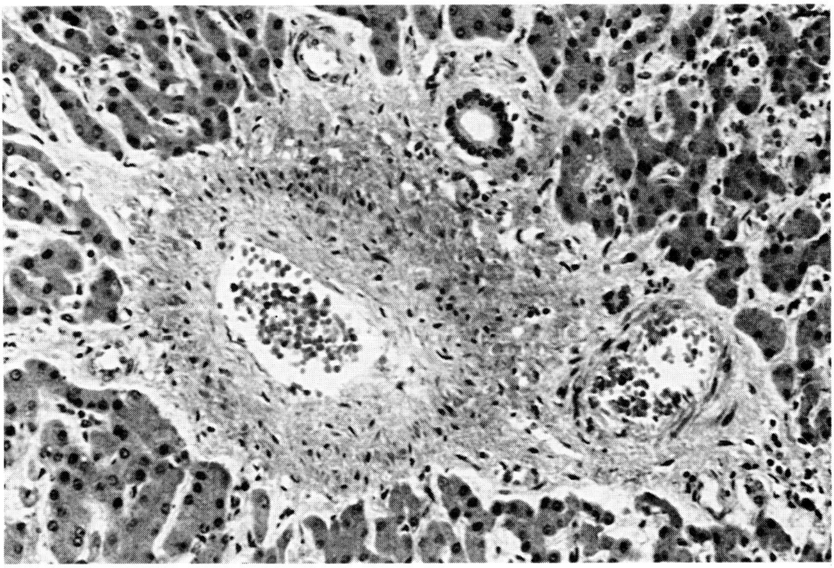

Fig. 10.4 Non-cirrhotic portal fibrosis. There is subintimal thickening of a portal vein (HE, × 220).

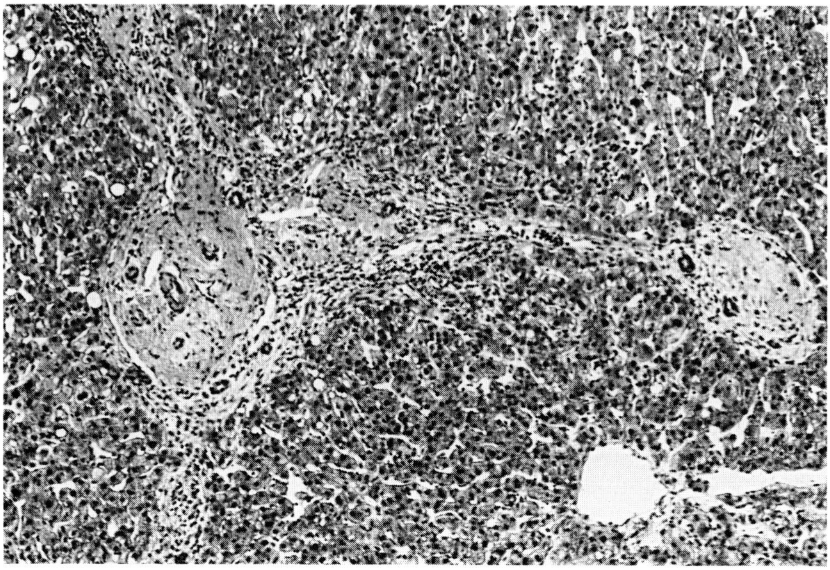

Fig. 10.5 Non-cirrhotic portal fibrosis. There is fibrous obliteration of portal venous radicles, and a septum bridges two portal tracts (HE, × 90).

There may also be striking hypertrophy of the walls of larger veins (Fig. 10.6). Organizing thrombi may be seen with development of small vascular channels to replace occluded vessels and these may be dilated and lined with swollen endothelial cells. All these vascular changes have a segmental distribution. Fibrosis affects portal tracts more widely and may extend as narrow septa into the parenchyma, often with inconspicuous inflammatory changes (Fig. 10.7), but it does not progress to cirrhosis (Okuda *et al.*, 1984). It is uncertain whether these are invariably manifestations of a single disease entity and whether terms such as 'chronic septal hepatitis', 'hepatoportal sclerosis' and 'Banti's syndrome' refer to the same condition. It seems possible that some cases are examples of healed or quiescent portal pyelophlebitis or persistent chronic hepatitis, especially after steroid therapy. Occasionally, lesions with similar histological features can be attributed to arsenic, thorotrast or vinyl chloride, all carcinogenic agents (see Fig. 13.22). In non-cirrhotic portal fibrosis liver biopsy is of value in excluding established cirrhosis and in indicating the need to search for other causes of portal hypertension. Unfortunately, at this time our knowledge of the condition is limited as many cases remain idiopathic while the extent of liver involvement is difficult to assess because of its irregular distribution.

Extrahepatic portal thrombosis may spread within the liver and be seen

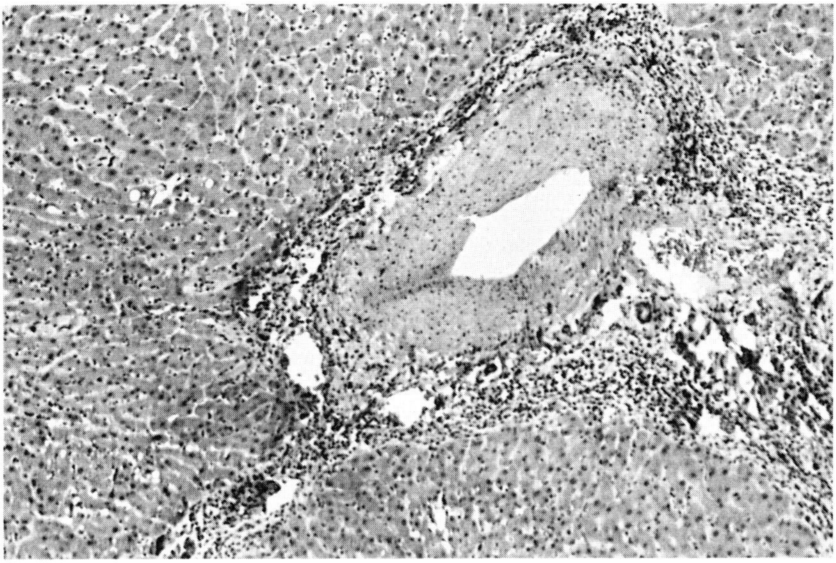

Fig. 10.6 Non-cirrhotic portal fibrosis. Thickening of the portal vein wall is evident (HE, × 90).

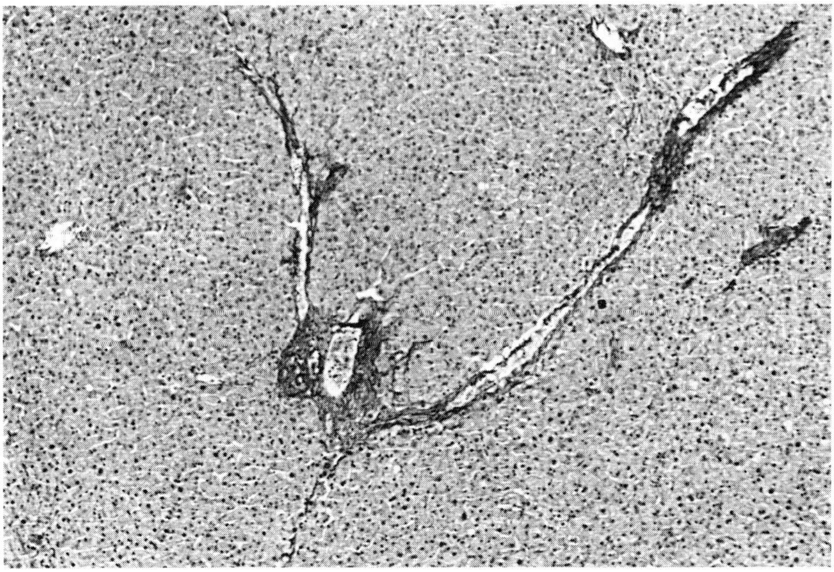

Fig. 10.7 'Chronic septal hepatitis' in a case of portal hypertension (H and van Gieson's stain, × 72).

in biopsies. Many of these cases are examples of thrombophlebitis arising as a complication of some intra-abdominal septic process; the example illustrated (Fig. 10.8) is from a case of empyema of gall bladder. Aseptic intrahepatic thrombosis may complicate viral hepatitis and cirrhosis. The portal venous system within and outwith the liver may be extensively invaded by carcinoma, especially primary liver cell cancer; this together with thrombosis can cause severe portal hypertension with death from oesophageal haemorrhage before the development of extensive metastases. The presence of ova and adult worms of *Schistosoma mansoni* and *japonicum* in branches of the portal vein within the liver causes thrombosis; a chronic granulomatous inflammatory reaction in and around the vessels, resulting in 'pipe-stem fibrosis' which is characteristic of this infestation may not be readily detected in needle biopsies. In late cases revascularization of portal tracts produces an angiomatoid appearance.

In congenital hepatic fibrosis portal hypertension is probably caused by maldevelopment of veins within the fibrous bands which traverse the liver. Partial nodular transformation of liver is another rare cause of portal hypertension. It comprises nodules of liver parenchyma measuring up to 8 cm in diameter but with little fibrosis. Being centrally located and maximal at the hilar region it usually goes undetected in needle biopsies (Sherlock *et al.*, 1966).

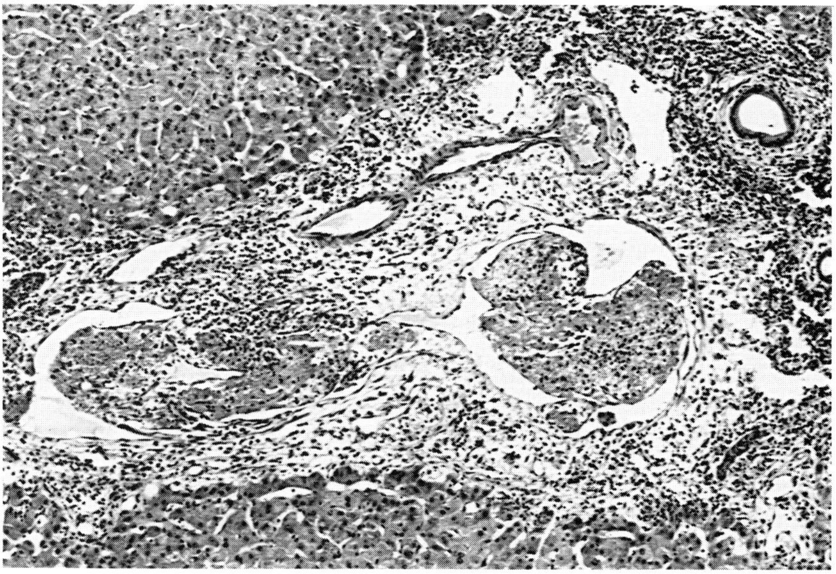

Fig. 10.8 Intrahepatic portal thrombophlebitis. Thrombi fill a dilated portal vein and the portal triad is infiltrated with acute inflammatory cells (HE, × 160).

In veno-occlusive disease some portal radicles show the same subintimal proliferation of loose connective tissue which is the characteristic lesion of the hepatic veins.

10.2.3 Portal venous infarct (Zahn's infarct)

Portal venous occlusion rarely causes submassive or centrilobular zonal necrosis, even in the absence of hepatic arterial obstruction. More frequently the ischaemic area is darkened by sinusoidal congestion, when there is atrophy of liver cell plates (Fig. 10.9). Occasionally the ischaemic area shows fatty change which should not be confused with lipoma (Fig. 10.10).

10.2.4 Portal-systemic shunts

Extensive haemorrhagic necrosis of liver has been reported rarely following operations to establish portacaval anastomosis. Gradual worsening of the primary hepatic disease despite the relief of portal hypertension sometimes occurs. These operations may also be followed by the conspicuous deposition of haemosiderin pigment in the liver.

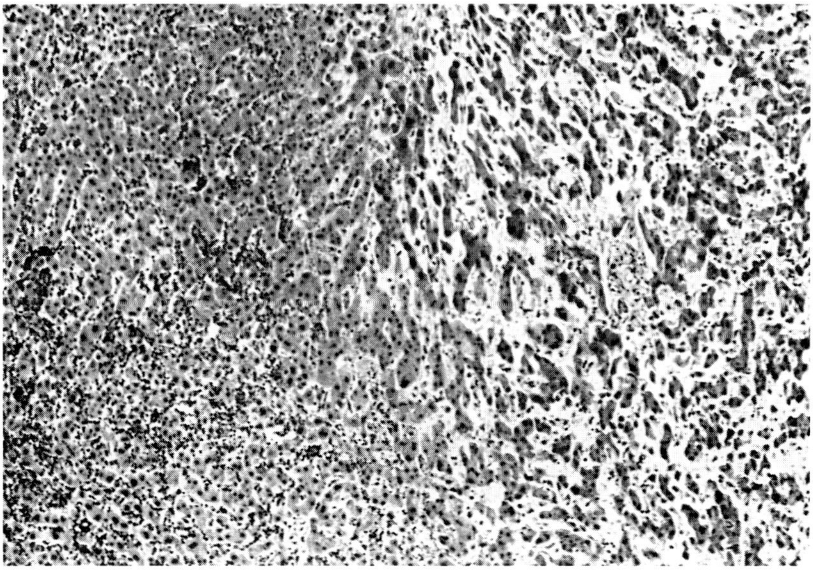

Fig. 10.9 Zahn's infarct (right) with atrophy of liver cell cords. The adjacent parenchyma shows sinusoidal congestion (HE, × 90).

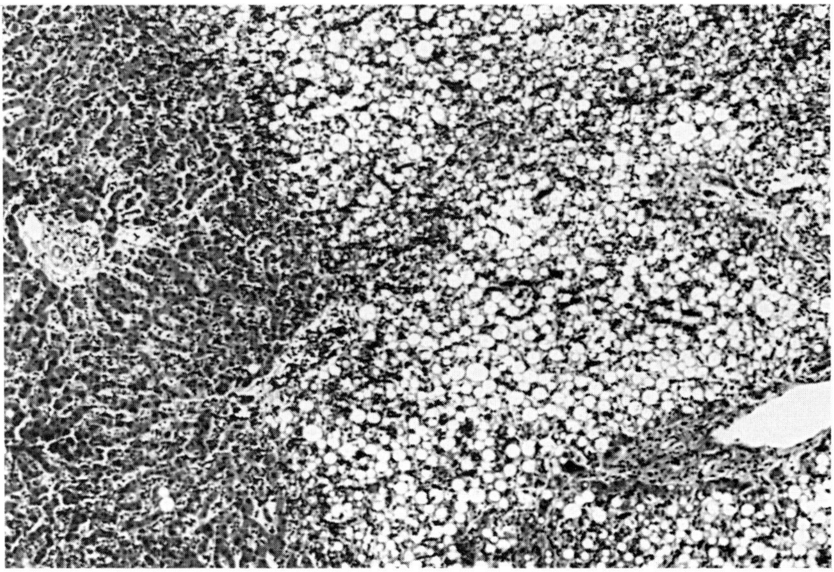

Fig. 10.10 Fatty change type of Zahn's infarct. Unlike lipoma and pseudolipomatous inclusion the fat is present in liver cells and the edge is irregular (HE, × 90). (See Fig. 13.50).

10.3 Lesions of the hepatic veins and venous congestion

10.3.1 Venous congestion and sinusoidal dilatation

Chronic venous congestion gives rise to centrilobular zonal atrophy or necrosis of hepatocytes. However, this is not commonly seen in routine liver biopsy examinations (Fig. 10.11) as, naturally, the procedure is rarely undertaken in patients with congestive heart failure. Unlike the centrilobular zonal necrosis induced by hepatotoxins, the sinusoids are congested and their openings into the central veins are wide (Fig. 10.12). Lipochrome pigment is frequently seen and there may be evidence of cholestasis. In a minority, steatosis is prominent in surviving hepatocytes. While these changes are sufficient to account for clinical jaundice in congestive cardiac failure this may be aggravated by the production of extra bilirubin in haemorrhagic infarcts, especially in the lungs. In the majority of cases of congestive cardiac failure, however, there is little clinical or biochemical evidence of liver damage despite hepatomegaly.

It should be noted that atrophy of centrilobular hepatocytes with sinusoidal dilatation can occur in left-sided cardiac failure without right-sided failure and systemic congestion. This is regarded as a consequence of hepatic hypoxia from diminished cardiac output and hepatic blood

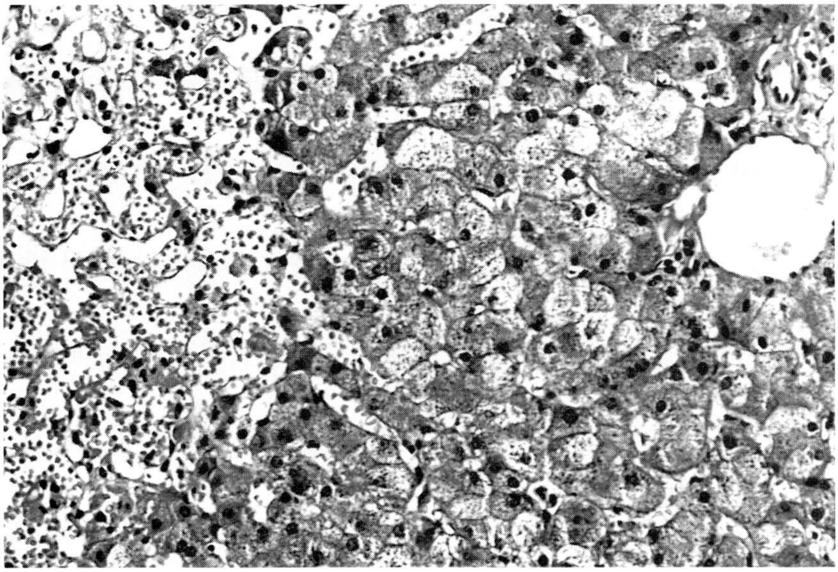

Fig. 10.11 Needle biopsy showing centrilobular loss of hepatocytes and sinusoidal congestion (left). From a case of mitral stenosis (HE, × 247).

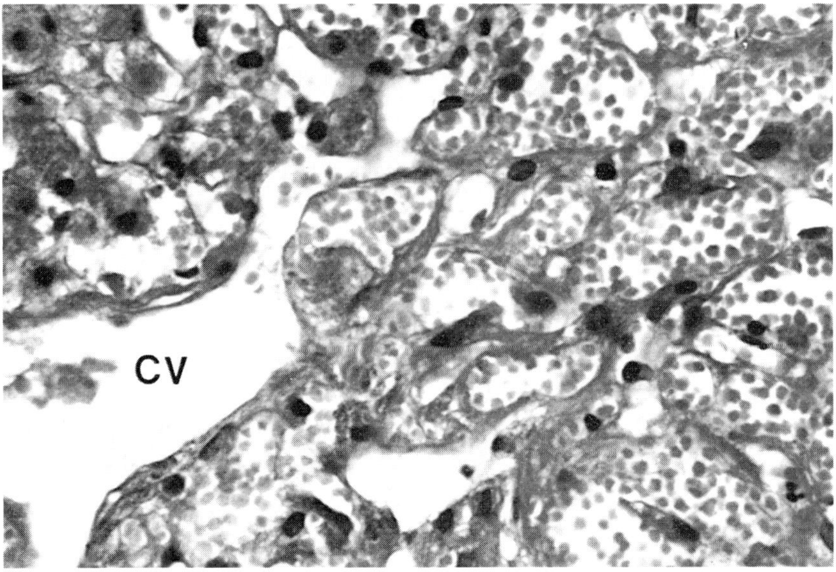

Fig. 10.12 Venous congestion of liver. The openings of sinusoids into the central vein (CV) are dilated (HE, × 560).

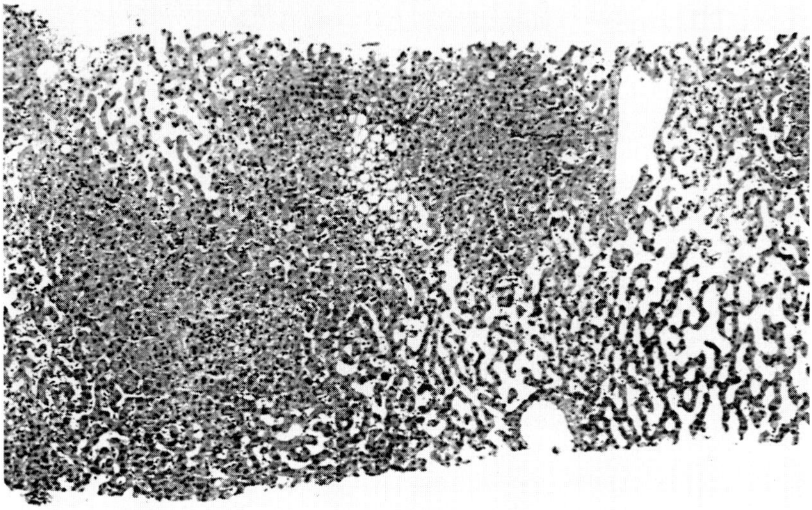

Fig. 10.13 Centrilobular and midzonal sinusoidal dilatation without congestion and without cardiac failure (HE, × 90).

flow (Ware, 1978). Sinusoidal dilatation is found also in biopsies from patients with normal cardiac function (Fig. 10.13); some of these suffer from tumours or from chronic inflammatory diseases, but the pathogenesis of the liver change is unknown (Bruguera *et al.*, 1978). Peripheral sinusoidal dilatation has been associated rarely with the intake of oral contraceptive steroids and may be a precursor of peliosis hepatis.

Very severe centrilobular congestion and haemorrhage are seen in acute cardiac failure, and in the acute form of the Budd–Chiari syndrome, which is due to occlusion of the ostia of the hepatic veins or the inferior vena cava, usually by thrombus, and sometimes related to contraceptive steroid intake, but it is unusual to find thrombus in biopsies from these cases. In the affected zones, blood cells apparently replace cords of atrophic hepatocytes while adjacent dilated sinusoids are empty (Fig. 10.14).

10.3.2 Cardiac 'cirrhosis'

In longstanding hepatic venous congestion fibrosis links adjacent centrilobular zones producing a 'reversed lobulation' pattern (Fig. 10.15). Nodular hyperplasia of surviving parenchyma may follow (true cardiac cirrhosis), but this is rarely seen. More commonly there is a rather diffuse fibrosis involving portal areas as well as centrilobular zones. Atrophic

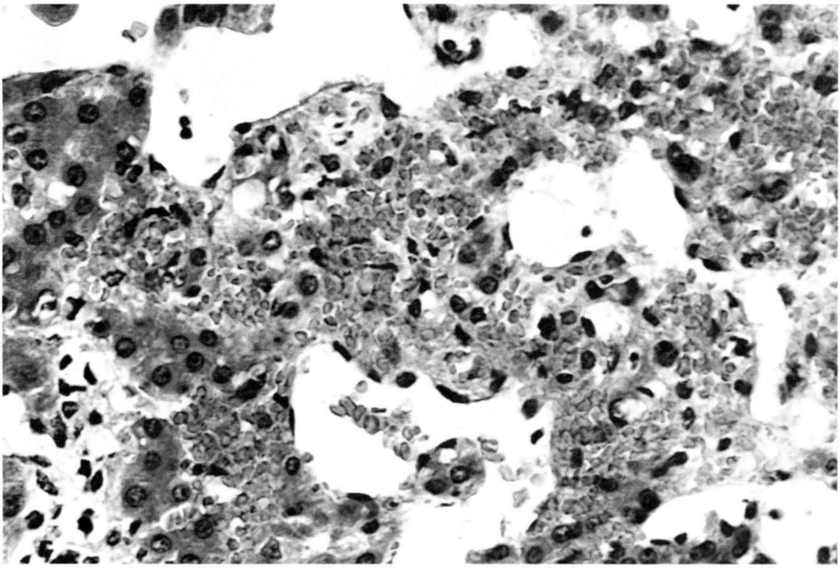

Fig. 10.14 Acute venous congestion. Erythrocytes are seen among cords of atrophic hepatocytes while the dilated sinusoids are mostly empty (HE, × 150).

hepatocytes can resemble bile ductular epithelium, giving an appearance of central cholangiolar proliferation (Fig. 10.16). These changes can complicate chronic cardiac disease such as mitral stenosis and constrictive pericarditis or the chronic form of Budd–Chiari syndrome due either to occlusion of the ostia of hepatic veins or inferior vena cava by thrombus or fibrous webs, or to compression of these vessels by some extrinsic lesion; venous drainage from the caudate lobe is often unaffected so that this part of the liver undergoes hypertrophy. Cardiac 'cirrhosis' can develop rapidly in infants with congenital heart disease. Some cases date from episodes of severe centrilobular hepatocyte damage in cardiogenic shock.

Occasionally nodular regenerative hyperplasia develops in the absence of fibrosis (Fig. 10.17), the parenchymal nodules being separated by bands of more normal liver showing sinusoidal dilatation (Steiner, 1959). Small nodules may have a periportal distribution and portal tracts may be seen within larger nodules, the intervening parenchyma being representative of congested centrilobular tissue. A similar lesion may be found in association with certain other chronic diseases such as rheumatoid arthritis, Felty's syndrome and myeloproliferative disorders, but in these the distribution of nodules is more haphazard. Nodular regenerative hyperplasia may be accompanied by portal hypertension. The pathogenesis is unknown (Stromeyer and Ishak, 1981; Weinbren and Mutum,

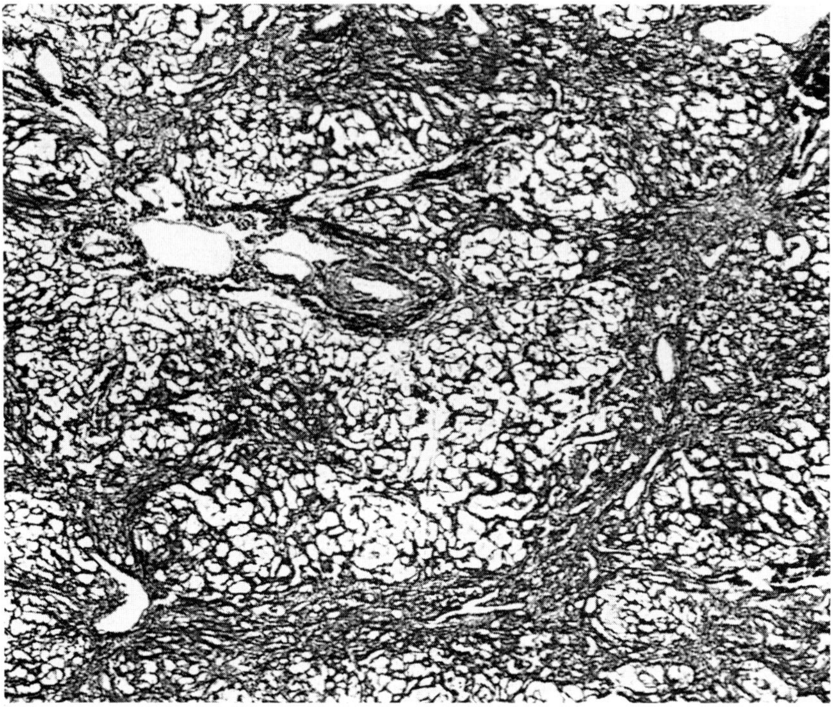

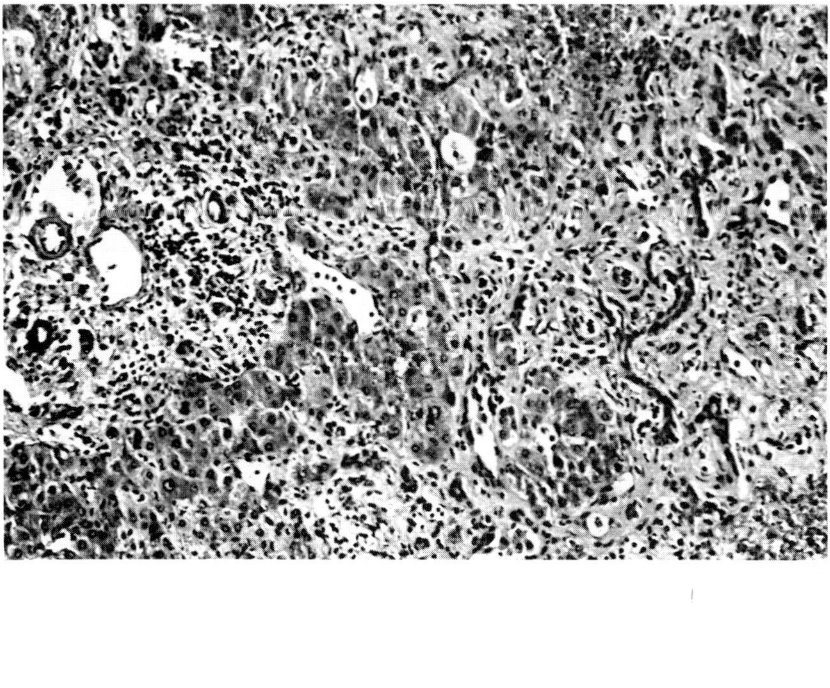

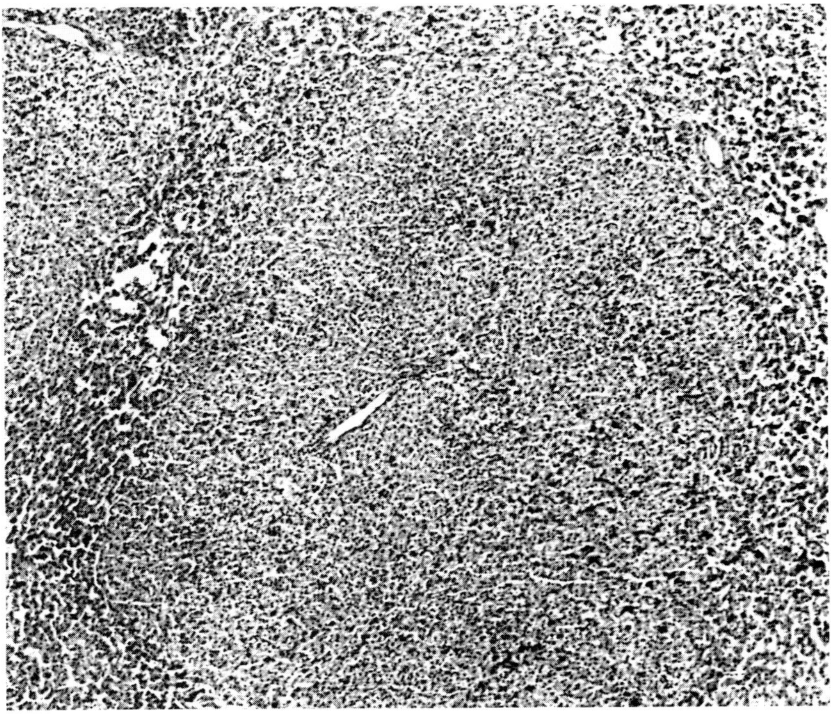

Fig. 10.17 Nodular regenerative hyperplasia in the nonfibrotic liver of a patient with chronic cardiac failure. Atrophic portal tracts are present within two nodules (HE, × 56).

1984), but in some cases it is associated with busulphan-thioguanine therapy for leukaemia (Key *et al.*, 1987).

10.3.3 *Veno-occlusive disease*

This produces a Budd–Chiari-like syndrome but the lesion is confined to the intrahepatic radicals of the hepatic veins. It is common in children in the Caribbean where it is due to the ingestion of pyrrolizidine alkaloids in bush teas and herbal medicines (Bras *et al.*, 1954). Small epidemics of unknown aetiology have been reported from other countries including

◀ **Fig. 10.15** Cardiac 'cirrhosis'. Parenchyma and centrally placed portal triad are surrounded by fibrous tissue linking adjacent centrilobular veins. This forms a 'reversed lobulation' pattern (reticulin stain, × 93).

◀ **Fig. 10.16** Cardiac 'cirrhosis'. The centrilobular zone in the right part of the field is fibrous and contains cords of atrophic parenchymal cells resembling bile ductules (HE, × 180).

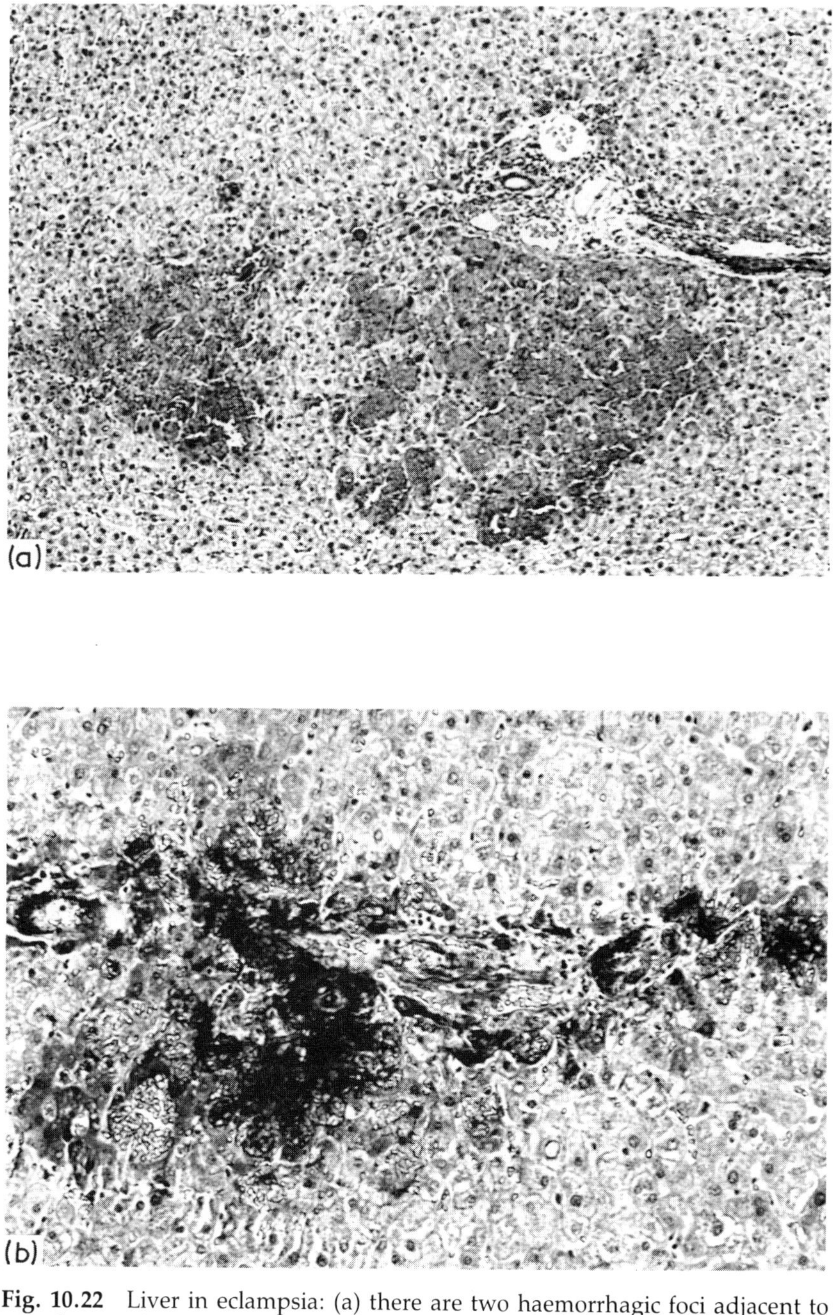

Fig. 10.22 Liver in eclampsia: (a) there are two haemorrhagic foci adjacent to portal tracts; (b) a zone of periportal haemorrhage stained to show dark fibrin. ((a) HE, × 90; (b) Mallory's stain × 180.)

10.4.2 *Altered sinusoidal permeability*

In eclampsia there is increased permeability of the sinusoidal walls and fibrin deposition in Disse's space. This is most marked in the sinusoids adjacent to the portal triads but the overall distribution in the liver can be very haphazard. Later there is haemorrhage in these areas and adjacent hepatocytes may show severe hydropic degeneration or necrosis (Fig. 10.22). Extensive subcapsular bleeding and spontaneous rupture of haemorrhagic necrotic liver are reported in eclampsia.

10.5 Liver ischaemia in general

10.5.1 *Shock*

This may be associated with steatosis or with centrilobular congestion and cholestasis. Cholestasis is sometimes prominent and can be confined within proliferating bile ductules in the portal areas, but not in normal ducts (see Fig. 3.25). There may be infiltration by acute inflammatory cells (Fig. 10.23) or necrosis in centrilobular zones, but the latter feature is usually conspicuous only in autopsy material. The mechanisms of liver damage due to endotoxin are discussed by Nolan (1981). In cases attributable to heat stroke, however, these changes can be severe and

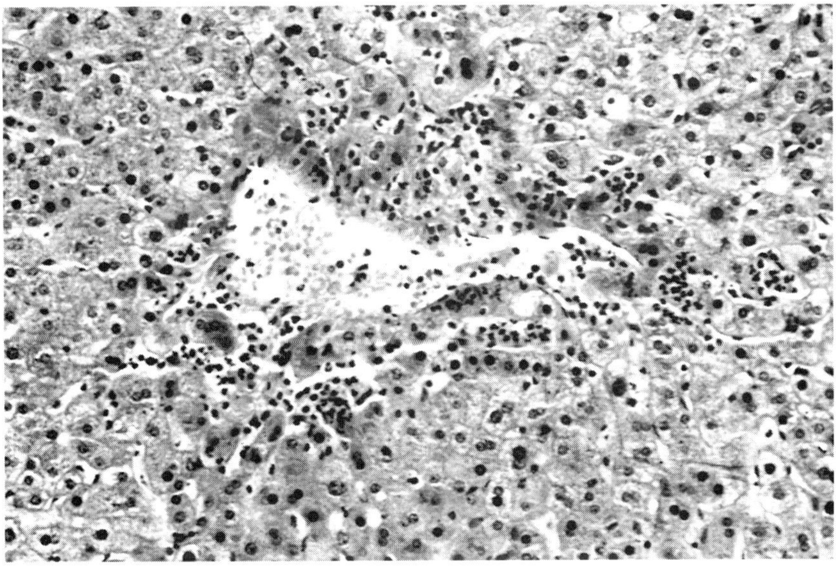

Fig. 10.23 Liver in shock. There is infiltration of the centrilobular zone with neutrophil polymorphs (HE, × 225).

bear some resemblance to cholestatic viral hepatitis, with which it may also be confused clinically.

10.5.2 *Cirrhosis*

In the parenchymal nodules of the cirrhotic liver there is a tendency to ischaemia because of the vascular distortion and shunting of much blood entering the liver directly into the hepatic venous radicles through anastomatic vessels lying in the bands of fibrous tissue. Accordingly, the hepatocytes at the centre of such nodules may show atrophy or fatty change as well as loss of glycogen (Fig. 10.24). If ischaemia is aggravated through severe gastrointestinal haemorrhage these areas can become necrotic and this should not be misinterpreted as a reactivation or recurrence of the original acute liver disease which led to cirrhosis.

10.5.3 *Hemilateral liver degeneration in infants*

Atrophy of the left lobe may be related to change in the hepatic blood supply at the time of birth when the amount of highly oxygenated blood reaching the left side of liver is reduced following closure of the ductus venosus. Hepatic fibrosis in infants and children is associated with generalized hypoxia from various causes and can be severe in the more susceptible left lobe.

Ischaemic damage to the right lobe is a consequence of prolongation of pregnancy beyond term, being found in about a quarter of post-mature neonates. The affected part is pale and well demarcated from the left side by an irregular line situated 1–2 cm to the right of the falciform ligament. The ischaemic hepatocytes are vacuolated or show fatty change. They may be affected by siderosis in babies who suffer also from rhesus incompatibility.

10.5.4 *Jaundice after cardiopulmonary bypass (post-pump jaundice)*

Jaundice with hyperbilirubinaemia of conjugated type occurs in a minority of patients by the second postoperative day. It is a serious complication, 25% in one large series being fatal (Collins *et al.*, 1983). Autopsies performed soon after death have failed to reveal any histological features of note other than mild congestion and portal inflammation. It seems appropriate to note the condition here, although there is no good clinical evidence for severe hypoxia, hypotension or hypothermia, and the pathogenesis is not known.

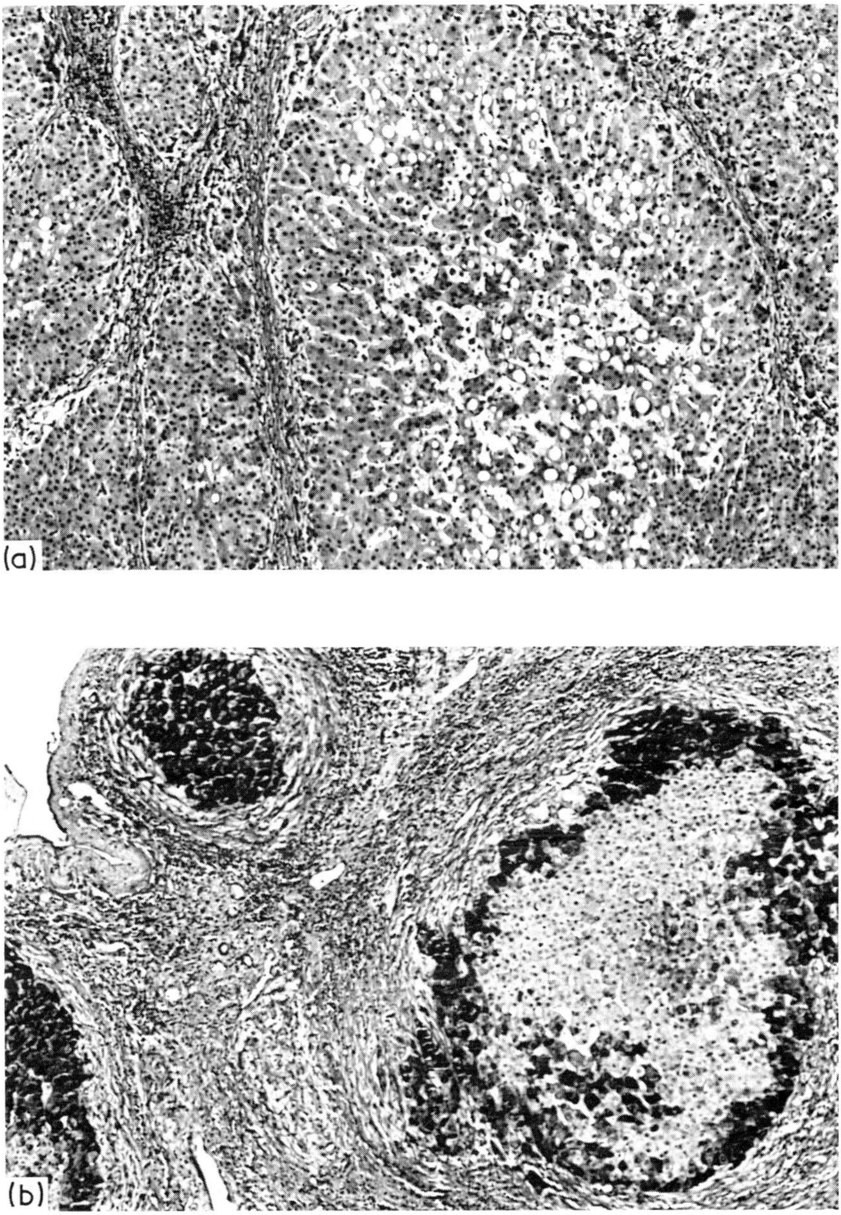

Fig. 10.24 (a) and (b), ischaemic cirrhotic nodules. These nodules show central hepatocyte atrophy, fatty change and loss of glycogen. ((a) HE, × 90; (b) PAS stain × 90.)

10.6 Vascular malformations and tumours

10.6.1 Peliosis hepatis

Throughout the liver there are numerous small blood-filled cystic spaces measuring up to several millimetres in diameter. They seem to be in continuity with the hepatic sinusoids but often show no endothelial lining. Some contain fluid blood and others thrombi. Small stellate scars mark the sites of organization (Fig. 10.25). This very rare condition usually occurs in association with a wasting disease or the administration of steroid hormones. Severe liver dysfunction is unusual but intraperitoneal haemorrhage can be fatal. Complete regression following cessation of androgen therapy is recorded.

Blood-filled cystic spaces may be seen adjacent to tumour deposits, especially primary liver carcinoma. Compared with peliosis they are more irregular in shape and merge more obviously with dilated sinusoids (Fig.

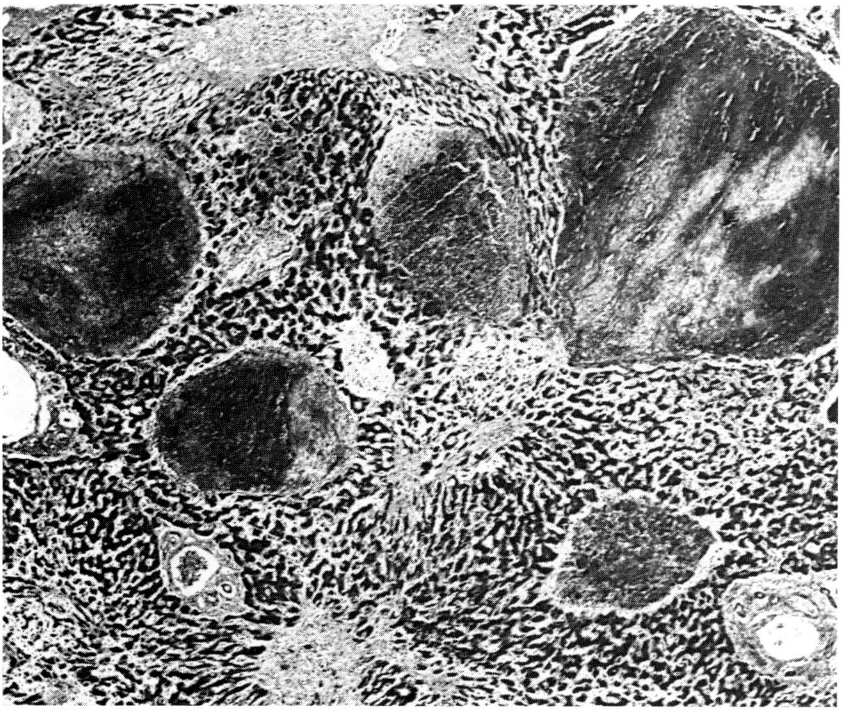

Fig. 10.25 Peliosis hepatis. There are multiple blood-filled cysts, some having undergone thrombosis. Stellate scars are also present (HE, × 45).

10.26). Grossly dilated vascular channels may occur also within these cancers.

10.6.2 *Hereditary haemorrhagic telangiectasia (Rendu–Osler–Weber syndrome)*

Abnormal dilated venous and capillary channels may be seen in portal triads and liver parenchyma (Figs 10.27 and 10.28). As the lesion expands in the liver it replaces many adjacent lobules. Cirrhosis may coexist but portal hypertension can also be caused by arteriovenous shunts which are part of the vascular abnormality. The diagnosis is often suspected at laparotomy from the presence of similar vascular abnormalities in other viscera, but the liver is involved in only a minority of cases.

10.6.3 *Tumours*

Angiosarcoma, vascular hamartoma, etc., are described in Chapter 13.

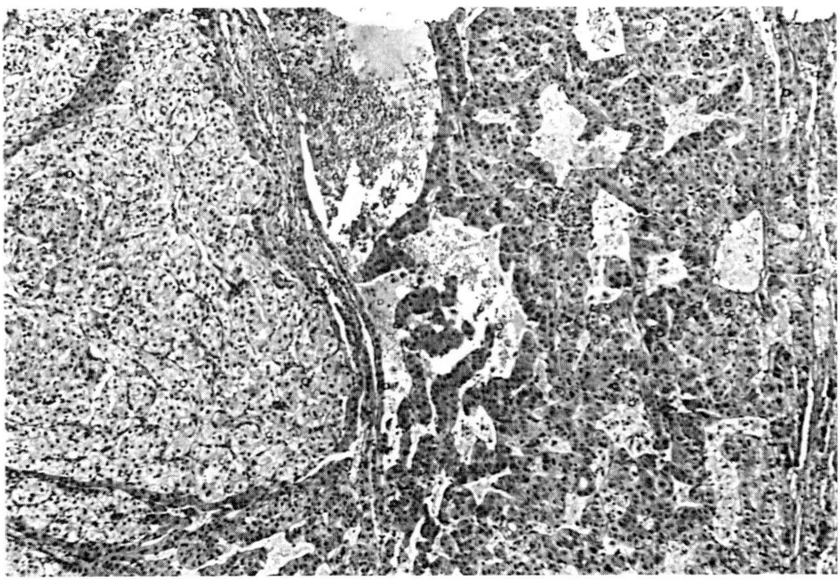

Fig. 10.26 Gross sinusoidal dilatation adjacent to nodules of hepatoma (H) (HE, × 72).

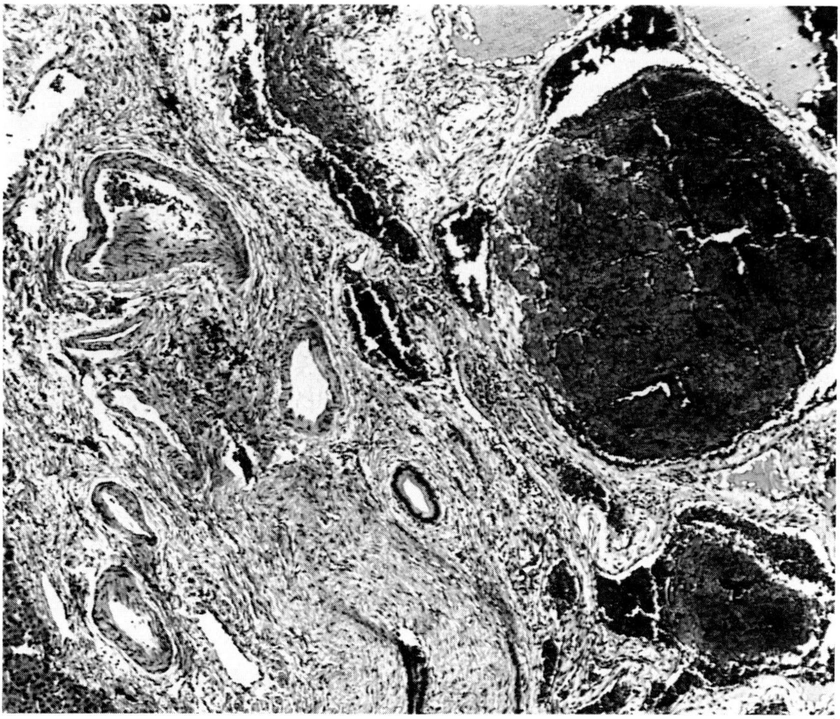

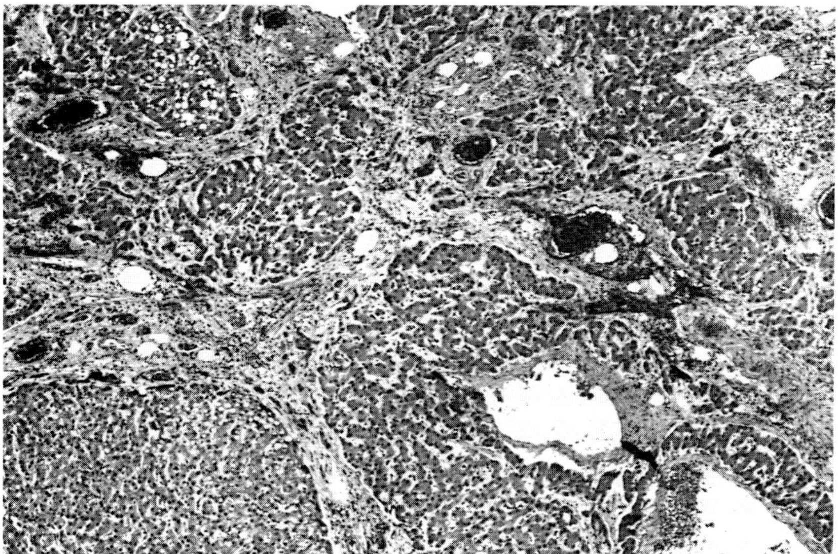

References

Aikat, B. K., Bhusnurmath, S. R., Chhuttani, P. N. *et al.* (1979), Pathology of non-cirrhotic portal fibrosis: a review of 32 autopsy cases. *Hum. Pathol.*, **10**, 405–18.

Berk, P. D., Popper, H., Krueger, G. R. F. *et al.* (1979), Veno-occlusive disease of the liver after allogeneic bone marrow transplantation: possible association with graft-versus-host disease. *Ann. Intern. Med.*, **90**, 158–64.

Boyer, J. L., Sen Gupta, K. P., Biswas, S. K. *et al.* (1967), Idiopathic portal hypertension: comparison with the portal hypertension of cirrhosis and extra-hepatic portal vein obstruction. *Ann. Intern. Med.*, **66**, 41–68.

Bras, G., Jelliffe, D. B. and Stuart, K. L. (1954), Veno-occlusive disease of liver with nonportal type of cirrhosis, occurring in Jamaica. *Arch. Pathol.*, **57**, 285–300.

Bruguera, M., Aranguibel, F., Ros, E. *et al.* (1978), Incidence and clinical significance of sinusoidal dilatation in liver biopsies. *Gastroenterology*, **75**, 474–8.

Collins, J. D., Bassendine, M. F., Ferner, R. *et al.* (1983), Incidence and prognostic importance of jaundice after cardiopulmonary bypass surgery. *Lancet*, **i**, 1119–23.

Dacie, J. V., Brain, M. C., Harrison, C. V. *et al.* (1969), Non-tropical idiopathic splenomegaly ('primary hypersplenism'): a review of ten cases and their relationship to malignant lymphomas. *Br. J. Haematol.*, **17**, 317–33.

Gottfried, M. R. and Sudilovsky, O. S. (1982), Hepatic veno-occlusive disease after high dose mitomycin C and autologous bone marrow transplantation therapy. *Hum. Pathol.*, **13**, 646–50.

Key, N. S., Kelly, P. M. A., Emerson, P. M. *et al.* (1987) Oesophageal varices associated with busulphan–thioguanine combination therapy for chronic myeloid leukaemia. *Lancet*, **ii**, 1050–2.

Kim, D. K., Kinne, D. W. and Fortner, J. G. (1973), Occlusion of the hepatic artery in man. *Surg. Gynecol. Obstet.*, **136**, 966–8.

McGee, J. O'D., Patrick, R. S., Wood, C. B. and Blumgart, L. H. (1976), A case of veno-occlusive disease of the liver in Britain associated with herbal tea consumption. *J. Clin. Pathol.*, **29**, 788–94.

Nanba, K., Soban, E. J., Bowling, M. C. *et al.* (1977), Splenic pseudosinuses and hepatic angiomatous lesions: distinctive features of hairy cell leukaemia. *Am. J. Clin. Pathol.*, **67**, 415–26.

Nolan, J. P. (1981), Endotoxin, reticulo-endothelial function and liver injury. *Hepatology*, **1**, 458–65.

Okuda, K., Kono, K., Ohnishi, K. *et al.* (1984), Clinical study of eighty-six cases of idiopathic portal hypertension and comparison with cirrhosis with splenomegaly. *Gastroenterology*, **86**, 600–10.

Sherlock, S., Feldman, C. A., Moran, B. *et al.* (1966), Partial nodular transformation of the liver with portal hypertension. *Am. J. Med.*, **40**, 195–203.

Steiner, P. E. (1959), Nodular regenerative hyperplasia of the liver. *Am. J. Pathol.*, **35**, 943–53.

◀ **Fig. 10.27** Telangiectasia involving liver. Prominent vascular channels from a portal tract lead into a cavernous angiomatous area. Multiple lesions of this type were present (HE, × 72).

◀ **Fig. 10.28** Telangiectasia involving liver. This is a focal cirrhotic area in which fibrous septa contain numerous abnormal vascular channels (HE, × 56).

Stromeyer, F. W. and Ishak, K. G. (1981), Nodular transformation (nodular 'regenerative' hyperplasia) of the liver. *Hum. Pathol.*, **12**, 60–71.
Vidins, E. I., Britton, R. S., Medline, A. *et al.* (1985), Sinusoidal caliber in alcoholic and non-alcoholic liver disease: diagnostic and pathogenic implications. *Hepatology*, **5**, 408–14.
Ware, A. J. (1978), The liver when the heart fails. *Gastroenterology*, **74**, 627–8.
Weinbren, K. and Mutum, S. S. (1984), Pathological aspects of diffuse nodular hyperplasia of the liver. *J. Pathol.*, **143**, 81–92.

11 Chronic hepatitis

In 1950 Waldenstrom described a liver disease in young women characterized by severe liver and endocrine dysfunction and hyperglobulinaemia. This probably constitutes the first clinical description of the entity now known as chronic hepatitis although the pathology of the disease was described 3 years earlier (Kalk, 1947). After Waldenstrom a number of authors described other clinical presentations of the same syndrome under different designations. In fact, in 1972 Desmet had collected from the literature no less than 49 terms which have been applied to chronic hepatitis. This confusion in terminology probably arose from a failure to appreciate that chronic hepatitis is not a single disease, but a syndrome with a number of causes such as viruses, drugs, etc. In 1968, 11 pathologists produced a classification of chronic hepatitis which has gone a considerable way towards resolving this confusion (De Groote *et al.*, 1968). This group proposed that there are essentially two types of chronic hepatitis, chronic persistent hepatitis and chronic aggressive hepatitis. This simple classification has been embellished but is still in general use, although the designation chronic active hepatitis is now preferred to chronic aggressive hepatitis.

Chronic hepatitis can be defined as an unresolving inflammatory disease of the liver which has persisted for at least 6 months as evidenced by clinical or biochemical findings. The 6-months time-interval is arbitrary; patients who have had subclinical disease for many months before presentation cannot easily be fitted into this scheme of things nor can patients with a prolonged but resolving illness following acute viral hepatitis.

Chronic hepatitis affects both sexes and all age groups and may present with few or very obvious signs of liver disease. It may, or may not be associated with non-organ specific autoantibodies (antinuclear factor, smooth muscle antibody) or liver 'specific' antibodies. Because of the wide spectrum of clinical, biochemical and immunological findings in this syndrome and the fact that similar clinical and laboratory findings are often present in other liver diseases, it is now accepted that a firm generic diagnosis of chronic hepatitis can only be made on liver biopsy. Even after liver biopsy doubt may remain about the aetiology and subgroup of chronic hepatitis into which the case fits. In such cases, all of the clinical, serological and biopsy data have to be considered and a balanced decision made on patient management and prognosis.

11.1 Chronic persistent hepatitis

This is usually a benign disease which persists for months, or years, before spontaneous remission occurs. Some cases, however, progress to CAH and cirrhosis (Fattovich *et al.*, 1986).

Patients with CPH are usually asymptomatic or present with vague symptoms and minimal elevations in liver function tests, particularly the transaminases. On liver biopsy the liver architecture is normal. Portal tracts are usually enlarged and infiltrated by small to moderate numbers of lymphocytes, together with smaller numbers of histiocytes and plasma cells (Fig. 11.1). There may be some 'spill over' of these inflammatory cells into the liver lobule but this is minimal and there is little or no piecemeal necrosis (Fig. 11.2). Fibrosis of portal tracts and septa formation are absent. Small areas of focal necrosis within lobules are present in some

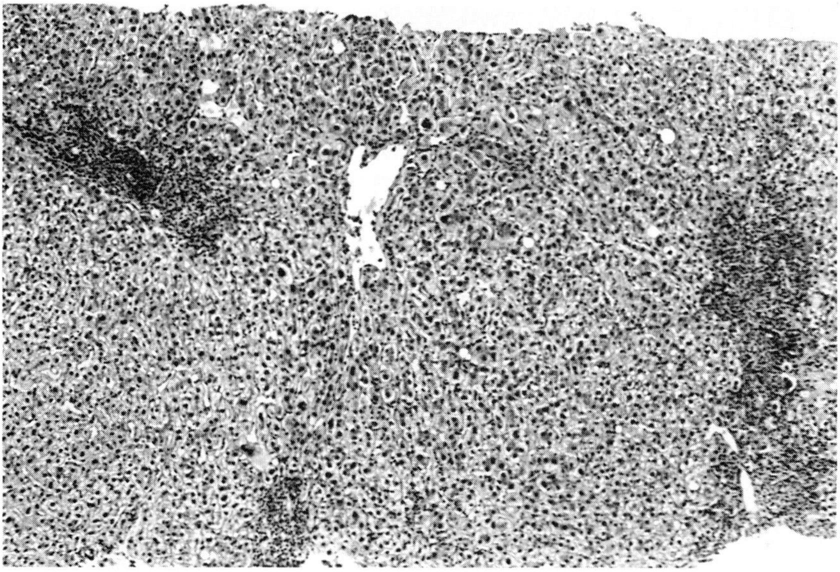

Fig. 11.1 CPH. The portal tracts are infiltrated by lymphocytes. The lobular architecture is almost normal (HE, × 72).

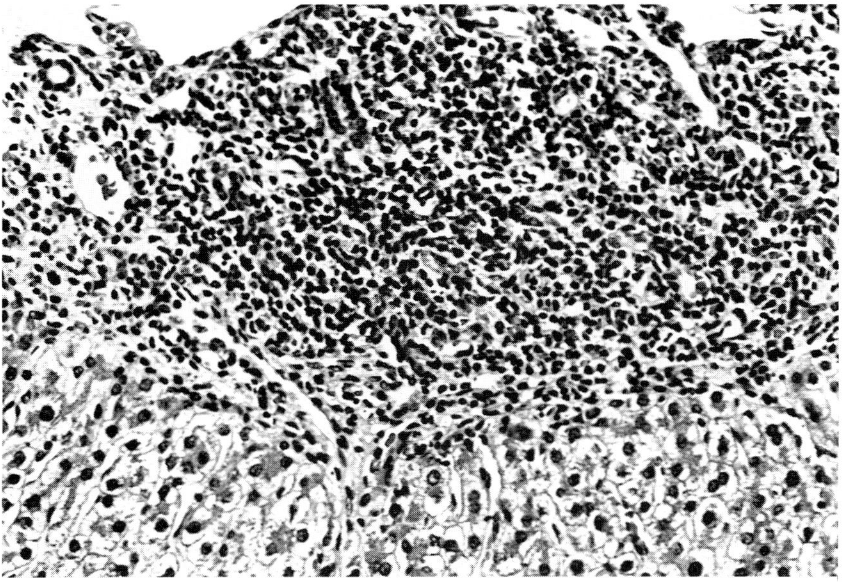

Fig. 11.2 CPH. The inflammatory infiltrate in the portal tract does not erode the limiting plate (HE, × 247).

cases. These morphological changes occur in CPH whether the disease is induced by drugs, a virus or of unknown aetiology.

There are several pitfalls in the diagnosis of CPH. A few inflammatory cells may spill over from the portal tract into the parenchyma but this is not associated with extension of fibrous tissue into the lobule. A confident diagnosis of CPH cannot be made on tiny fragmented biopsies because an adequate biopsy (2 cm long) may reveal CAH. It should also be remembered that patients with CAH who have had steroid therapy often show liver changes similar to CPH. Similar histological changes occur in the liver in non-specific reactive hepatitis (p. 41) and in early primary biliary cirrhosis.

Since the 1968 classification (De Groote *et al.*, 1968) of chronic hepatitis appeared, two other relatively benign forms of the disease have been described – chronic septal hepatitis and chronic lobular hepatitis.

11.1.1 *Chronic lobular hepatitis*

There is no general agreement on the nomenclature of this entity. On biopsy the inflammatory changes in the liver are most prominent in the lobules with mild portal tract inflammation. The lobular changes range from areas of focal necrosis to widespread ballooning of hepatocytes, acidophil body formation and marked inflammatory cell infiltration typical of viral hepatitis (see p. 80). It is thought that this form of chronic hepatitis is the result of a viral infection and hence other terms, such as unresolved viral hepatitis and persistent viral hepatitis, have been applied. Chronic lobular hepatitis may persist for years but at this time it is not considered a precursor of cirrhosis unless superimposed on a lesion with piecemeal necrosis (Scheuer, 1986).

11.1.2 Chronic septal hepatitis

This may be a variant of CPH or, alternatively, a quiescent stage of CAH which has occurred spontaneously or after therapy. Some may be examples of non-cirrhotic portal fibrosis (see Chapter 10). On biopsy it is characterized by the presence of long thin collagenous septa which extend from portal tracts into the hepatic parenchyma (Fig. 11.3). There is usually a mild inflammatory reaction in the portal tracts and septa which is similar in character to that found in CPH. Piecemeal necrosis is not a feature of this condition.

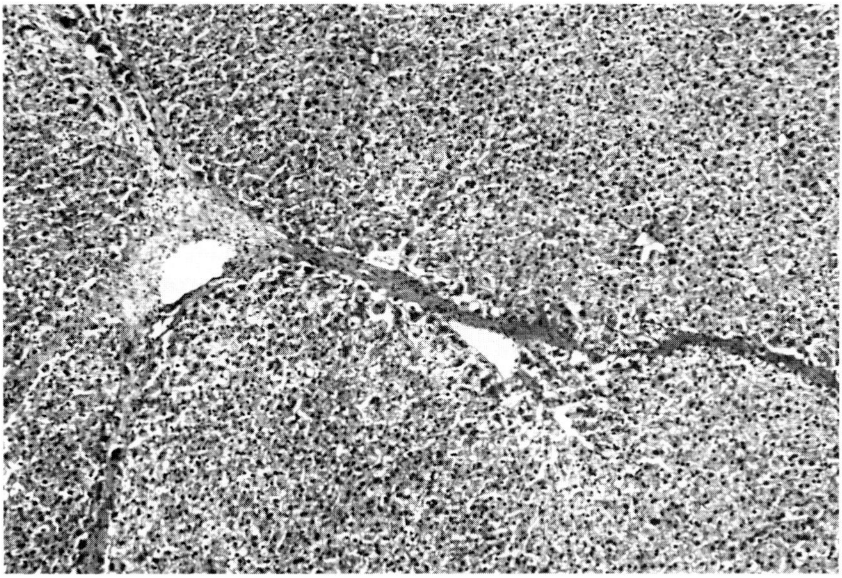

Fig. 11.3 Chronic septal hepatitis. Thin fibrous bands extend from a portal tract into the liver parenchyma. There are a few lymphocytes in the septa but piecemeal necrosis is absent (HE, × 60).

11.2 Chronic active hepatitis

11.2.1 Aetiology

CAH is not a single disease but a syndrome. It can be initiated by a viral infection. The incidence of HBsAg in CAH varies from 10% to 67% but the incidence of antigenaemia in British and Australian cases is low. However, liver markers of HBV may be present in 20% of cases when serology is negative; it is important, therefore, to do immunohistochemistry on all cases of CAH to firmly exclude an HBV aetiology (Cuccurallo *et al.*, 1987). NANB infection is also a frequent cause of CAH. Drugs, such as oxyphenisatin, α-methyldopa, and isoniazid are established causes of CAH. Some cases of Wilson's disease present initially as CAH and this disease should always be actively excluded in young patients since it is amenable to specific therapy. Alpha-1-antitrypsin deficiency may also present as CAH but on biopsy the typical PAS-positive granules in periportal hepatocytes serve to distinguish this form of CAH from the idiopathic ('autoimmune') variety. Alcohol abuse may also be associated with CAH.

When no aetiological agent can be identified it is assumed that the disease has an autoimmune basis but there is not conclusive evidence to support, or refute, this contention. In those cases of supposed autoimmune origin, although liver signs and symptoms dominate the clinical picture, in addition there may be multisystem involvement with fever, arthralgia, skin rashes and hypergammaglobulinaemia. Three subgroups of *autoimmune CAH* have been proposed on the basis of serological and clinical features. Classic 'lupoid' CAH is characterized by hypergammaglobulinaemia, female preponderance, positive antinuclear factor, smooth muscle and liver membrane antibodies, and a high frequency of HLA phenotypes 8 and DR3. The second group is characterized by serum liver-kidney microsomal antibodies (Manns *et al.*, 1984). The third group does not have antinuclear factor, or antibodies to liver-kidney microsomes, but does have antibodies to soluble liver antigens (Manns *et al.*, 1987). This latter group comprises females mainly with hypergammaglobulinaemia who respond well to immunosuppressive therapy. The nature of the soluble liver antigen has not been defined but it is not species or organ specific.

11.2.2 Clinical and biopsy presentation

In contrast to CPH, patients with CAH usually, but not always, have overt signs and symptoms of liver disease such as jaundice, hepatomegaly, palmar erythema and spider naevi. These clinical features, however, are common to other forms of chronic liver disease such as chronic alcoholic liver injury and primary biliary cirrhosis. For these reasons, therefore, liver biopsy is necessary to establish a definitive diagnosis of CAH. Serological tests (smooth muscle antibody and anti-nuclear factor) are positive in some, but not all, cases of CAH. Patients with CAH usually have elevated transaminases and, particularly in women, a serum immunoglobulin level greater than 0.3 g/l.

The four cardinal histopathological features of CAH are: portal and periportal hepatitis, piecemeal necrosis, fibrosis, and lobular damage and inflammation. The portal tracts are enlarged and infiltrated in decreasing order of frequency by lymphocytes, plasma cells and histiocytes (Fig. 11.4). These inflammatory cells spill over into the parenchyma and this is associated with death of hepatocytes in the limiting plate which becomes eroded and ill defined; this complex of features is referred to as piecemeal necrosis (Fig. 11.5). When prominent, piecemeal necrosis is usually associated with a polymorphonuclear cell infiltrate (Fig. 11.5). Simultaneously with the development of portal tract inflammation there is an increase in portal tract collagen which extends into the liver lobule as septa, and collagen fibres frequently surround individual hepatocytes or groups of hepatocytes which then come to lie in the portal tracts or septa (Fig. 11.6). In addition, there is regeneration of hepatocytes and this is

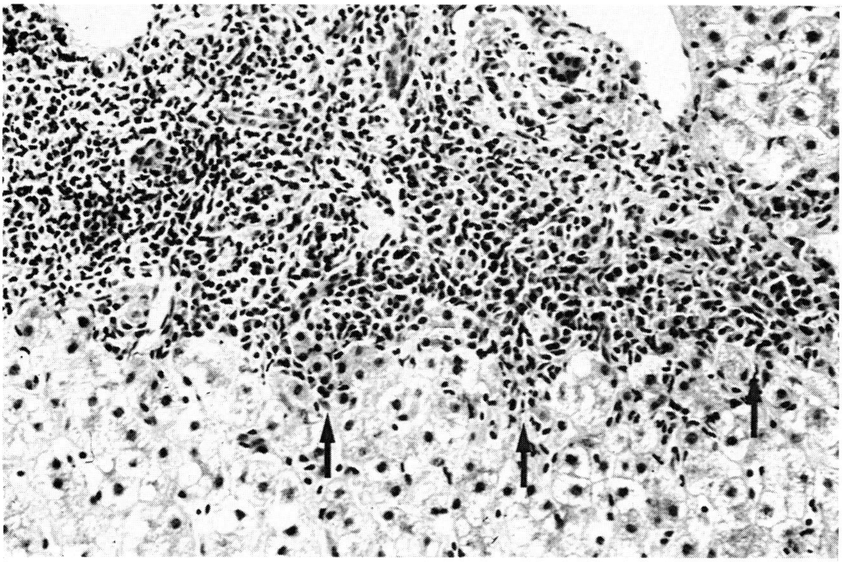

Fig. 11.4 CAH. A portal tract is heavily infiltrated by lymphocytes and plasma cells which have eroded the limiting plate (arrows) and spilled over into the lobule (HE, × 360).

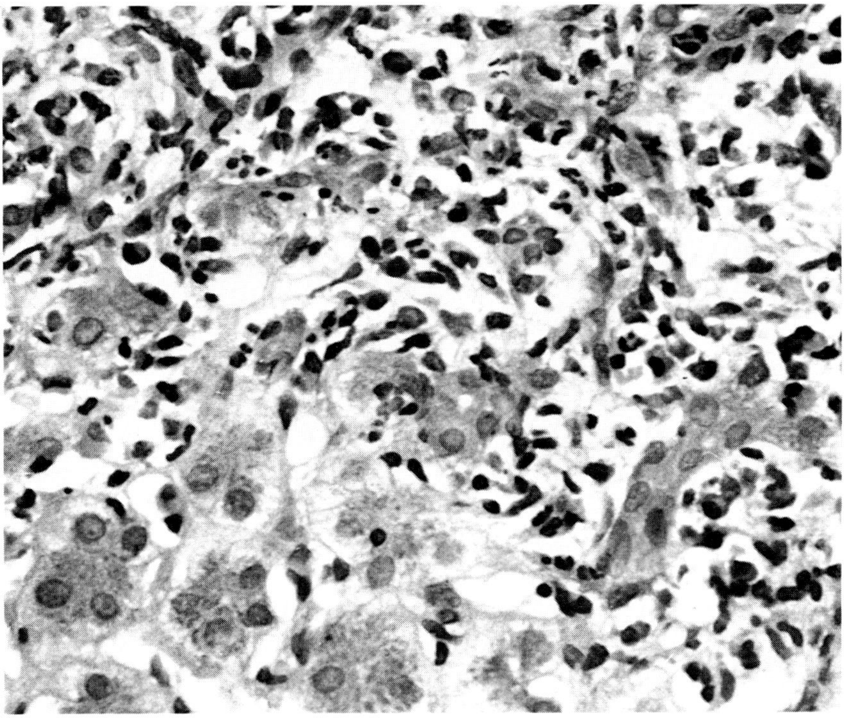

Fig. 11.5 CAH. Lymphocytes, plasma cells and connective tissue have extended from the portal tract through the limiting plate. This process of piecemeal necrosis is associated with polymorphonuclear leucocyte infiltration (HE, × 720).

identified by double-cell thick hepatocyte plates and by the formation of hepatocyte rosettes (Fig. 11.7). Together with portal and periportal changes there are often changes within the liver lobule. These range from activation of sinusoidal cells and focal necrosis to extensive ballooning of hepatocytes, particularly at the junction of hepatocytes with collagen septa; acidophil bodies may also be found (Figs 11.8 and 11.9).

It is evident, therefore, that all of the changes of typical viral hepatitis can be found in some cases of CAH and these occur in the absence of viral markers (e.g. autoimmune or NANB types). In hepatitis B carriers ground-glass cells and viral antigens (p. 91) can be demonstrated on liver biopsy (Fig. 11.10). HBsAg may be cytoplasmic or associated with cell membranes, while HBcAg is nuclear. HBcAG-containing nuclei have a sanded appearance. In patients with a concomitant HDV infection, liver nuclei, and sometimes cytoplasm, contain delta antigen. Lobular damage is prominent in 'autoimmune' CAH in the absence of viral markers.

Occasionally in CAH there is piling up of swollen hydropic bile duct

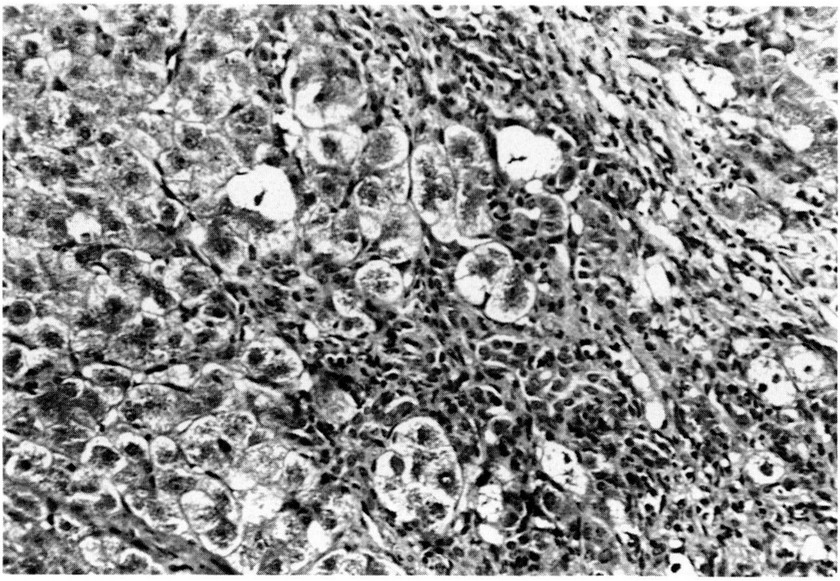

Fig. 11.6 CAH. Inflammatory cells and connective tissue have destroyed the limiting plate and groups of hepatocytes are isolated in connective tissue (HE, × 560).

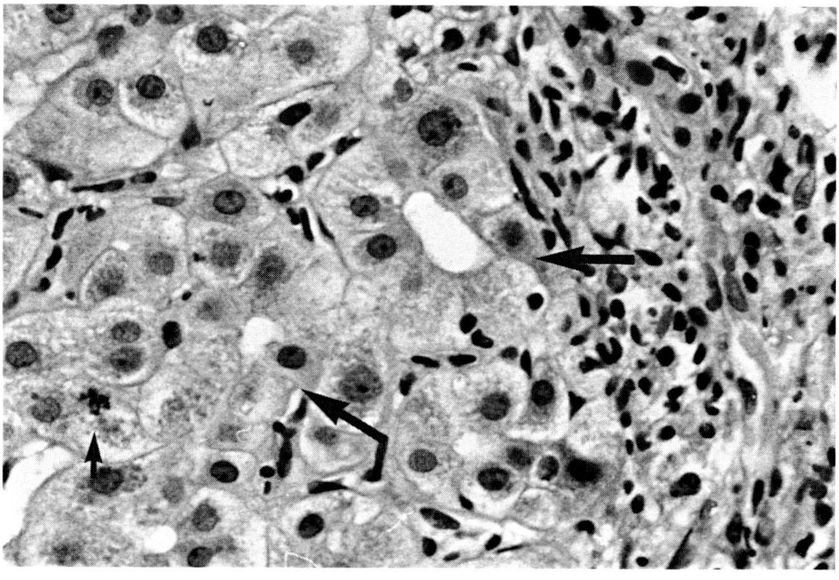

Fig. 11.7 CAH. There is piecemeal necrosis and hepatocyte regeneration; there is a mitotic figure (small arrow) in one hepatocyte and others have formed rosettes (large arrows) (HE, × 720).

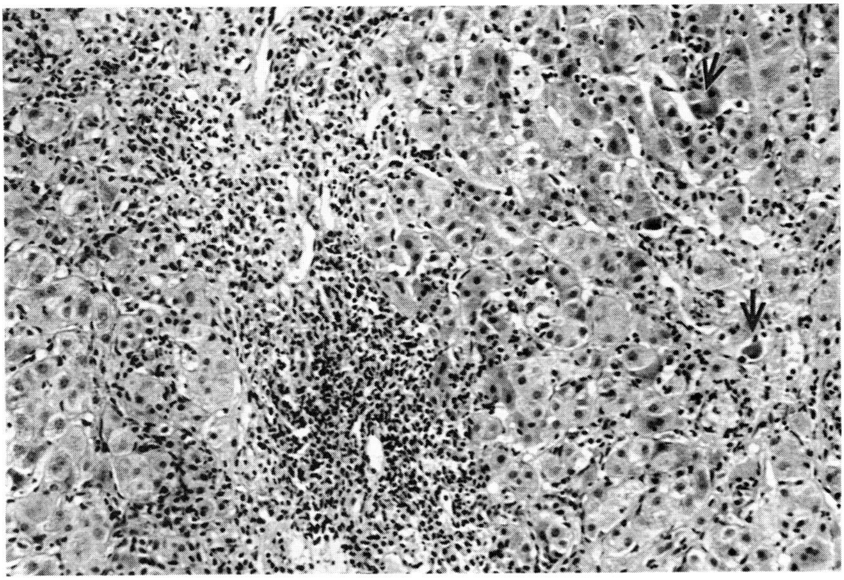

Fig. 11.8 CAH. This is a florid CAH in which there is piecemeal necrosis and intralobular focal necrosis. Acidophil cells (arrows) are also present in the right half of the field (HE, × 160).

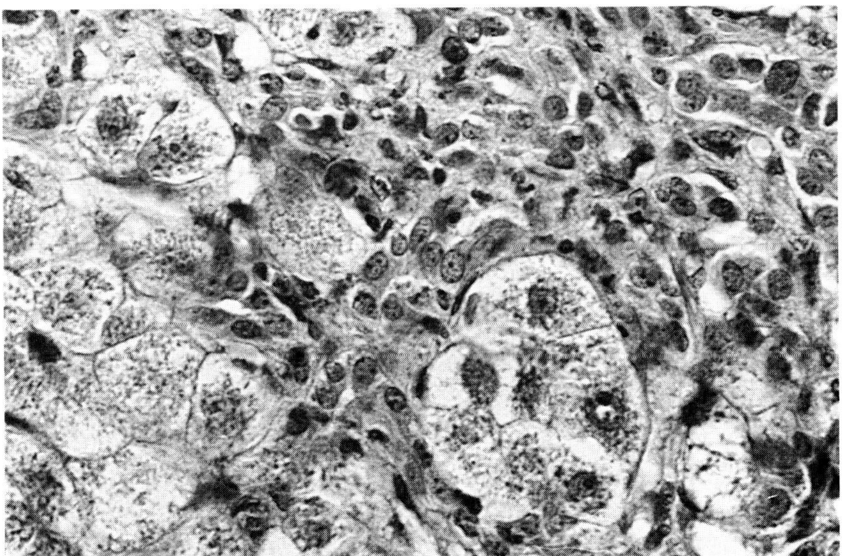

Fig. 11.9 CAH. There is piecemeal necrosis and 'balloon' degeneration of hepatocytes at the portal tract–parenchymal junction (HE, × 720).

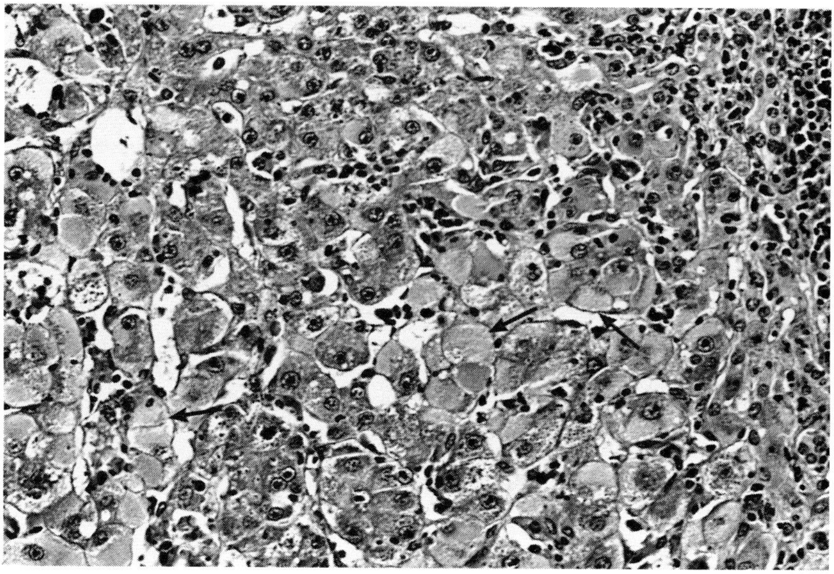

Fig. 11.10 CAH. There are numerous ground-glass hepatocytes in this field (arrows) (HE, × 247).

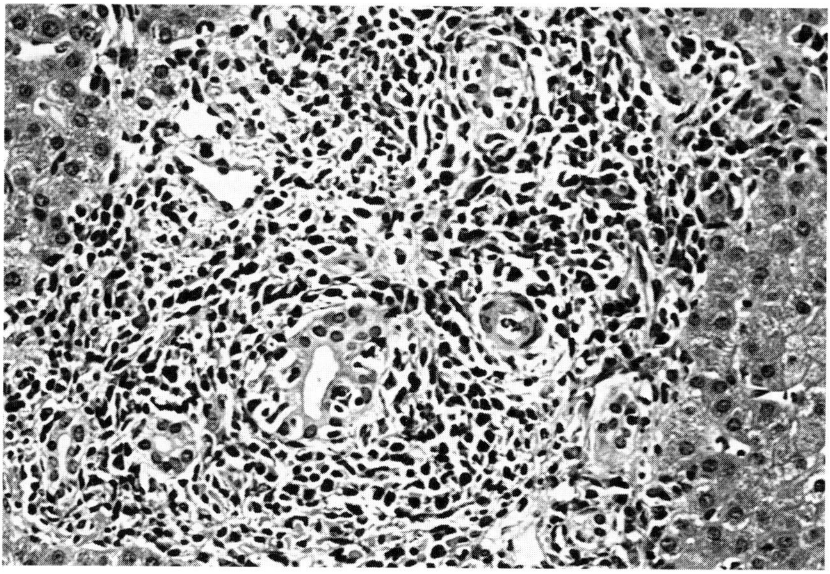

Fig. 11.11 CAH. An interlobular bile duct is damaged. There is mild piecemeal necrosis (HE, × 274).

epithelial cells which may be infiltrated by inflammatory cells (Fig. 11.11) and necrosis of the lining of larger ducts can also occur. These changes have a patchy distribution and can pose difficulty in the differential diagnosis of primary biliary cirrhosis. Unlike PBC, granulomas are not seen, piecemeal necrosis is often more prominent and circulating anti-mitochondrial antibody (AMA) is absent. In our experience, this is an uncommon feature of CAH but has been recorded frequently by some observers, especially from Scandinavia (Poulsen and Christoffersen, 1972).

These morphological features vary in intensity within the liver. Small biopsies, therefore, may give an erroneous impression of severity.

11.2.3 *Prognostic features*

Although there is general agreement that CAH has a potential to progress to cirrhosis, there is debate about which morphological features in a liver bioposy are of importance in assessing prognosis. Recently, great emphasis has been placed on the presence of lobular necrosis and particularly bridging hepatic necrosis as morphological features which connote a bad prognosis. Bridging hepatic necrosis is recognized as areas of necrosis which link portal tracts and portal tracts, portal tracts and central veins, or central veins with each other. Central portal bridges are probably the most important in determining a cirrhotic outcome (Scheuer, 1986). In these areas there is cell dropout but necrotic hepatocyte debris has usually been removed at the time of biopsy and the necrotic zones are replaced by lymphocytes, mononuclear cells and fine reticulin fibres (Fig. 11.12); the histiocytes often contain abundant ceroid pigment. In one study it was shown that of the patients with bridging hepatic necrosis on first biopsy, 27% developeed cirrhosis (Baggenstoss *et al.*, 1974). The same authors have described a more extensive necrotizing process referred to as subacute hepatitis with multilobular necrosis; this is associated with a 40% incidence of progression to cirrhosis. As its name implies this lesion is characterized by complete necrosis of several adjacent lobules and is usually associated with a vigorous chronic inflammatory cell infiltrate and evidence of CAH elsewhere in the biopsy (Fig. 11.13). More recently, a Scandinavian group have questioned the ominous significance of these necrotizing lesions because they found no correlation between confluent necrosis and the subsequent development of cirrhosis (Dietrichson and Christoffersen, 1977). In this study the degree of portal tract inflammation and abnormal bile duct proliferation were the only biopsy features which showed a significant correlation with the development of cirrhosis. In neither study was bridging defined anatomically (see p. 85). It is probable that central portal bridging

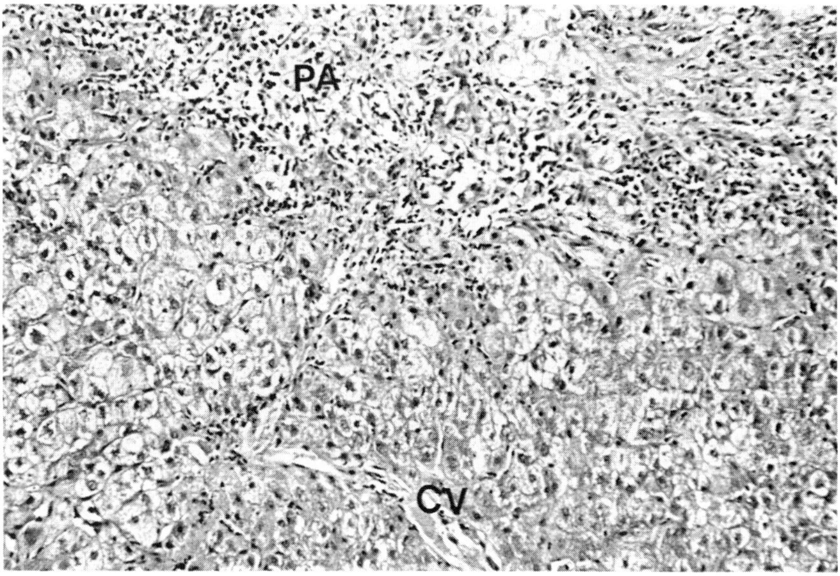
PA
CV

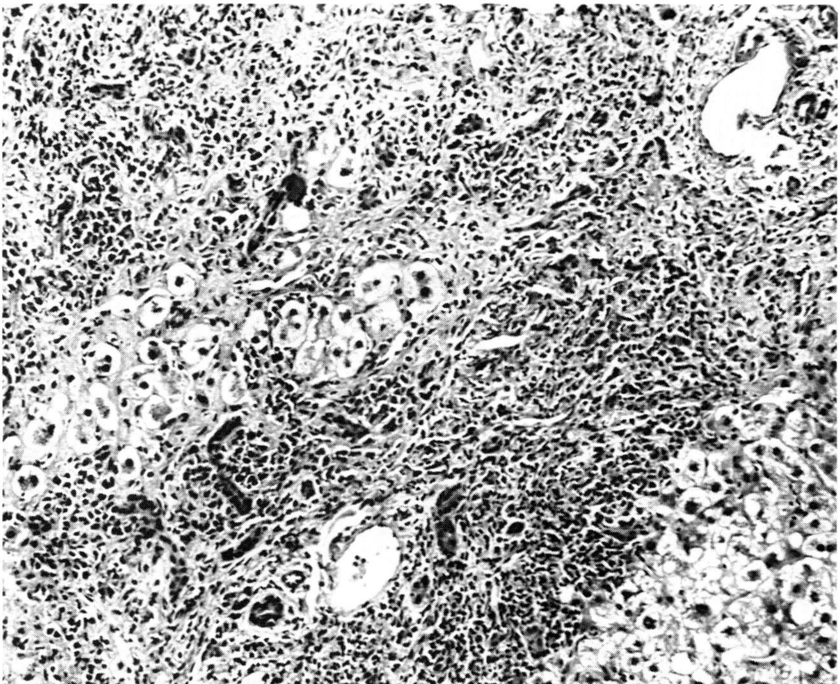

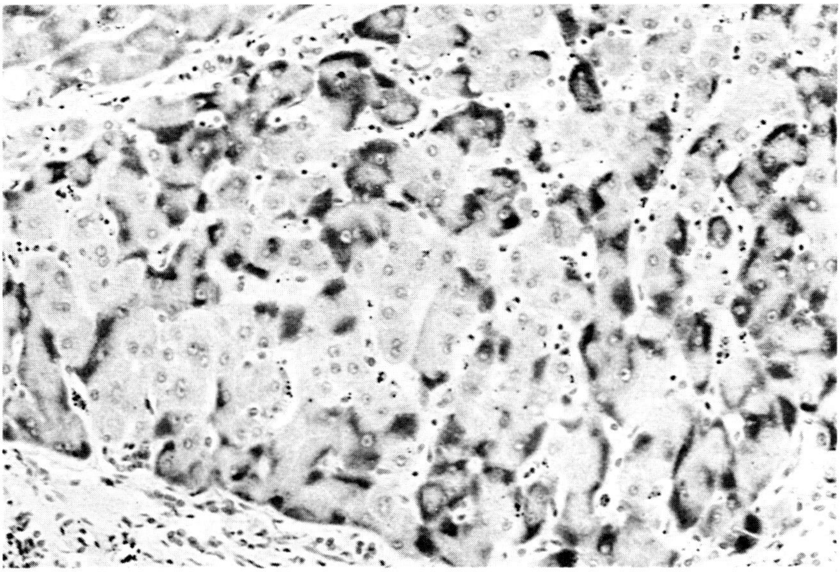

Fig. 11.14 CAH. HBsAg is present on the surface of many hepatocytes (immunoperoxidase using anti-HBsAg, × 225).

correlates with prognosis. Assessment of CAH biopsies should therefore include a statement about lobular necrosis.

In patients with CAH of presumed viral B origin, it has been suggested that localization of viral antigen in the hepatocyte cell membrane or cell periphery may be an important factor in determining chronicity of hepatitis (Fig. 11.14) (Ray *et al.*, 1976). Probably of more importance in prognosis is whether the patient still has active viral replication as indicated by persisting serum HBV-DNA or HBeAg. If this is the case, firstly the patient can obviously continue to deteriorate because of immune damage to hepatocytes. Secondly, even when such patients seroconvert to anti-HBe with diminution in viral replication, this may also be associated with deterioration in liver function due to elimination of virus-containing hepatocytes. Thirdly, patients who have seroconverted to anti-HBe sometimes have reactivation of viral replication with associated flare in their liver disease. Fourthly, if the patient becomes super-

◀ **Fig. 11.12** CAH. There is bridging necrosis between a portal area (PA) and a central vein (CV) (HE, × 160).

◀ **Fig. 11.13** CAH. There is multilobular necrosis; between two portal areas there are only a few surviving hepatocytes, many of which show balloon degeneration (HE × 180).

infected with HDV, liver function again deteriorates. Superinfection with HAV or nonA, nonB viruses also leads to deterioration. Finally, the viral type which has induced CAH also influences outcome. NonA, nonB chronic infections are frequently punctuated by exacerbation of disease activity (Berman *et al.*, 1979).

11.2.4 Differential diagnosis

On biopsy CAH has to be distinguished from CPH and primary biliary cirrhosis. This distinction is not only of academic interest but of practical significance to the patient, as the therapy for each condition is different. CPH differs from CAH in that piecemeael necrosis and fibrosis are absent in the former. Biopsies from patients with CAH being treated with steroids may show a pattern similar to CPH but fibrosis is usually still evident. PBC can usually be distinguished from CAH because in the former the lobular architecture remains intact at least early in the disease, there is marked bile ductular proliferation and the typical interlobular bile duct lesion (p. 268) may be present; further helpful points are that the portal tract lymphocytes in PBC frequently form focal aggregates, and cholestasis when present is more marked on the periphery of the lobule. Where piecemeal necrosis is florid this is much more in favour of CAH. Antimitochondrial antibody is positive in 90% or more of patients with PBC and uncommon in CAH; while in CAH high titres of smooth muscle antibody and antinuclear factor are typical. It has to be acknowledged, however, that even after careful analysis of the pathological, clinical, biochemical and serological data a firm diagnosis of CAH or PBC may still not be possible, and in particular when any of these two sets of data are at variance. In these cases the designation chronic active liver disease should be applied and the patient followed up. The true nature of the condition may be revealed as the natural history of the disease unfolds.

The morphological features of CAH may be present also in cirrhosis from many different causes (see p. 289). In this instance some authors use the term 'CAH with cirrhosis' while others prefer the designation 'active cirrhosis'. In a small biopsy it is sometimes difficult to exclude the possibility that cirrhosis has already developed in CAH.

CAH may develop in intravenous drug abusers. These patients show the same features on biopsy as described above but in addition there may be evidence of hepatitis B virus infection in the form of ballooned hepatocytes, focal necrosis, etc. In North American addicts double refractile material is also a frequent finding in portal tracts and small granulomas may develop in the liver. These features are probably related to the material used to dilute the drug before it is administered; in the UK, however, the authors have rarely seen this on liver biopsy.

In summary, chronic hepatitis is a syndrome which may be induced by viruses, drugs, metabolic disorders, alcohol or unknown agents. The syndrome is associated with overt signs of liver disease, and may or may not be associated with serological abnormalities. Biopsy is the only way in which a definitive diagnosis of chronic hepatitis can be made. It is useful to divide these cases up into CPH and CAH, because it is generally true that the former has a better prognosis than the latter. In each case, lobular necrosis and inflammation, and the degree of piecemeal necrosis, should be recorded since these also reflect a worse prognosis. It should be realized that in any given patient these broad prognostic indicators may not exactly predict disease evolution. It is important, therefore, when assessing evolution that the aetiology of the condition be taken into account. In chronic HBV infections, the status of viral replication and the patient's immune response to HBV are other data that bear on prognosis.

11.3 Primary biliary cirrhosis

PBC is primarily a disease of bile ducts. It is also associated with chronic hepatitis.

'Primary biliary cirrhosis' is an inaccurate term because at first presentation patients with this disease do not have cirrhosis. Of the alternative terms suggested the most accurate designation for this disease in its early stages of development is 'chronic non-suppurative destructive cholangitis', but the term 'primary biliary cirrhosis' is so entrenched in the literature that it will be retained here.

11.3.1 Presentation

PBC presents in the fourth to sixth decade and is about eight times more common in females than males. The initial symptoms are usually pruritus and fatigue and although jaundice and pale stools may be complained of these signs may not develop until late in the course of the disease. Xanthelasma on the hands and feet and in skin flexures present early in the disease. Steatorrhoea, with poor absorption of vitamins D and K, is also a feature. These defects in absorption give rise to osteomalacia and a prolonged prothrombin time respectively. Classically, there is a markedly elevated alkaline phosphatase with a normal bilirubin early in the history of the disease, although the bilirubin eventually rises. Elevated serum cholesterol and bile acids also occur at an early stage but serum transaminases usually rise later in the disease. The serum IgM is characteristically elevated and antimitochondrial antibody is present in 80–95% of cases in a titre of 1 : 10 to 1 : 6000. The latter is also occasionally positive in CAH, systemic lupus erythematosis and rheumatoid arthritis. Antinuclear factor and smooth muscle antibody is found in 20–50% of cases of PBC. This condition may be fatal, with a natural history of 1–20 years and the mean survival from initial diagnosis is 5 years (Foulk and Baggenstoss, 1975). The disease is not infrequently associated with other disorders in which there are immunological abnormalities such as Sjögren's syndrome, thyroid disease, renal tubular acidosis, fibrosing alveolitis and CRST syndrome (Calcinosis, Raynaud's phenomenon, Scleroderma and Telangiectasia). In spite of associated immunological abnormalities and the finding of cell-mediated immunity to a bile duct antigen, PBC does not fulfil all of Witebsky's postulates for an autoimmune disease.

11.3.2 Biopsy features

On the basis of liver biopsy features, PBC has been divided into four stages: stage 1 (cholangitis) showing the typical duct lesion; stage 2 (ductular) characterized by atypical ductular proliferation; stage 3 (scarring) typified by connective tissue deposition; and stage 4 showing a fully developed cirrhosis. Although this staging terminology is used here for descriptive convenience, it should be realized that this division is artificial in that features of the first three stages may be seen in any single biopsy and the duct lesion is often still seen in fully established cirrhosis. Furthermore, the dominant pathological feature sometimes correlates poorly with the duration of the patient's symptoms.

A lesion of interlobular and septal bile ducts is typical of stage 1. This is characterized by degeneration of the bile duct epithelium which takes the form of cytoplasmic vacuolation, irregularity in size and shape of the cells and nucleus pyknosis (Figs 11.15 and 11.16). This may progress to frank necrosis of the bile duct epithelium and this is pathognomonic of PBC. Concomitant with these degenerative changes evidence of regeneration is also present – the bile duct epithelium may be stratified and have as many as six layers. There is a florid inflammatory reaction around these damaged ducts, in which the inflammatory cells are present in focal aggregates; this and the bile duct lesion may only be evident when serial sections are examined. These aggregates are composed of mainly

---

Fig. 11.15 Primary biliary cirrhosis, stage 1. The epithelium of an interlobular bile duct has been destroyed (arrows) and lymphocytes have eroded into the duct ▶ lumen. There are numerous lymphocytes and some plasma cells around the duct (HE, × 450).

Fig. 11.16 Primary biliary cirrhosis, stage 1. An abnormal interlobular bile duct is surrounded by lymphocytes and mononuclear cells. The bile duct epithelial cells are taller than normal, some are vacuolated, and the polarity of some cells is ▶ disturbed (small arrows); lymphocytes are also present in the epithelium (large arrows) (HE, × 640).

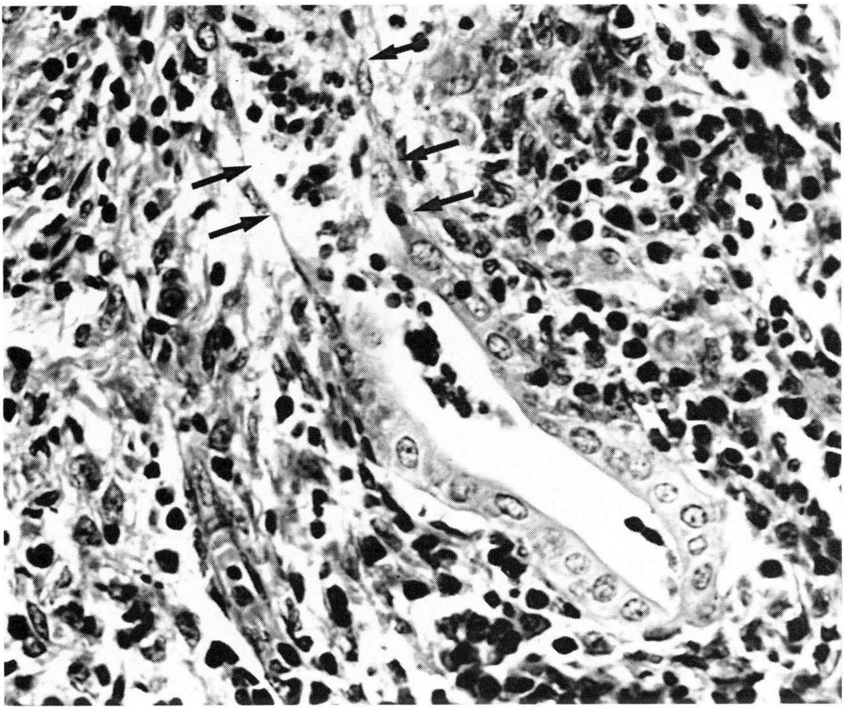

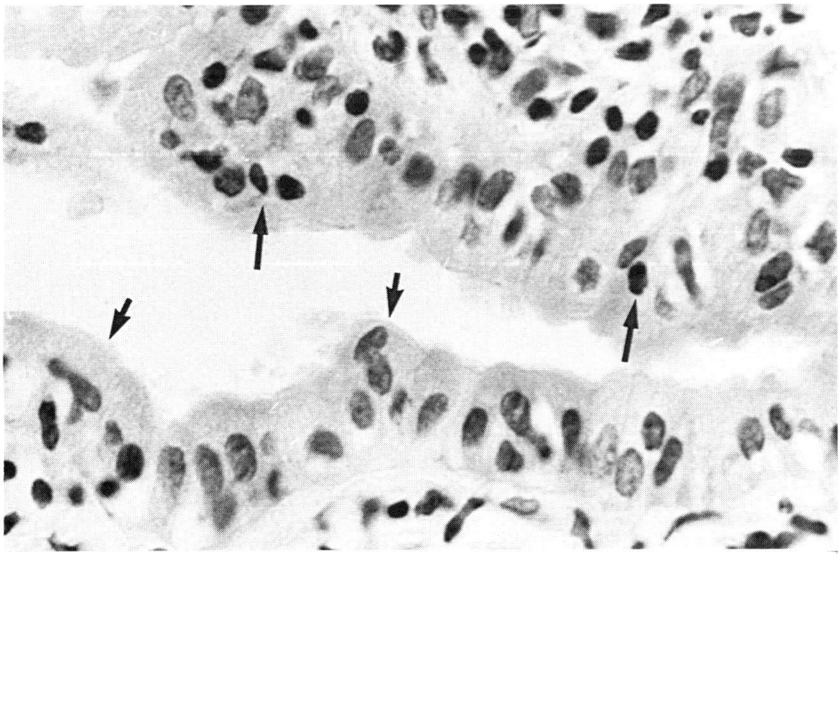

lymphocytes and plasma cells and sometimes have a germinal centre (Fig. 11.15). Degenerate bile duct epithelium in the centre of lymphoid aggregates can masquerade as germinal centres. Lymphocytes and plasma cells erode the damaged bile duct and are found in their lumina (Fig. 11.15) together with mucin and other debris. Similar immunocytes are also found diffusely in portal tracts and may be associated with variable degrees of erosion of the limiting plate. Small granulomas develop in about a third of cases in relation to damaged bile ducts and more uncommonly in the liver lobule. These granulomas are composed of macrophages, which are usually PAS positive, together with multinucleate giant cells (Fig. 11.17); typically, however, these granulomas are not as well defined as the one illustrated.

Within the parenchyma sinusoidal cells are prominent and a few acidophil cells and hydropic hepatocytes are infrequently present. Early in the disease cholestasis is usually absent but its presence correlates with the serum bilirubin level. When present cholestasis is diffuse or more typically periportal in distribution (Fig. 11.18). Copper-binding protein

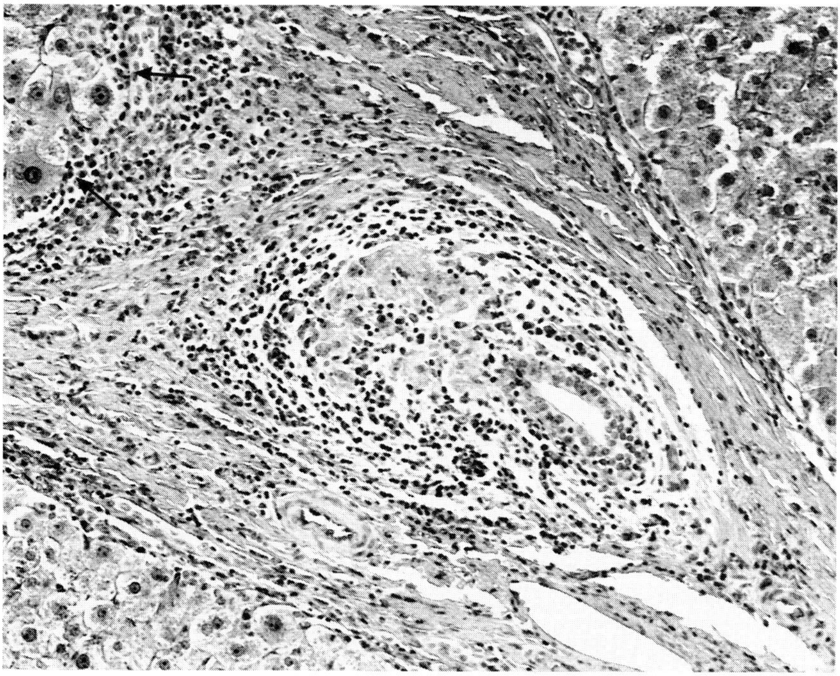

Fig. 11.17 Primary biliary cirrhosis, stage 1. There is a well-formed granuloma at one end of an interlobular bile duct which has presumably ruptured. The bile duct epithelium is hyperplastic. Piecemeal necrosis is also evident (arrows) (HE, × 225).

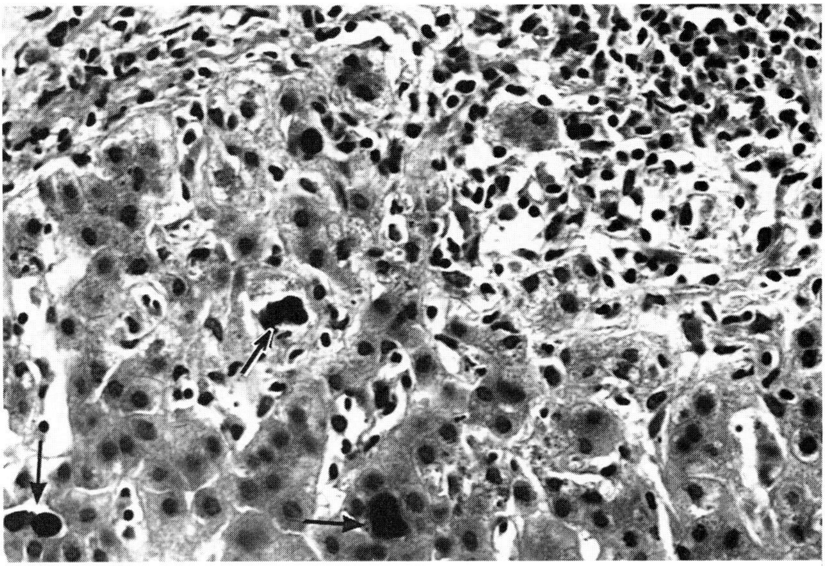

Fig. 11.18 Primary biliary cirrhosis, stage 2. Peripheral cholestasis is evident as bile thrombi (arrows) in cells around a portal tract. There is piecemeal necrosis (HE, × 458).

accumulates in peripheral hepatocytes and is evidence of biliary damage. When cholestasis is severe feathery degeneration and bile infarcts may be found rarely.

Stage 2 is characterized by a decrease in the number of interlobular bile ducts and proliferation of bile ductules. In normal liver the number of mature ducts in any portal tract is one or more, and only 15% of tracts contain no bile ducts. In PBC, however, the mean number of bile ducts in any one portal tract is less than one and 60% of tracts contain no ducts. The ductules are lined by a single layer of cells ranging from flat to columnar in shape (Fig. 11.19). The lumen is small and on light microscopy may not be visible. Lymphocytes, plasma cells and few polymorphs, surround these structures. Macrophages are also present and these may contain lipid and form small xanthomas (Fig. 11.20).

In parallel with the changes described above, connective tissue is deposited in portal tracts which become broadened. This fibrotic process is typical of stage 3. The connective tissue in portal tracts spreads out as septa linking adjacent portal tracts but often avoids central veins (Fig. 11.21). In this way large masses of parenchyma are isolated and, although partially surrounded by fibrous tissue, they still have a normal vascular architecture. Eventually, however, the lobular parenchyma is dissected by fibrous tissue, the vascular architecture is destroyed and a true

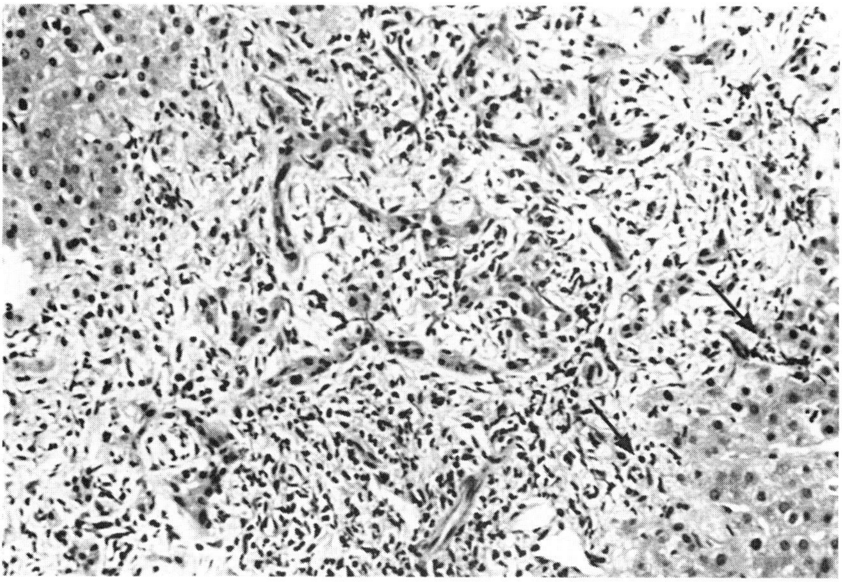

Fig. 11.19 Primary biliary cirrhosis, stage 2. There is exuberant bile duct proliferation in a portal area which is expanded by connective tissue. Piecemeal necrosis is also evident (arrows) (HE, × 240).

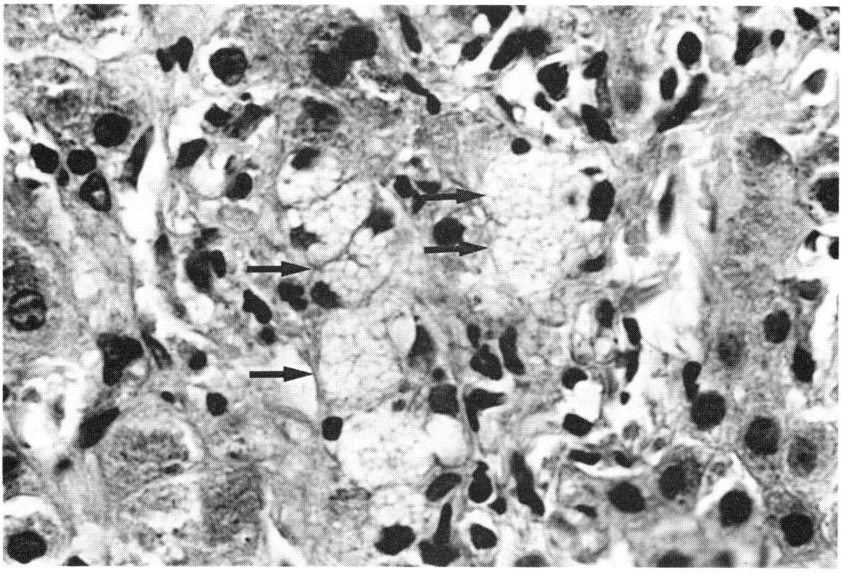

Fig. 11.20 Primary biliary cirrhosis, stage 2. Large foamy histiocytes (arrows) are intermixed with hepatocytes at the portal tract parenchymal junction. These histiocytes have presumably engulfed material which has leaked from a damaged bile duct (not shown) (HE, × 720).

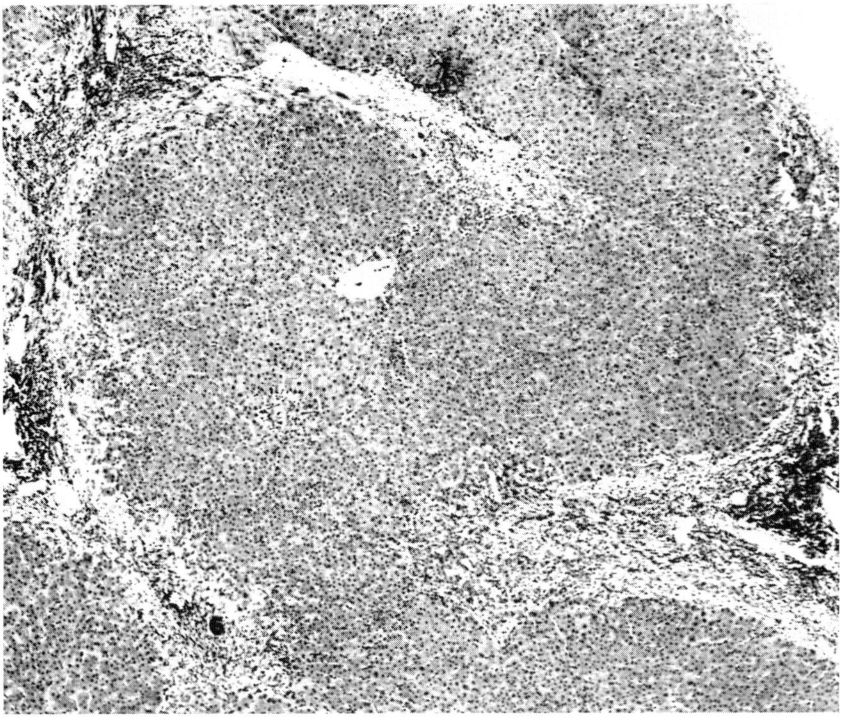

Fig. 11.21 Primary biliary cirrhosis, stage 3. There is extensive fibrosis, focal collections of lymphocytes in septa and nodular regeneration but fibrous septa do not extend to the central vein (HE, × 56).

cirrhosis (stage 4) results. At this stage bile ductular proliferation is still present but is unhelpful in diagnosis since it is found in almost all aetiological forms of cirrhosis. An occasional florid duct lesion or granuloma may still be present in the cirrhotic stage but of more help in diagnosis is the decreased number of mature bile ducts and the presence of aggregates of lymphoid cells in connective tissue septa (Fig. 11.22). Cholestasis is usually marked at this stage as is CBP.

Mallory bodies are found in about one-fifth of the cases of PBC and may be present at any stage. They are most common in periportal hepatocytes. Copper and copper-binding protein in peripheral hepatocytes increases with time and is most prominent when cirrhosis has developed.

In the histopathological assessment of PBC it should be remembered that the disease in its early stages is focal in the liver and medium-sized bile ducts showing the characteristic lesion may not be present in needle biopsies. All of the other features described, however, do show themselves in needle biopsy.

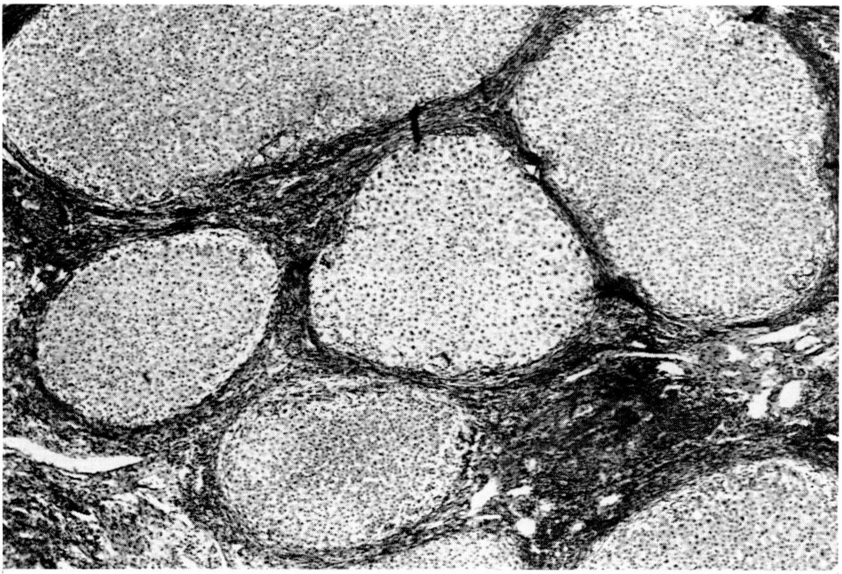

Fig. 11.22 Primary biliary cirrhosis, stage 4. This liver is truly cirrhotic. The architecture is totally destroyed by fibrous tissue and nodular regeneration. Focal collections of lymphocytes are characteristic of this stage of the disease. Note also the absence of interlobular bile ducts in the septa (HE, × 60).

In a recent large study, Portmann *et al.* (1985) have looked at the changes in biopsy appearances in PBC and have defined four forms of piecemeal necrosis. These have been correlated with stages as defined above. Biliary piecemeal necrosis is associated with erosion of the limiting plate and a variable degree of polymorph infiltration. The peripheral hepatocytes are prominent. This form occurs at all stages except stage 1 but increases from stages 2 to 4. Lymphocytic piecemeal necrosis is characterized by lymphocytic and plasma cell erosion of the limiting plate and the adjacent hepatocytes are atrophic or ballooned. This type of necrosis with ballooned hepatocytes can be identical to autoimmune CAH; it is most common in stages 2 and 3. Ductular piecemeal necrosis is characterized by ductules which extend into the parenchyma and are typically associated with polymorph infiltration; it is equally common in stages 2–4. Fibrotic piecemeal necrosis appears predominantly in stages 3 and 4. Fibrosis accompanies all forms of piecemeal necrosis but in this variety the characteristic feature is a hypocellular halo at the junction of portal tracts with hepatic parenchyma and extension of fibrosis into the parenchyma. The 'halo' lesion occurs predominantly in stage 4. These authors also stressed the diagnostic significance of hypocellular scars which occur predominantly in stages 3 and 4. The scars are $\geqslant$ 1 mm thick, and composed of loose collagen in which there are many thick-walled arteries (suggesting tortuosity). Larger bile ducts (300 μm) than those damaged in PBC are found in some areas of these scars and frequently show micropapillary projections. These ducts do not have a normal relation with arteries. In some areas of these scars, plexiform arrangements of ductules are found associated with polymorph infiltration. Mallory bodies and copper-binding protein occur predominantly in stages 3–4 and 2–4 respectively. These patterns of piecemeal necrosis were maintained throughout the evolution of PBC in individual patients. Those features indicating a poor prognosis were cirrhosis, increasing cholestasis with Mallory body formation, and 'halo' formation at the margins of fibrous septa. Patients with stage 1 lesions may be asymptomatic and may have a normal life-span without deterioration of liver function (Beswick *et al.*, 1985). These recent studies re-emphasize that no one morphological feature predicts duration or prognosis of PBC except cirrhotic transformation.

11.3.3 Differential diagnosis

Five main conditions should be considered in the differential diagnosis of PBC and these are CAH, ulcerative colitis with pericholangitis or sclerosing cholangitis, large duct obstruction, drug-induced cholestasis, and granulomatous hepatitis; rarely lymphomatous infiltration of liver may present diagnostic difficulty. In ascending order of diagnostic importance in arriving at a diagnosis in PBC are: (1) sinusoidal cell hypertrophy and hyperplasia; (2) epithelioid granulomas; (3) peripheral bile stasis; (4) no cholestasis in the presence of an elevated alkaline phosphatase; (5) destruction of medium-sized bile ducts and (6) absence of bile ducts in more than 50% of portal tracts (Foulk and Baggenstoss, 1975). In some cases it is difficult to differentiate between PBC and CAH because, in both, piecemeal necrosis and bile duct abnormalities can occur. In CAH, the interlobular bile ducts (Poulsen and Christoffersen, 1972) involved are the same size as in PBC but the inflammatory reaction associated with this is not usually focal but diffuse throughout portal tracts and piecemeal necrosis is more prominent. In addition, granulomas are uncommon in CAH. In both conditions there may be lobular hepatitis with swelling of hepatocyte and focal necrosis; in the absence of bile duct lesions, copper-binding protein and granulomas, it is difficult to distinguish these PBC lesions from CAH. In sclerosing cholangitis, the classic lesion is 'onion-skin' fibrosis around interlobular and larger bile ducts together with piecemeal necrosis (Barbatis *et al.*, 1985). Large bile duct obstruction may occasionally be difficult to differentiate from PBC. The most helpful feature is the presence of polymorphs within and around ducts in large duct obstruction. Polymorphs around ductules may be part of the piecemeal necrosis process in PBC and are not a helpful discriminating criteria. A positive feature indicating drug-induced cholestasis is the presence of cholestasis as the only abnormality or in association with a portal tract inflammatory infiltrate in which eosinophils are present. Apart from septrin, few drugs produce bile duct damage or loss. Granulomatous hepatitis is distinguished from PBC on the basis that granulomas only are present in the former. In PBC, granulomas may occur anywhere in the lobule and should be considered in the differential diagnosis of granulomatous hepatitis.

In graft-versus-host disease and in transplant rejection, bile duct damage is the rule (p. 68, 70); piecemeal necrosis is absent and vascular injury is frequently present. A bile duct lesion very similar to PBC may be seen in liver transplants years after transplantation (Neuberger *et al.*, 1982). This lesion may be the result of low-grade chronic rejection rather than recurrence of PBC as originally reported. Lymphomatous infiltrates in portal tracts are usually easily distinguished from PBC because of the monomorphous cell type and these are not associated with piecemeal necrosis, bile duct destruction or granulomas. Granulomas may occur in Hodgkin's disease on their own or in combination with a Hodgkin's infiltrate. When the latter is present there is no diagnostic difficulty, but when granulomas only are evident PBC should be excluded.

11.4 Chronic hepatitis in systemic 'autoimmune' disease

In systemic 'autoimmune' diseases, serious liver dysfunction is uncommon and is usually overshadowed by other organ pathology. In spite of the fact that systemic lupus erythematosis and CAH have in common some clinical features such as arthralgia, skin rashes, positive serum antinuclear factor, etc., liver pathology is usually normal or unimpressive in the former condition. There may be non-specific reactive hepatitis, focal aggregates of lymphocytes with germinal centre formation in portal tracts (Fig. 11.23), or fatty change due to steroid therapy. Rarely, CAH may be found at biopsy.

Rheumatoid arthritis may cause secondary amyloid deposition in the liver and diffuse nodular regenerative hyperplasia can occur in Felty's syndrome.

Progressive systemic sclerosis is not associated with overt liver disease but scleroderma localized to the digits together with calcinosis, Raynaud's phenomenon and telangiectases (CRST syndrome) may be present in PBC. In progressive systemic sclerosis, megamitochondria unassociated with other liver pathology are frequently found in hepatocytes.

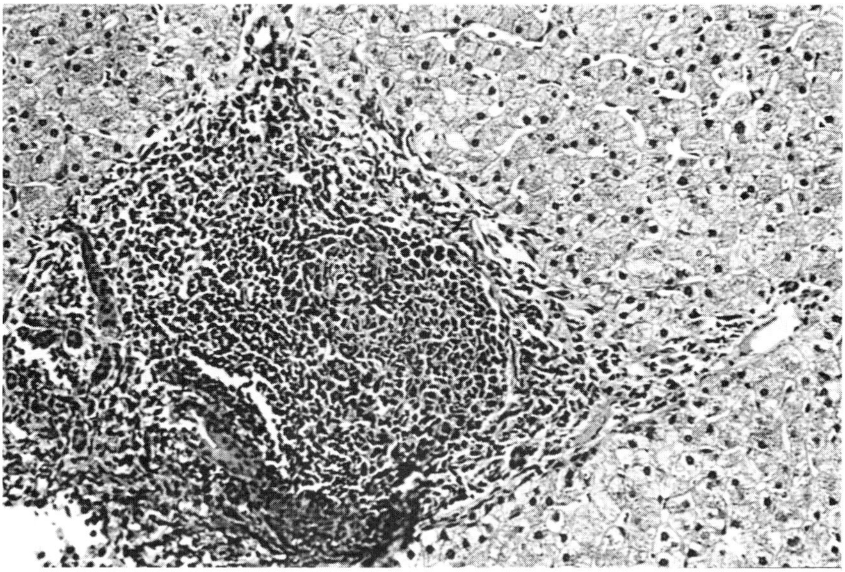

Fig. 11.23 Systemic lupus erythematosis. The characteristic lesion in this condition is the presence of focal aggregation of lymphocytes in portal tracts; this aggregate contains a germinal centre. The parenchyma is normal (HE, × 160).

11.5 Liver disease in AIDS

Hepatitis in patients with human immunodeficiency virus (HIV) infection is secondary to the immune deficiency caused by the virus and to viruses which may be expected in association with HIV, such as, hepatitis B. Of 82 consecutive cases in the UK (Boylston *et al.*, 1987), five showed acute viral hepatitis, three non-specific reactive hepatitis and three CMV infection. Mycobacterial (including atypical forms), malarial, and *Histoplasma* infections and granulomas without organisms were each seen in one patient. There are no liver findings specific for AIDS and Kaposi's sarcoma is found in liver only when present elsewhere in the patient (Glasgow *et al.*, 1985).

11.6 Indian childhood cirrhosis

This uncommon disease affects young children in India and has occurred in immigrants to other countries including the UK. Most cases occur in the age group 6 months to 5 years. There is hepatomegaly and gradually progressive hepatic failure with a high mortality rate. It tends to occur in families and among the upper social classes but there is no convincing evidence for a genetic or environmental cause. During its development, there is hepatitis with numerous lymphocytes, histiocytes and neutrophil polymorphs, but few plasma cells. This is accompanied by severe hydropic degeneration and Mallory bodies in hepatocytes, although fatty change is not present (Fig. 11.24). The liver cells are rich in copper but, unlike Wilson's disease, there is no deficiency of serum ceruloplasmin. Diffuse fine pericellular fibrosis is another prominent feature which leads to a very fine micronodular cirrhosis with little evidence of parenchymal regenerative activity. Copper chelation therapy may be useful.

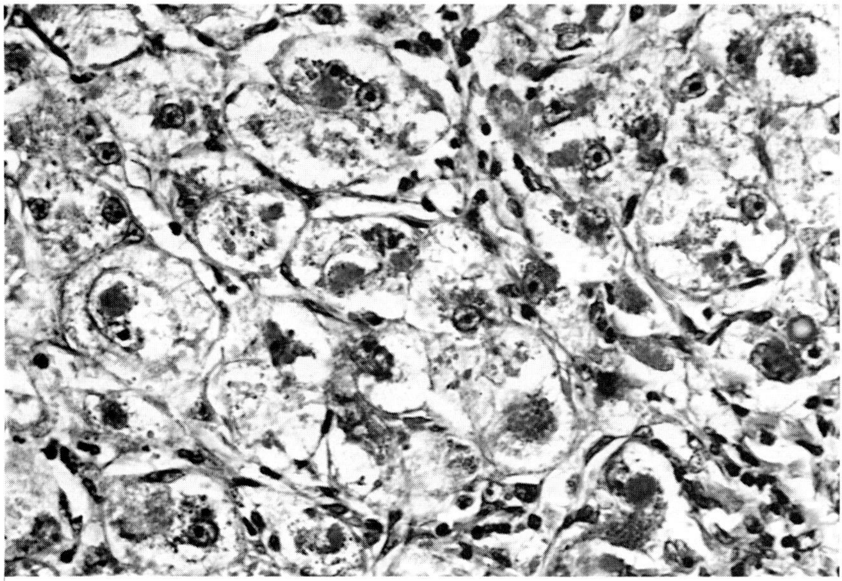

Fig. 11.24 Indian childhood cirrhosis. The hepatocytes are swollen and contain Mallory bodies (HE, × 450).

11.7 Non-specific reactive hepatitis

The liver reacts to a number of disparate stimuli by producing the picture of non-specific reactive hepatitis, the most common cause being a febrile illness. It is a frequent finding on biopsy and may be provoked by any systemic or severe local infection and by a variety of drugs (p. 176). It is characterized by prominent sinusoidal cells and a mild infiltrate of lymphocytes and mononuclear cells in portal tracts which have an irregular distribution in the liver (Fig. 11.25). There may also be small foci of necrosis and mild fatty change in hepatocytes. It is important to emphasize to the clinician that this is not a primary liver disease but is a reaction of the liver to extrahepatic stimuli. The portal tract infiltrate is not

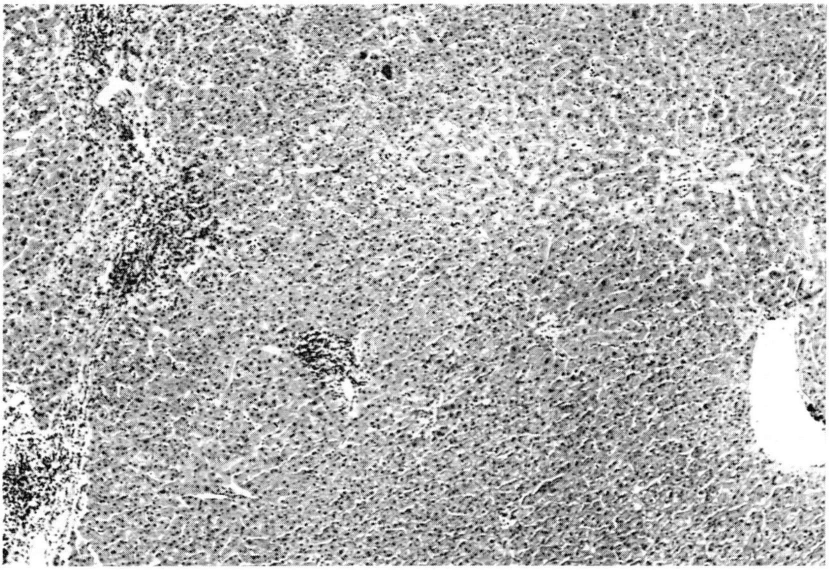

Fig. 11.25 Nonspecific reactive hepatitis. Small collections of chronic inflammatory cells have an irregular distribution in parenchyma and portal tracts (HE, × 56).

associated with piecemeal necrosis and this should not be confused with CAH but may be difficult to distinguish from CPH.

References

Baggenstoss, A. H., Summerskill, W. H. J. and Ammon, H. V. (1974), The morphology of chronic hepatitis, in *The Liver and its Diseases* (eds. F. Schaffner, S. Sherlock and C. M. Leevy), Intercontinental Medical Book Corp., New York, p. 199.

Barbatis, C., Grases, P., Shepherd, H. A. *et al.* (1985), Histological features of sclerosing cholangitis in patients with ulcerative colitis. *J. Clin. Pathol.*, **38,** 778–83.

Berman, M., Alter, H. J., Ishak, K. G. *et al.* (1979), The chronic sequelae of non-A, non-B hepatitis. *Ann., Intern. Med.*, **91,** 1–6.

Beswick, D. R., Klatskin, G. and Boyer, J. L. (1985), Asymptomatic primary biliary cirrhosis: a progress report on long-term follow-up and natural history. *Gastroenterology*, **89,** 267–71.

Boylston, A. W., Cook, H. T., Francis, N. D. *et al.* (1987), Biopsy pathology of acquired immune deficiency syndrome (AIDS). *J. Clin. Pathol.*, **40,** 1–8.

Cuccurallo, G., Rimbaldi, M., Iaquinto, G. *et al.* (1987), Importance of showing HBsAg and HBcAg positivity in the liver for better aetiological definition of chronic liver disease. *J. Clin. Pathol.*, **40,** 167–71.

De Groote, J., Desmet, V. J. *et al.* (1968), A classification of chronic hepatitis. *Lancet*, **ii,** 626.

Desmet, V. J. (1972), Chronic hepatitis (including primary biliary cirrhosis), in *The Liver* (eds E. A. Gall and F. K. Mostofi), Williams and Wilkins, Baltimore, p. 286.

Dietrichson, O. and Christoffersen, P. (1977), The prognosis of chronic aggressive hepatitis. A clinical and morphological follow up study. *Scand. J. Gastroenterol.*, **12**, 289.

Fattovich, G., Rugge, M., Brollo, L. *et al.* (1986), Clinical, virologic and histologic outcome following seroconversion from HBeAg to anti-HBe in chronic hepatitis type B. *Hepatology*, **6**, 167–72.

Foulk, W. T. and Baggenstoss, A. H. (1975), Biliary cirrhosis, in *Diseases of the Liver* (ed. L. Schiff), Lippincott, Philadelphia, p. 94.

Glasgow, B. J., Anders, K., Layfield, L. J. *et al.* (1985), Clinical and pathologic findings in the liver in the acquired immune deficiency syndrome (AIDS). *Am. J. Clin. Pathol.*, **83**, 582–8.

Kalk, H. (1947), Die chronischen Verlangsformen der Hepatitis epidemica im Hinblick auf ihre klinische Symptomatologie. *Dtsch. Med. Wochenschr.*, **72**, 471.

Manns, M., Gerken, G., Kyriatsoulis, A. *et al.* (1987), Characterisation of a new subgroup of autoimmune chronic active hepatitis by autoantibodies against soluble liver antigens. *Lancet*, **i**, 292–4.

Manns, M., Meyer zum Buschenfelde, K. H., Slusarczyk, J. *et al.* (1984), Detection of kidney–liver microsomal autoantibodies by radioimmunoassay and their relation to antimitochondrial antibodies in inflammatory liver diseases. *Clin. Exp. Immunol.*, **57**, 600–8.

Neuberger, J., Portmann, B., MacDougall, B. R. D. *et al.* (1982), Recurrence of primary biliary cirrhosis after liver transplantation. *N. Engl. J. Med.*, **306**, 1–4.

Portmann, B., Popper, H., Neuberger, J. *et al.* (1985), Sequential and diagnostic features in primary biliary cirrhosis based on serial histologic study in 209 patients. *Gastroenterology*, **88**, 1777–90.

Poulsen, H. and Christoffersen, P. (1972), Abnormal bile duct epithelium in chronic aggressive hepatitis and cirrhosis. *Hum. Pathol.*, **3**, 217.

Ray, M. B., Desmet, V. J., Fevery, J. *et al.* (1976), Distribution patterns of hepatitis B surface antigen (HBsAg) in the liver of hepatitis patients. *J. Clin. Pathol.*, **29**, 94.

Scheuer, P. J. (1986), Changing views on chronic hepatitis. *Histopathology*, **10**, 1–4.

Waldenstrom, J. (1950), *Leber, Blutproteine und Nahrungseiweiss Stoffwechsel* (ed., Kr. Sonderband, XV), Tagung, Bad Kisingen, p. 8.

12 End-stage liver disease and developmental abnormalities

Although the term 'cirrhosis' is entrenched in the literature, the entity which it describes is more properly called 'end-stage liver disease'. By analogy with 'granular contracted kidney', it connotes a condition which is the end result of many disease processes and is irreversible. Further, it admits that in some instances the foregoing aetiology is no longer discernible on pathological or other form of examination. Almost any disease which produces liver damage can, on occasion, give rise to end-stage liver disease.

Cirrhosis is best defined as a chronic irreversible condition in which the normal parenchymal and vascular architecture of the entire liver is destroyed by the formation of regenerating nodules of hepatic parenchyma intersected by bands of connective tissue. Although hepatocyte necrosis may be a feature during the development of end-stage liver disease it is not mandatory for the diagnosis, and is frequently absent (except in autopsy material) in the established condition.

Whatever the initial cause of cirrhosis there are essentially two mechanisms involved in its pathogenesis: deposition of large amounts of connective tissue in an organ which, in normality, is remarkable for its paucity of this tissue; regeneration of hepatocytes and increase in number of mesenchymal cells which populate regenerated nodules. It is probable that both mechanisms are activated simultaneously during the development of end-stage liver disease. The molecular mechanisms regulating the connective tissue productive and regenerative responses are reviewed elsewhere (Popper, 1977; Fleming and McGee, 1984).

End-stage liver disease presents clinically with evidence of hepatocellular failure, portal hypertension, hepatocellular cancer or a combination of these. An appreciable number of cases, however, are asymptomatic and are found incidentally during examination for other reasons or are discovered at autopsy.

12.1 Cirrhosis

12.1.1 Morphological types of cirrhosis

Many classifications of cirrhosis have been introduced in an attempt to correlate morphology with aetiology. None of these classifications, however, are entirely satisfactory because different aetiological agents can result in cirrhoses which on routine morphological criteria are indistinguishable and the same agent can produce different morphological types. Identification of aetiology, therefore, rests on morphology and other ancillary forms of pathological and clinical examination. Nevertheless, since most aetiological agents are usually associated with the generation of one or other morphological variant of cirrhosis a simple classification is still worthwhile.

The most useful macroscopic classification of cirrhosis is based on the size of regenerating nodules (Anthony *et al.*, 1978). There is a micronodular and macronodular form which are distinguished by the fact that the nodules in the former are generally less than 3 mm in diameter and in the latter generally greater than 3 mm in diameter (Figs 12.1 and 12.2). In the macronodular form, nodules may be as large as 1–3 cm in diameter on macroscopic examination but on microscopy these may be further subdivided by thin septa of connective tissue; it is this which makes it difficult to adhere rigidly to any classification of cirrhosis. In the micronodular form the nodules are usually separated by thin bands of connective tissue (Fig. 12.1), while in macronodular cirrhosis the fibrous bands take the form of thick scars (Fig. 12.2). There is a subtype of macronodular cirrhosis in which the fibrous bands are thin and delicate and this type has

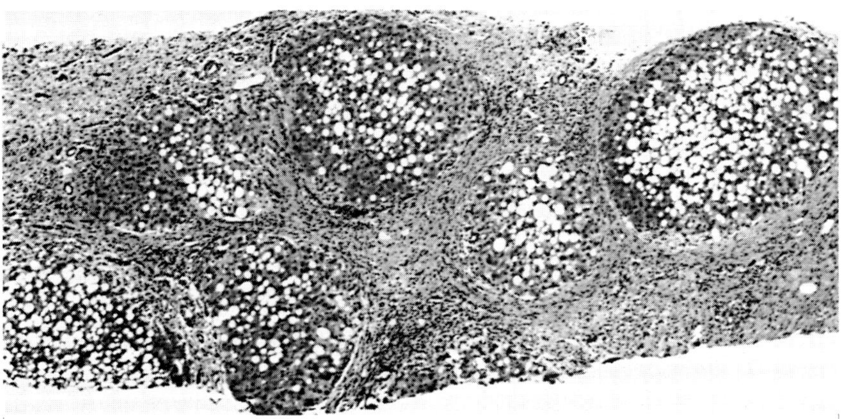

Fig. 12.1 Micronodular cirrhosis. The small regenerating nodules show fatty change and are delineated by connective tissue septa (HE, × 45).

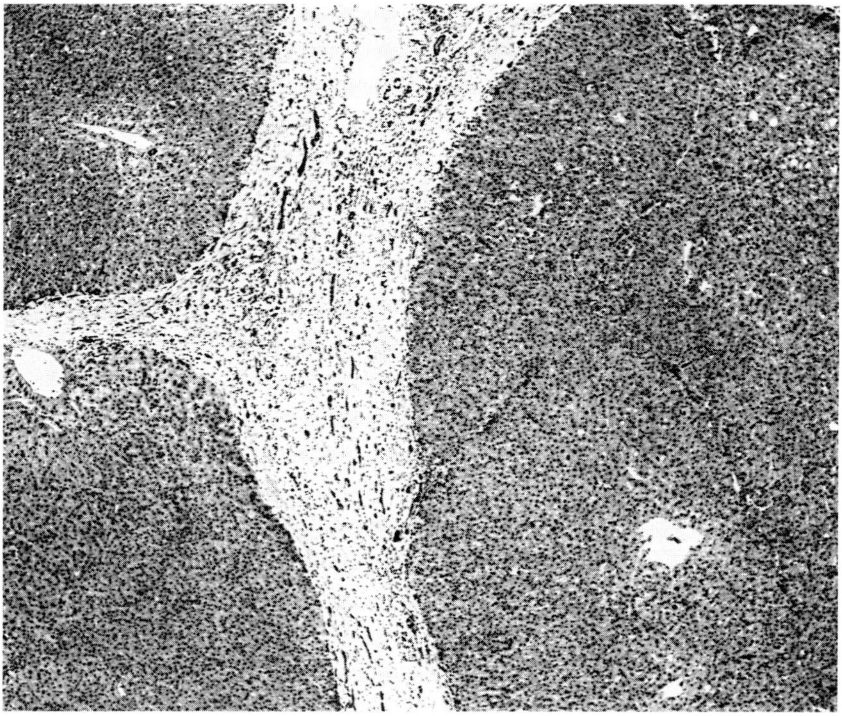

Fig. 12.2 Macronodular cirrhosis. Parts of three large nodules are seen separated by connective tissue septa. The nodules contain small blood vessels and a poorly developed portal tract. Note that the magnification is the same as that of Fig. 12.1 (HE, × 45).

been referred to as post-hepatitic cirrhosis (Fig. 12.3). The third morphological variant is a mixed type which shows the features of micronodular and macronodular cirrhosis. Classification is relatively easy at autopsy but on biopsy this may present difficulties because of sampling problems – a needle biopsy at best yields 100 mg of tissue, i.e. 1 : 15 000 of the entire organ. Furthermore, it is physically impossible for a biopsy needle (1 mm in diameter) to include an entire nodule from a macronodular cirrhosis.

The micronodular cirrhosis of biliary obstruction and venous outflow obstruction presents distinguishing features. In 'biliary' cirrhosis the nodule contours fit into each other as in a jig-saw (Fig. 12.4), while in 'cardiac' cirrhosis fibrosis starts around central veins giving rise to reversed lobulation of liver architecture (Fig. 12.5).

Micronodular cirrhosis is associated with alcohol abuse, haemochromatosis, biliary obstruction, obstruction of hepatic venous outflow, intestinal bypass, mucoviscidosis and Indian childhood cirrhosis. A

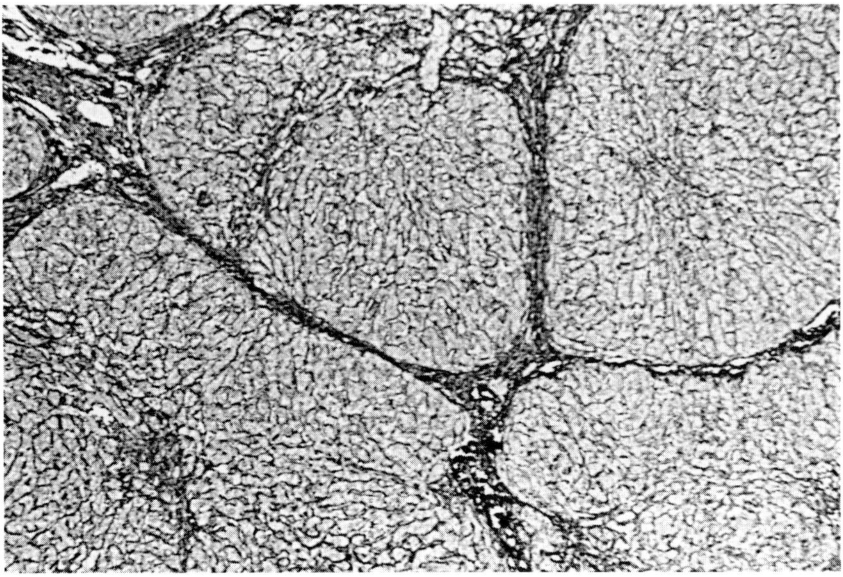

Fig. 12.3 Macronodular cirrhosis. The fibrous septa which separate nodules of regenerating parenchyma are thin and acellular. This morphological pattern is sometimes referred to as post-hepatitic cirrhosis (reticulin, × 45).

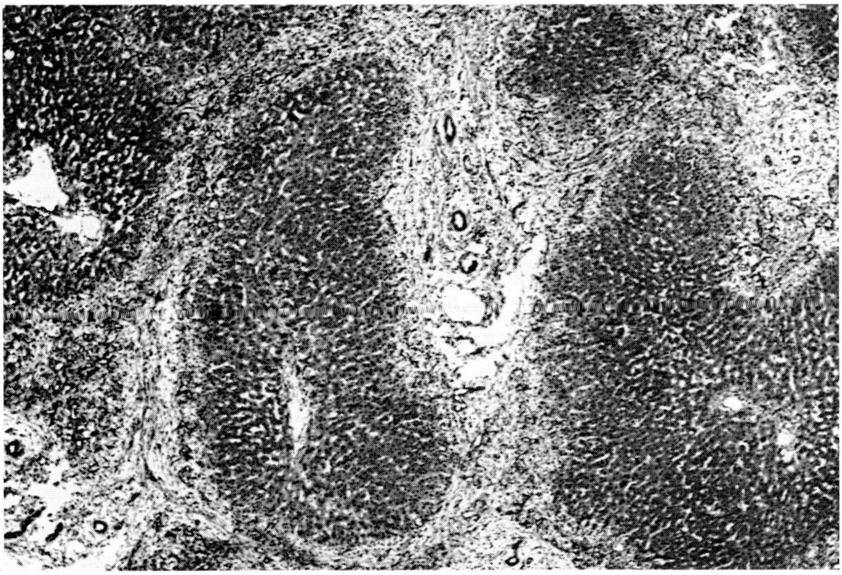

Fig. 12.4 Biliary cirrhosis secondary to large duct obstruction. The pattern tends to be monolobular, each lobule having a central vein uninvolved in the fibrotic process (HE, × 45).

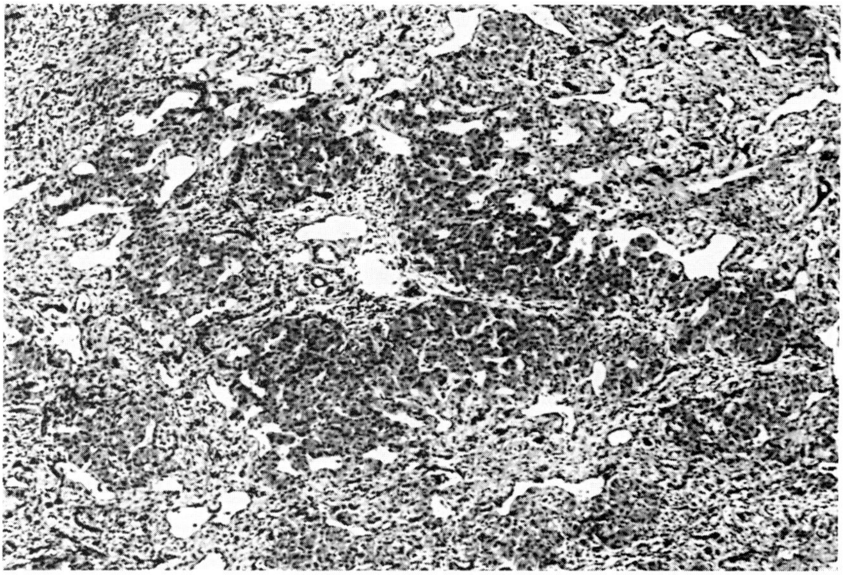

Fig. 12.5 Cardiac 'cirrhosis'. There is marked central fibrosis which has almost completely surrounded an area of parenchyma which has a portal area at its centre producing a pattern of reversed lobulation (HE, × 72).

macronodular pattern is characteristic of viral hepatitis and Wilson's disease and idiopathic (cryptogenic) cirrhosis in the UK. A mixed variant may be found in alcoholism and alpha-1-antitrypsin deficiency.

Conversion of one cirrhotic type to another occurs. In alcoholics, micronodular cirrhosis may eventually be converted to a macronodular type as these patients now survive much longer (Fauerholdt *et al.*, 1983). Similarly, macronodular cirrhosis may be converted to micronodular cirrhosis possibly by ischaemic injury within macronodules.

12.1.2 Histological diagnosis of cirrhosis

The histopathological diagnosis of cirrhosis is based on three features: the presence of excessive fibrous tissue, regenerating nodules and complete distortion of the normal relationship of hepatic venous outflow radicles and portal veins. Excessive fibrous tissue is usually readily identified on HE sections (Fig. 12.2), but it is unwise to make this assessment without a reticulin stain which demonstrates better than any other procedure the connective tissue stroma of the liver (Fig. 12.3). Fibrous septa may be thin delicate structures or wide thick scars containing large venous radicles, arteries and sizeable bile ducts. In fragmented biopsies fibrous septa may

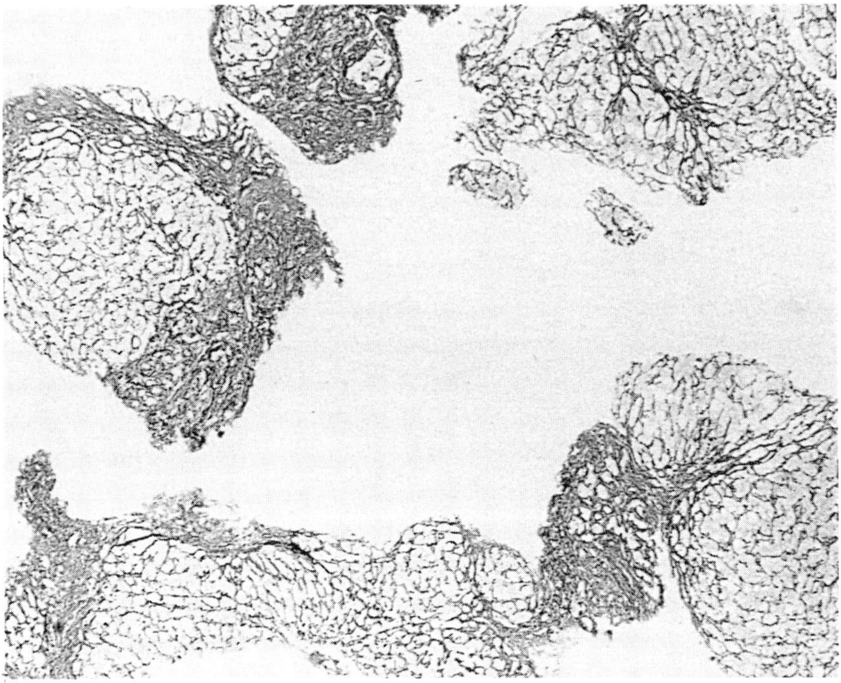

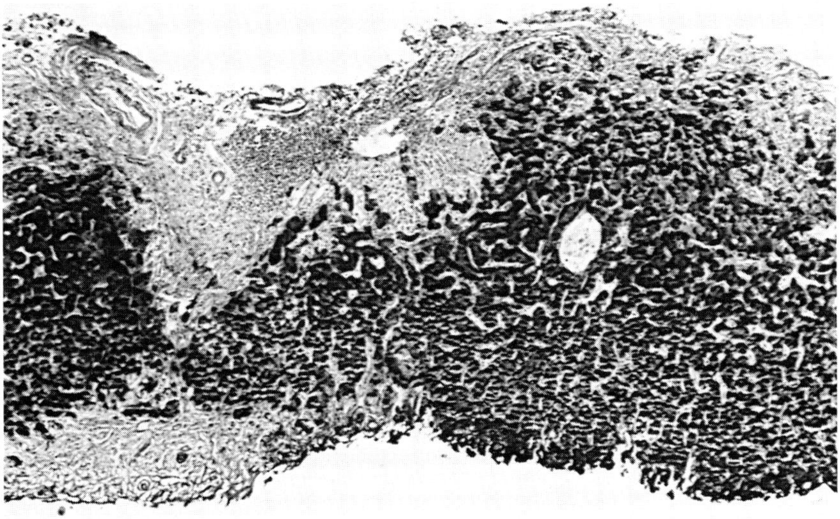

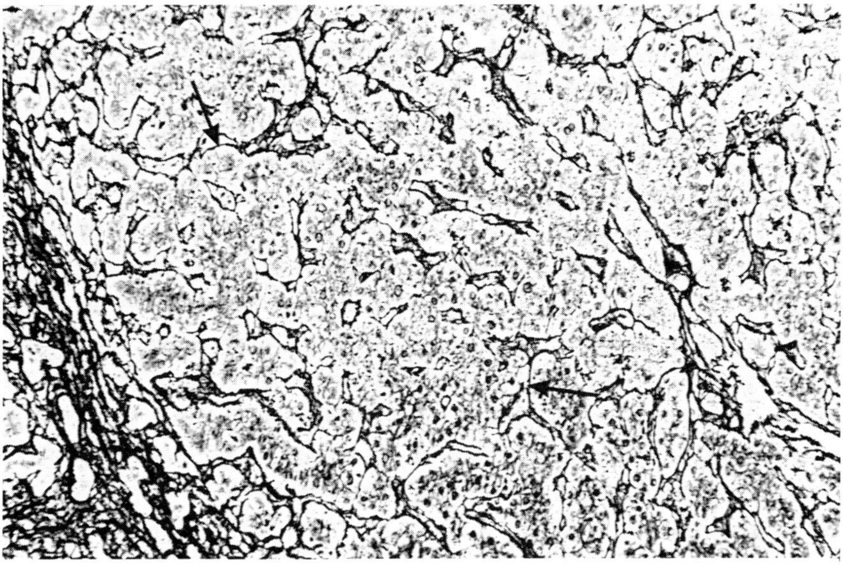

Fig. 12.8 Regeneration in a nodule in a cirrhotic liver. The liver cell plates are three to four cells thick in most areas (arrows) (reticulin, × 500).

delineate only the surface of the fragments and this phenomenon is frequently only recognizable in reticulin stains (Fig. 12.6) or PAS preparations which distinguish glycogen-containing hepatocytes from fibrous tissue (Fig. 12.7). Sometimes, in macronodular cirrhosis, a biopsy is completely devoid of fibrous tissue and the diagnosis depends on the absence of normal vascular architecture and the presence of regeneration.

Regenerating nodules may be self-evident as nodular areas of hepatic parenchyma circumscribed by connective tissue. The hepatocyte sheets forming these nodules are two or more cells thick (Fig. 12.8) and in general do not contain portal tracts except occasionally in the macronodular form of end-stage liver disease; when portal tracts are found within nodules they are hypoplastic. In assessing the extent of regeneration it should be realized that some plates within a normal liver may appear two cells thick because of tangential sectioning. In addition, even in a proven cirrhotic liver, many plates are one cell thick so that this criterion is not absolute. In a cirrhotic liver it is difficult to discern a normal

---

◀ **Fig. 12.6** Fragmented needle biopsy of a micronodular cirrhosis. The reticulin stain outlines the fibrous tissue on the surface of regenerating nodules (reticulin, × 60).

◀ **Fig. 12.7** Micronodular cirrhosis. The fibrous tissue is negatively outlined against the glycogen-containing hepatic parenchyma (PAS, × 72).

relationship between hepatic outflow and portal veins with a regularly orientated series of hepatocyte cords linking each venous structure. In developing cirrhosis, however, there is a stage at which these normal arrangements have not yet been completely destroyed (Fig. 12.4). In addition, portal tracts and effluent veins may be found in macronodules but they do not bear a normal relationship to each other (Fig. 12.2); these may represent residual normal or newly formed structures. In 'cardiac' cirrhosis portal tracts are evident in nodules (Fig. 12.5), while in 'biliary' cirrhosis effluent veins are present (Fig. 12.4).

Morphological features helpful in establishing the aetiology of cirrhosis are given in Table 12.1.

Table 12.1 Morphological correlates of aetiology in cirrhosis.

Aetiology	*Predominant type*	*Characteristic morphological features*
Alcohol	Micronodular (macronodular in advanced stages)	Fatty change; Mallory bodies; hydropic swelling; siderosis
Viral hepatitis	Macronodular, mixed	Acidophilic cells; ground-glass changes; HBV antigens in hepatocytes
Large bile duct obstruction	Micronodular	Jigsaw pattern; cholestasis; bile ductular proliferation; CBP
Primary biliary cirrhosis	Micronodular	Absence or reduction of medium-sized bile ducts; lymphoid aggregates; granulomas; peripheral cholestasis; Mallory bodies sometimes present, excess copper and CBP
Haemochromatosis	Micronodular	Haemosiderin in hepatocytes, ductular epithelium, Kupffer cells, macrophages and in septa
Alpha-1-antitrypsin	Micronodular, macronodular or mixed	PAS-positive diastase-resistant granules and globules in hepatocytes
Wilson's disease	Macronodular	Excess copper in hepatocytes; glycogenic vacuolation of nuclei, fatty change; Mallory bodies sometimes present
Toxins and drugs	Macronodular or mixed	Variable

Table 12.1 *cont.*

Aetiology	*Predominant type*	*Characteristic morphological features*
Intestinal bypass	Micronodular	Fatty change; Mallory bodies may be present
Venous outflow impairment	Micronodular	Reversed lobulation, passive congestion
Other inborn errors of metabolism*	Micronodular	Accumulation of carbohydrates and lipids, pseudoglandular appearance of liver cell cords, ductular proliferation, giant cell transformation, intrahepatic inclusions
Mucoviscidosis*	Micronodular	PAS-positive secretion in proliferating ductules: biliary pattern may be only focal
Hereditary haemorrhagic telangiectasia	Macronodular	Dilated vessels in septa
Indian childhood cirrhosis	Micronodular	Severe hepatocellular degeneration; Mallory bodies; creeping fibrosis

* Glycogen storage disease (types III and IV), galactosaemia, hereditary fructose intolerance, Wolman's disease, cholesterol ester storage disease, hereditary tyrosinaemia (see Chapter 9).

12.1.3 *Activity in cirrhosis*

In any cirrhotic liver there is a variable number of inflammatory cells in fibrous septa. The degree to which this is present and its association with parenchymal destruction is regarded as an index of activity of the cirrhotic process. The most severe degree of activity is reflected in heavy infiltration of septa by lymphocytes and plasma cells which, together with new fibrous tissue, overflow into the parenchyma destroying the usually clear junction between septa and regenerating nodules (Fig. 12.9). This process is identical in type to piecemeal necrosis which characterizes CAH. In aetiological terms it is preferable in a cirrhotic liver to equate this process with activity since a designation of CAH superimposed on cirrhosis may connote that the cirrhosis has arisen in a pre-existing CAH syndrome; in many instances this conclusion would be erroneous. As in CAH, however, steroid therapy may improve patient well-being when activity is marked.

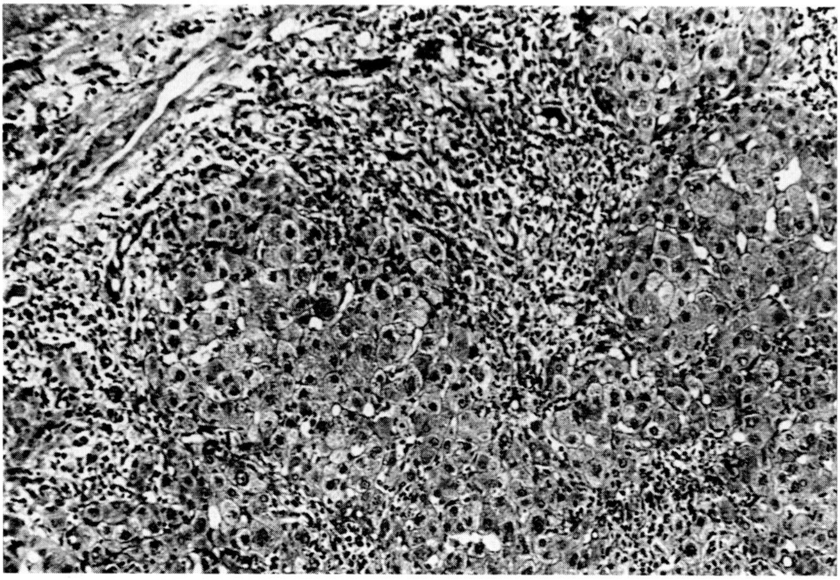

Fig. 12.9 End-stage liver disease in which there is continuing activity as evidenced by the heavy infiltrate of lymphocytes and plasma cells in the fibrous septum. The inflammatory cells and fibrous tissue have eroded into the parenchyma producing piecemeal necrosis (HE, × 400).

12.1.4 Secondary phenomena in cirrhosis

Secondary phenomena may be present in any cirrhosis and need not have aetiological significance. Of these, bile ductular proliferation and ischaemic necrosis are the most common. Ischaemic necrosis is almost restricted in its occurrence to autopsy material (Fig. 12.10). Cholestasis may also occur within regenerating nodules in any variety of cirrhosis. Siderosis is found not only in iron storage disorders but also after portocaval shunting and in alcoholic cirrhosis. Dysplasia, another secondary feature, is identified by enlargement and hyperchromatism of hepatocyte nuclei (Fig. 12.11). It is associated with the development of primary hepatic cancer in African cirrhotics and may precede the development of neoplasia (Anthony *et al.*, 1973). The concept that dysplasia is a premalignant forerunner of hepatocellular cancer has been questioned (Cohen and Berson, 1986).

Adenomatous hyperplasia in cirrhosis is characterized by disproportionately large nodules composed of liver plates several cells thick in which the nuclei show atypia. Hepatocellular cancer sometimes develops as nodules within these larger nodules (Arakawa *et al.*, 1986).

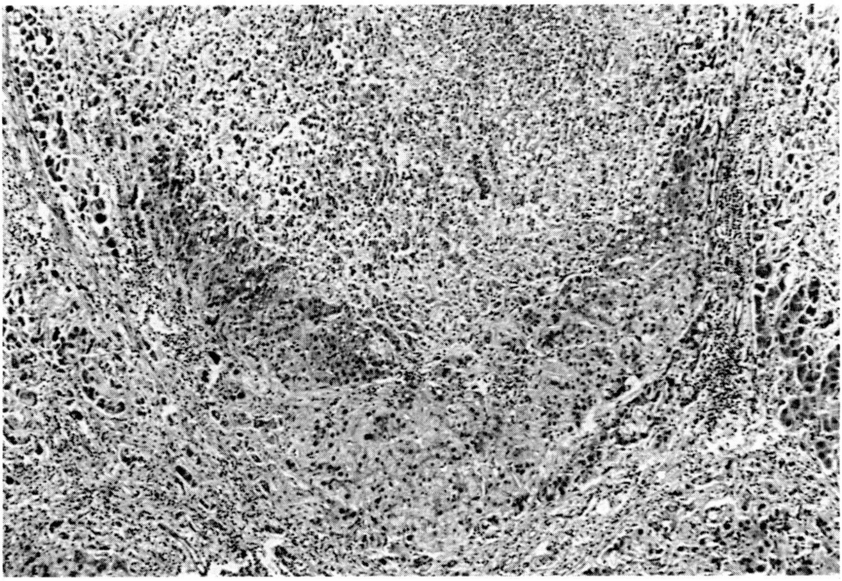

Fig. 12.10 Cirrhotic liver at autopsy. There is necrosis in the centre of several nodules (HE, × 60).

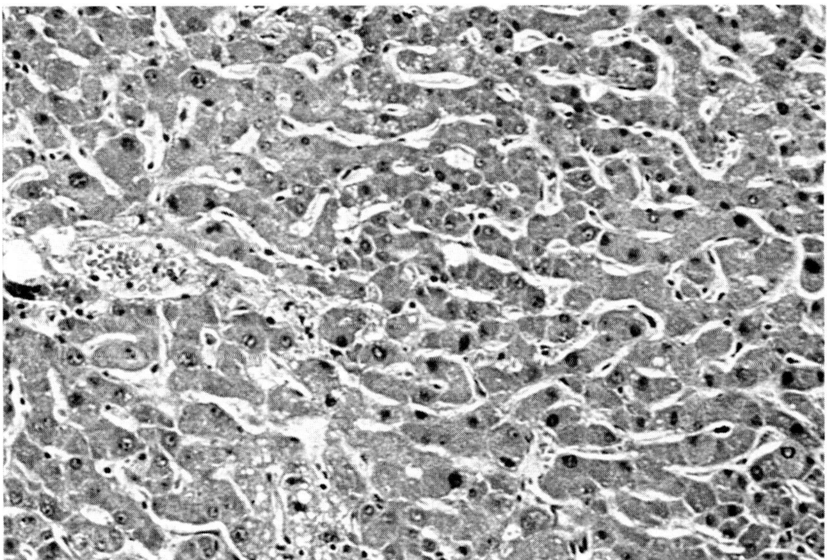

Fig. 12.11 Dysplasia in a cirrhotic nodule. The hepatocyte nuclei show nuclear pleomorphism and hyperchromatism (HE, × 400).

12.1.5 *Differential diagnosis*

End-stage liver disease should be suspected in any biopsy which is fragmented, provided the clinician who has performed it is experienced. The diagnosis is dependent on the demonstration of excessive connective tissue, nodular regeneration of hepatocytes and the absence of normal vascular relationships; all three criteria must be fulfilled.

Cirrhosis should be differentiated from diseases producing diffuse hepatic fibrosis such as early schistosomiasis, congenital hepatic fibrosis, radiation damage (p. 240) and veno-occlusive disease. Subcapsular fibrosis is frequently seen in surgical wedge biopsies and this should not be misinterpreted as cirrhosis (p. 13); the fibrosis in this condition is restricted to the subcapsular region and is not associated with hepatocyte regeneration. Diseases in which nodular regeneration of hepatocytes is prominent also should not be confused with cirrhosis; nodular formations of hepatocytes are found in diffuse nodular regenerative hyperplasia (p. 237), focal nodular hyperplasia (see p. 239) and partial nodular transformation of the liver (p. 232).

Central sclerosing hyaline necrosis in alcoholics presents with minor to massive fibrosis around central veins but the architecture is initially preserved. Veno-occlusive disease presents a similar pattern of fibrosis but is not associated with Mallory body formation.

In India and Japan especially non-cirrhotic portal fibrosis may pose a differential diagnostic problem. In this condition although nodule formation appears in subcapsular areas, the vascular architecture of the liver is preserved. The typical lesions are broad fibrotic scars and there is absence or partial obliteration of portal venous radicles in portal tracts (see p. 229). Because of the preservation of liver architecture these patients who survive after variceal bleeding do not develop a true cirrhosis on long-term follow up.

12.2 Developmental abnormalities

12.2.1 *Congenital hepatic fibrosis*

Congenital hepatic fibrosis is a condition in which there is both diffuse hepatic fibrosis and nodule formation; this condition can be differentiated from cirrhosis. Characteristically, the liver is transected by bands of mature connective tissue containing numerous well-formed bile ducts particularly at the fibrous tissue–parenchymal junction (Figs 12.12 and 12.13). The disease is distinguished from cirrhosis by preservation of normal vascular architecture in the parenchyma between fibrous bands

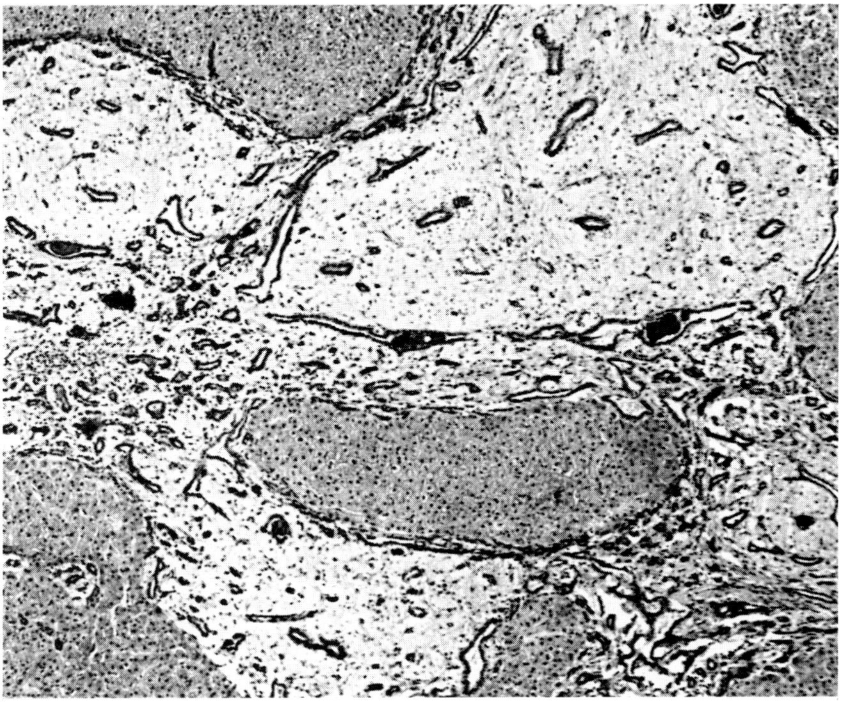

Fig. 12.12 Congenital hepatic fibrosis. The prominent bands of fibrous tissue contain numerous bile ducts especially at their periphery. Some ducts contain inspissated bile (HE, × 72).

(Fig. 12.12); there is no regeneration of liver cell plates. After abdominal surgery these patients may develop severe cholangitis (Fig. 12.14). Congenital hepatic fibrosis may occur in families and like cirrhosis often presents with signs or symptoms of portal hypertension. Unlike cirrhosis, however, hepatocyte function is normal.

12.2.2 *Von Meyenburg complexes*

Von Meyenburg complexes, which may be single or multiple, are focal collections of fibrous tissue containing numerous well-formed bile ducts (Fig. 12.15). In isolation, these bile duct-containing lesions can be mistaken for cirrhosis. Von Meyenburg complexes produce no symptoms but may be biopsied out of curiosity in the process of a laparotomy for other reasons. These complexes, congenital hepatic fibrosis, Caroli's syndrome and polycystic disease of the liver may be different manifestations of the same basic disease process.

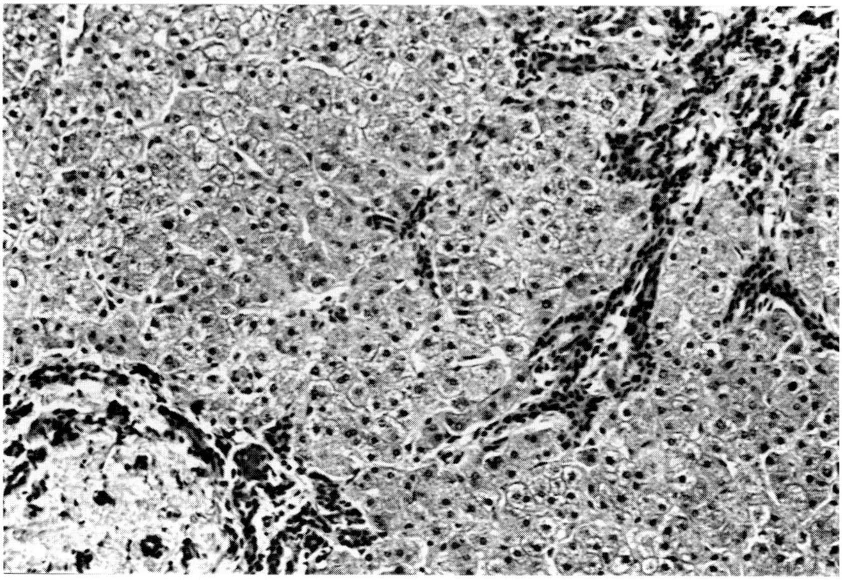

Fig. 12.13 Congenital hepatic fibrosis. In this field there are prominent bile ductules, some with little or no fibrous tissue (HE, × 180).

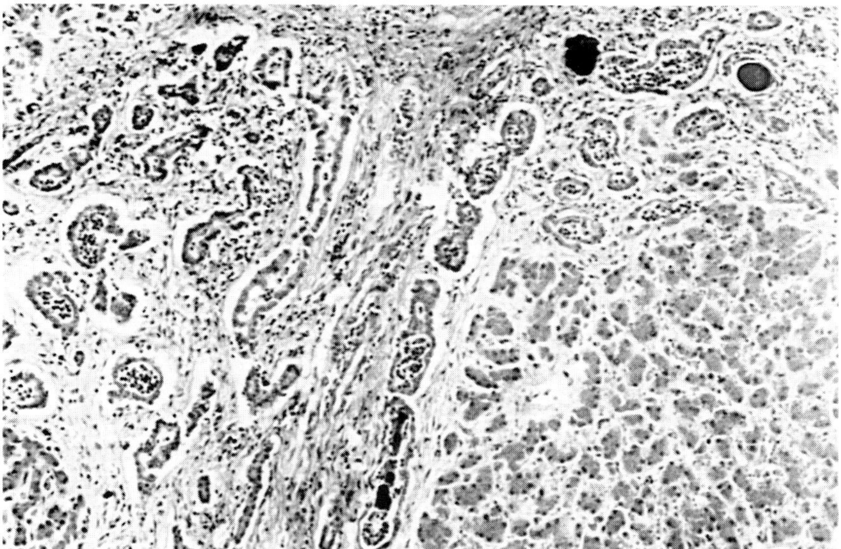

Fig. 12.14 Congenital hepatic fibrosis with superimposed acute cholangitis; there are numerous polymorphs within bile ducts. Cholangitis is a not infrequent complication of congenital hepatic fibrosis following surgery (HE, × 128).

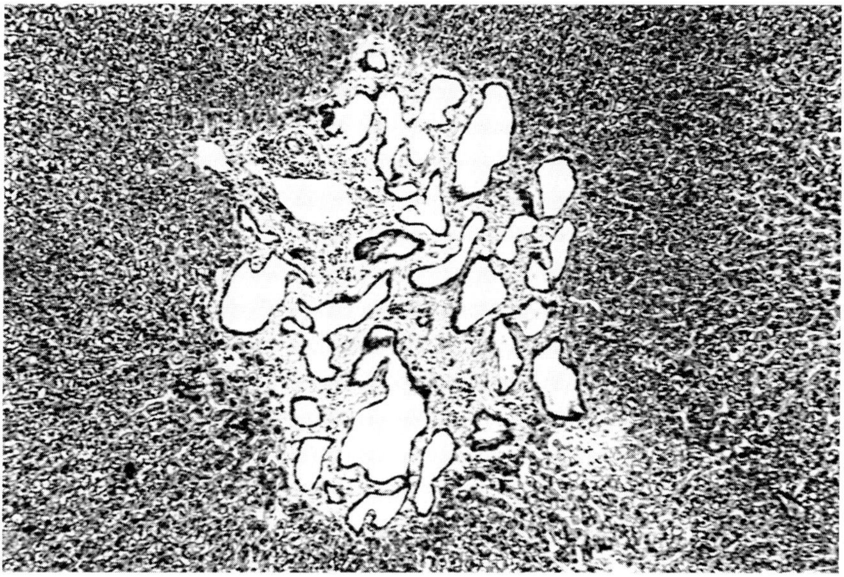

Fig. 12.15 von Meyenburg complex. This consists of numerous dilated bile ducts in a connective tissue stroma (HE, × 72).

12.2.3 *Adult polycystic disease*

Polycystic disease of the liver is frequently associated with polycystic disease of the kidneys and less commonly with cysts in the lungs, spleen, pancreas and ovary. The cysts vary in size from a few millimetres to 20 cm and may be present throughout the whole liver or restricted to one lobe. The cysts are lined by a cuboidal epithelium of biliary type or in large cysts by a flattened epithelium; the cysts are surrounded by varying amounts of fibrous tissue (Fig. 12.16). Single unilocular or multilocular cysts are relatively common at autopsy and do not indicate polycystic disease. They may cause clinical problems because of their size or present a differential diagnostic problem on ultrasonography.

12.2.4 *Infantile polycystic disease*

This condition is inherited as an autosomal recessive and presents with cysts in kidney and liver. The kidney cysts are grossly obvious, whereas microscopic examination is usually required for cyst recognition in the liver. The portal tracts are enlarged and contain numerous bile ducts which communicate with each other. The biopsy appearances in this condition are similar to congenital hepatic fibrosis with the exception that the thick fibrotic scars of the adult disease are absent in these babies.

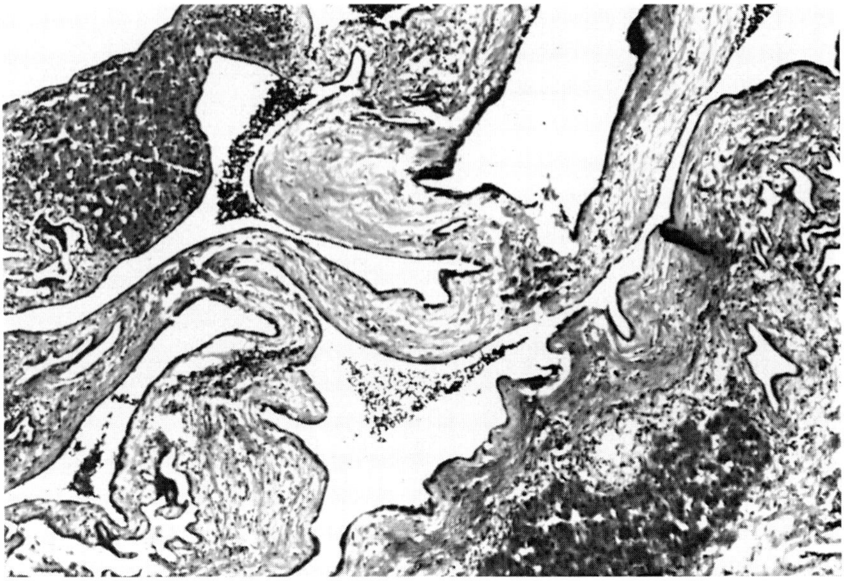

Fig. 12.16 Polycystic liver. This cyst has collapsed. It is lined by a cuboidal epithelium; there are numerous well-formed bile ducts in the connective tissue surrounding the cyst (HE, × 90).

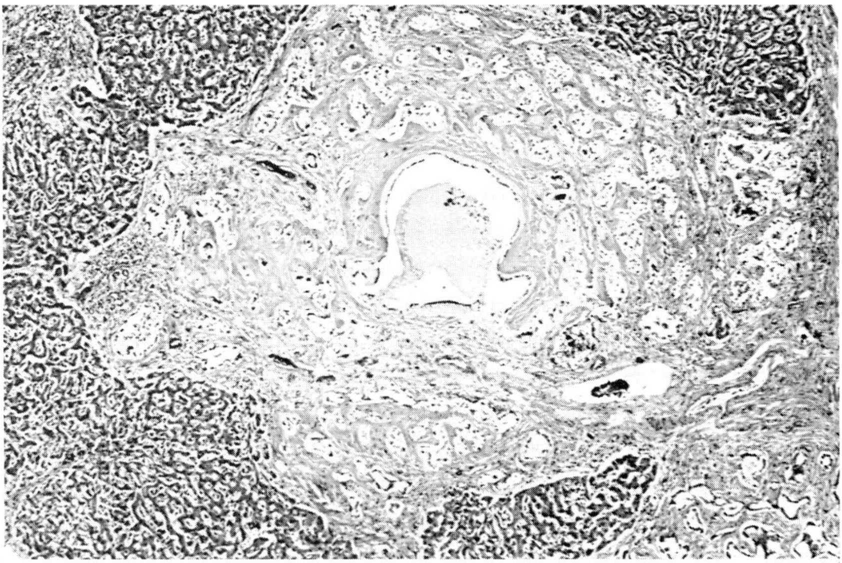

Fig. 12.17 Infantile polycystic liver. The fibrotic area contains a grossly dilated bile duct surrounded by numerous smaller biliary channels (HE, × 45).

Additionally, there may be prominent dilatation of interlobular bile ducts surrounded by numerous ductules (Fig. 12.17). If the infants survive the neonatal period, portal hypertension develops. Infantile polycystic disease may be associated with various other developmental anomalies. These are rare (see Landing *et al.*, 1980).

12.2.5 *Caroli's disease*

In Caroli's disease there is dilatation of intrahepatic bile ducts which communicate with the main bile ducts. These dilated ducts are not associated with fibrosis but they may become infected and contain calculi. The liver may also show features of congenital hepatic fibrosis. In addition, the kidneys may be cystic in Caroli's disease and cholangiocarcinoma has been reported as a complication (Gallacher *et al.*, 1972).

12.2.6 *Hamartomatous lesions*

See Chapter 13.

References

Anthony, P. P., Ishak, K. G., Nayak, N. C. *et al.* (1978), The morphology of cirrhosis. *J. Clin. Pathol.*, **31**, 395.

Anthony, P. P., Vogel, C. L. and Barker, L. F. (1973), Liver cell dysplasia: a premalignant condition. *J. Clin. Pathol.*, **26**, 217.

Arakawa, M., Kage, M., Sugihara, S. *et al.* (1986), Emergence of malignant lesions within an adenomatous hyperplastic nodule in a cirrhotic liver: observations in 5 cases. *Gastroenterology*, **91**, 198–208.

Cohen, C., and Berson, S. D. (1986), Liver cell dysplasia in normal, cirrhotic and hepatocellular carcinoma patients. *Cancer*, **57**, 1535–8.

Fauerholdt, L. *et al.* (1983), Conversion of micronodular cirrhosis into macronodular cirrhosis. *Hepatology*, **3**, 928–31.

Fleming, K. A., and McGee, J. O'D. (1984), Alcohol induced liver disease. *J. Clin. Pathol.*, **31**, 150.

Gallacher, P. J., Mills, R. R. and Mitchinson, M. J. (1972), Congenital dilatation of intrahepatic bile ducts with cholangiocarcinoma. *J. Clin. Pathol.*, **25**, 804.

Landing, B. H., Wells, T. R. and Claireaux, A. E. (1980), Morphometric analysis of liver lesions in cystic diseases of childhood. *Hum. Pathol.*, **11**, 549–60.

Popper, H. (1977), Pathologic aspects of cirrhosis: a review. *Am. J. Pathol.*, **87**, 228.

13 Tumours of the liver

This Chapter is concerned with neoplastic conditions of liver and intrahepatic bile ducts. Hamartomatous malformations are included but more diffuse developmental anomalies and non-neoplastic cystic conditions are considered elsewhere (see Chapter 12).

Some of these tumours are described under various names, while the same name may be applied to more than one entity. The classification adopted in this Chapter uses those names which are least likely to cause confusion. Synonyms are given in brackets.

13.1 Tumours derived from hepatocytes

13.1.1 Liver cell adenoma (benign hepatoma; hepatoma)

These are very uncommon and occur mostly in women who have been taking oral contraceptives for a number of years (Edmondson *et al.*, 1976). The majority are only a few centimetres in diameter and may be found accidentally at laparotomy or be the source of serious intra-abdominal bleeding. Large specimens measuring up to 30 cm in diameter have been reported which cause abdominal pain and swelling.

Liver cell adenomas are usually single and well demarcated from darker normal liver tissue. They may be distinctly yellow due to lipochrome pigment, fat or, occasionally, bile. There may be areas of necrosis and haemorrhage. Microscopically the tumour cells closely resemble normal hepatocytes apart from their slightly larger size and increased nuclear : cytoplasmic ratio (Fig. 13.1). Mitotic activity is inconspicuous or absent. The cytoplasm may be vacuolated due to fat or glycogen. The cells are arranged characteristically in trabeculae between compressed sinusoids where Kupffer cells are scanty. There is usually a distinct fibrous capsule which may be deficient in places but there is no invasion of surrounding tissue. Fibrous bands and scars may indicate the sites of old haemorrhage but there are no portal triads or well-formed vessels so

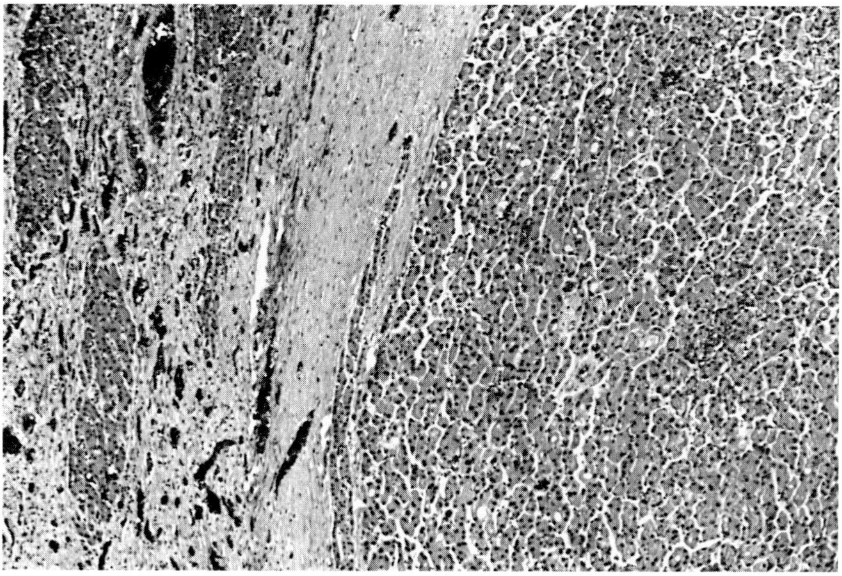

Fig. 13.1 Benign liver cell adenoma. There is a distinct capsule and compression with fibrosis of hepatic tissue on the left (HE, × 72).

that these tumours can be relatively hypovascular. By contrast, some adenomas contain many thin-walled vascular channels which are sources of serious haemorrhage and are attributed to oral contraceptive intake.

In multiple adenomatosis the liver may contain 50 or more such tumours of small size. Probably this is an entity distinct from single liver cell adenoma and from focal nodular hyperplasia as the incidence is equal in both sexes and is not related to hormone administration (Flejou *et al.*, 1985); in some there is associated glycogen storage disease.

13.1.2 *Focal nodular hyperplasia of liver (focal cirrhosis; mixed adenoma; mixed hamartoma; hamartomatous cholangiohepatitis)* (Gold *et al.*, 1978; Knowles *et al.* 1978) (Fig. 13.2)

This hamartomatous lesion typically is not encapsulated but has a central stellate scar. Multiple nodules which can measure up to 10 cm in diameter may be found close to the liver surface and may be pedunculated. Hepatocytes tend to be arranged around cores of vascular connective tissue (Fig. 13.3) which contain bile ductular structures (Fig. 13.4). Fibrous septa extend from these cores between ill-defined lobules. It has been suggested that focal hyperplasia may be a sequel to enhanced blood supply from a primary vascular developmental anomaly (Wanless *et al.*, 1985).

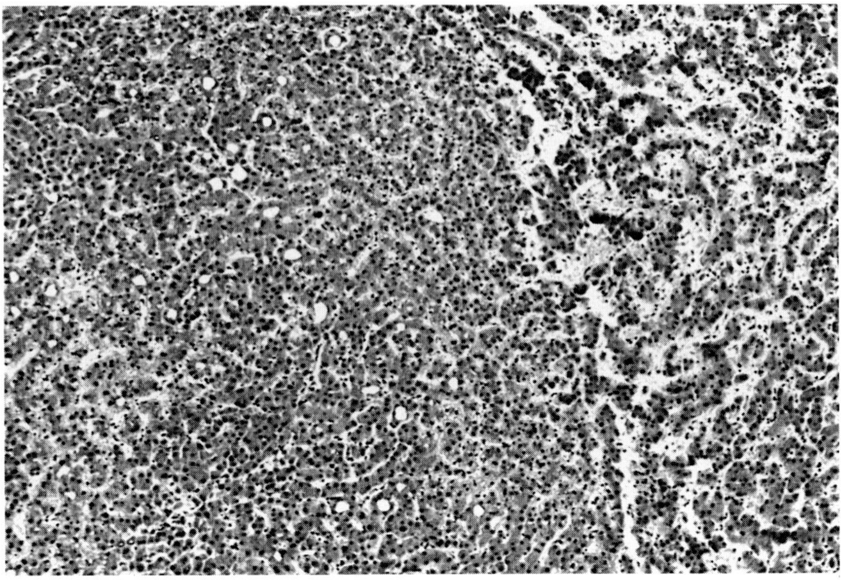

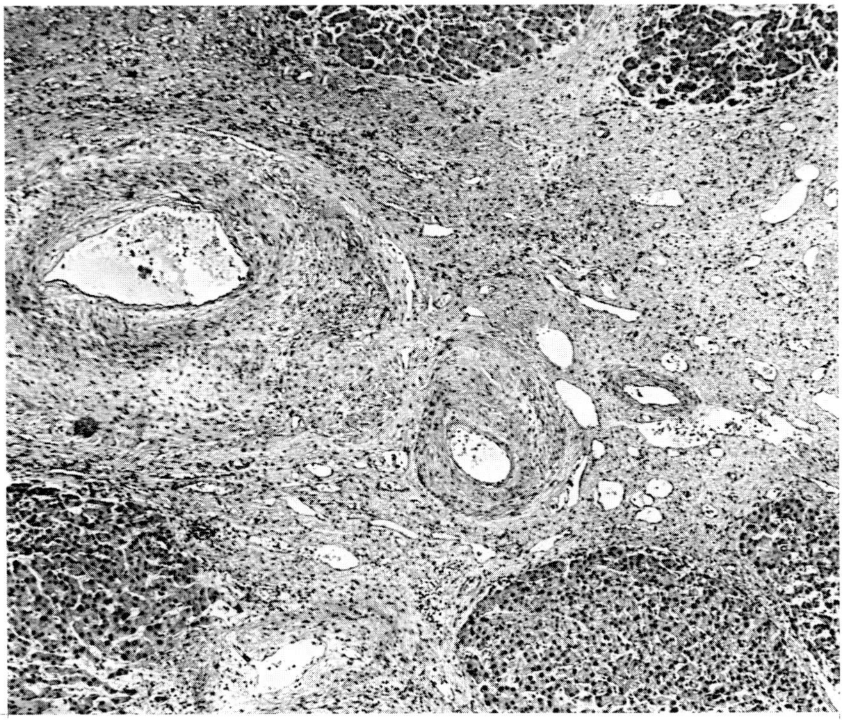

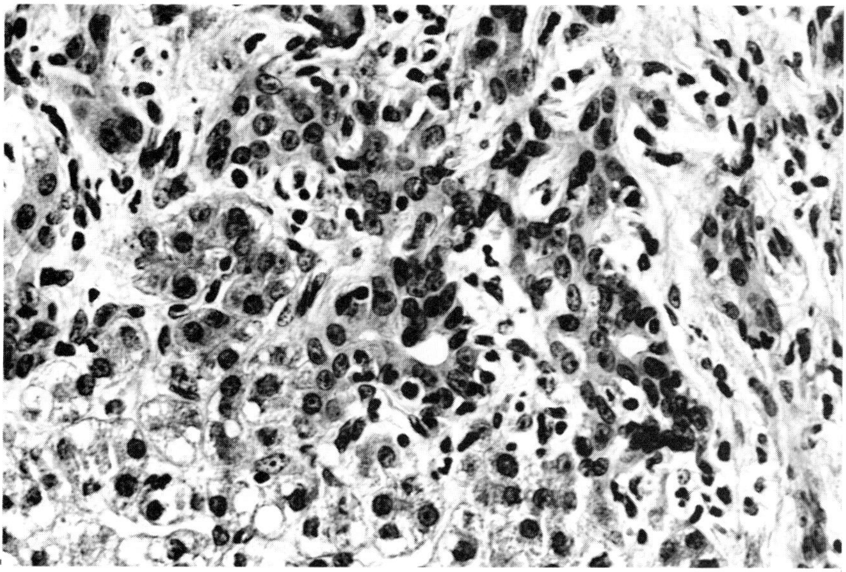

Fig. 13.4 Focal nodular hyperplasia. Part of the central scar showing bile ductules which seem to blend with hepatocytes (lower left) (HE, × 450).

Focal nodular hyperplasia, like true adenoma, is relatively common in young women and again there is an association with oral contraceptive intake and the risk of severe haemorrhage. Proliferation of fibroblasts, bile ductules, vascular subintimal cells and lymphocytic infiltration of the central scar may be stimulated, if not initiated, by steroid hormones.

The structural differences between focal nodular hyperplasia and liver cell adenoma are not always as distinct as described above; thus, a fibrous capsule may delineate part of a hyperplastic lesion and be absent from part of an adenoma. It seems possible, therefore, that both may be structural variants of the same disease (Klatskin, 1977). Certainly it is important to remember the features they have in common – their benign nature apart from the risk of serious bleeding and the frequent association with oral contraceptive intake. Regression, at least of focal nodular hyperplasia, is recorded following discontinuation of oral contraceptives (Edmondson *et al.*, 1977), and this should be done before resection of large lesions is contemplated. Malignant change in either condition is a

---

◀ **Fig. 13.2** Focal nodular hyperplasia. There is no fibrous capsule and the hyperplastic cells resemble normal liver cells very closely (nodules contained bile ductules but none are present in this field) (HE, × 72).

◀ **Fig. 13.3** Focal nodular hyperplasia. Part of the central scar containing numerous vascular channels (HE, × 72).

doubtful possibility but the rare association of hepatocellular carcinoma and oral contraceptive intake suggests that this can occur.

It is important to distinguish focal nodular hyperplasia and liver cell adenoma from:

(a) Hepatocellular carcinoma

The cells are usually more pleomorphic but well-differentiated tumours in non-cirrhotic liver may be impossible to distinguish from liver cell adenoma when the biopsy shows no evidence of invasion.

(b) Adenomatous hyperplasia in macronodular cirrhosis (multiple nodular hyperplasia)

This may be recognized by the pseudocapsule of compressed hepatic tissue and fibrous septa containing vessels and bile ducts which separate these nodular lesions.

(c) Nodular regenerative hyperplasia of liver (Chapter 10)

In this condition the liver is not fibrotic and consists of small nodules of hepatocytes separated by narrow bands of normal liver which may contain portal tracts or central veins. There is an association with congestive cardiac failure and certain chronic systemic diseases.

13.1.3 *Hepatocellular carcinoma (hepatoma; malignant hepatoma; liver cell carcinoma)*

Hepatocellular carcinoma, which is so common in many parts of Africa and Asia, is seen less frequently in the Western world. However, it is by no means a rare tumour and the slowly rising incidence in Europe and America may be related to an increase in chronic liver disease. Various factors have been implicated in the aetiology of primary liver cancer and, of these, hepatitis B has the most convincing association (Sumithran and MacSween, 1979; London, 1981); this is based on the immunohistochemical and serological demonstration of the infection in established cases of tumour and on the frequency of tumour development in carriers of the virus and in cases of chronic hepatitis attributed to virus B infection. Hepatocellular carcinoma frequently occurs in association with and possibly as a consequence of cirrhosis usually of the macronodular variety but also in some micronodular cases with haemochromatosis or chronic alcoholism. It is possible that the association with hepatitis B could be explained simply by the fact that the virus is a known cause of macronodular cirrhosis and need not be primarily oncogenic (Zaman *et al.*, 1985). In the non-cirrhotic liver there may be a single discrete nodule or a central large lesion surrounded by smaller satellites which are also well delineated. Other more massive tumours infiltrate the liver in an irregular fashion. A diffuse rather scirrhous type is also described which may be mistaken for micronodular cirrhosis on naked-eye examination. In some cases bile production is evident in the gross specimen.

These patients often present with painful hepatomegaly, but in non-cirrhotic cases there may be no liver dysfunction until the organ is extensively invaded by tumour. In chronic liver disease malignancy is suspected from clinical deterioration or by the presence of blood-stained ascitic fluid, cytological examination of which is often diagnostic. The use of peritoneoscopy in directing the needle to suspicious areas of liver is valuable in reducing the risk of false-negative results from aspiration biopsy.

(a) Histopathology

In the majority of cases the tumour is well enough differentiated to make the histological diagnosis fairly easy (Fig. 13.5), and this becomes certain if bile production by tumour cells can be detected. Indeed, the cytological characteristics and arrangement of cells in plates can resemble normal liver so closely that the true diagnosis may be missed by cursory examination of small fragments which have no normal tissue for comparison, but closer inspection should reveal some irregularity of hepatocyte arrangement and absence of prominent Kupffer cells. As in benign adenoma, tumour nodules contain no portal triads and are often separated from surrounding tissues by a fibrous capsule; this can be prominent in slowly expanding growths and is relatively common in Japanese cases (Okuda *et al.*, 1977). The invasive nature of the tumour is often apparent however (Fig. 13.6), or there may be an abrupt transition between normal and neoplastic cells (Fig. 13.7). In cirrhosis the arrangement of tumour nodules often mimics the cirrhotic pattern (Fig. 13.8); individual nodules may consist partly of hyperplastic and partly of neoplastic cells or malignancy may appear to develop in the centres of nodules. In haemochromatosis the tumour contains no iron (Fig. 13.9) (this is true also of secondary cancer in a siderotic liver).

While some of these tumours are not highly differentiated the histogenesis can be recognized fairly readily from the quantity of well-stained cytoplasm, which is often rather basophilic, with distinct nuclear membranes and prominent nucleoli. Nucleoli are nearly always present. Apart from bile pigment, some cells contain prominent sharply defined

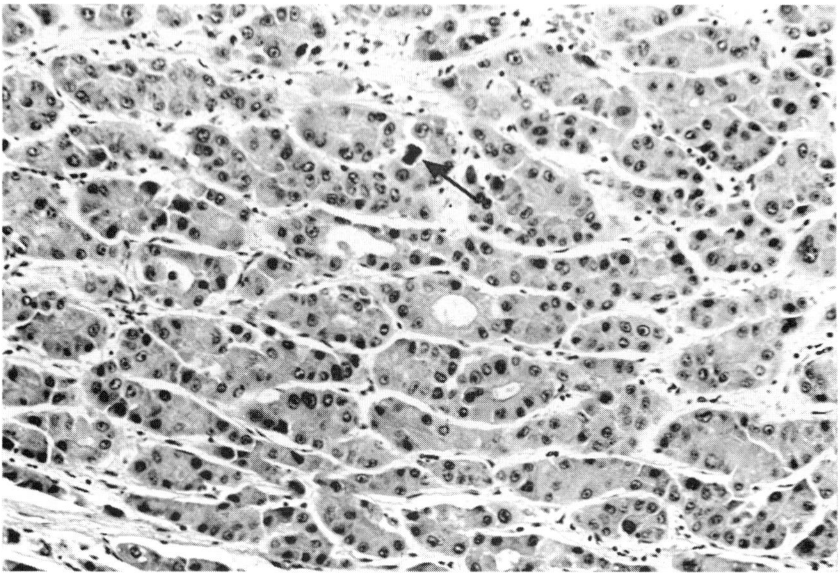

Fig. 13.5 Hepatocellular carcinoma showing a typical trabecular arrangement of moderately well differentiated tumour cells. There is a tendency to acinar formation and one small bile plug is present (arrow) (HE, × 180).

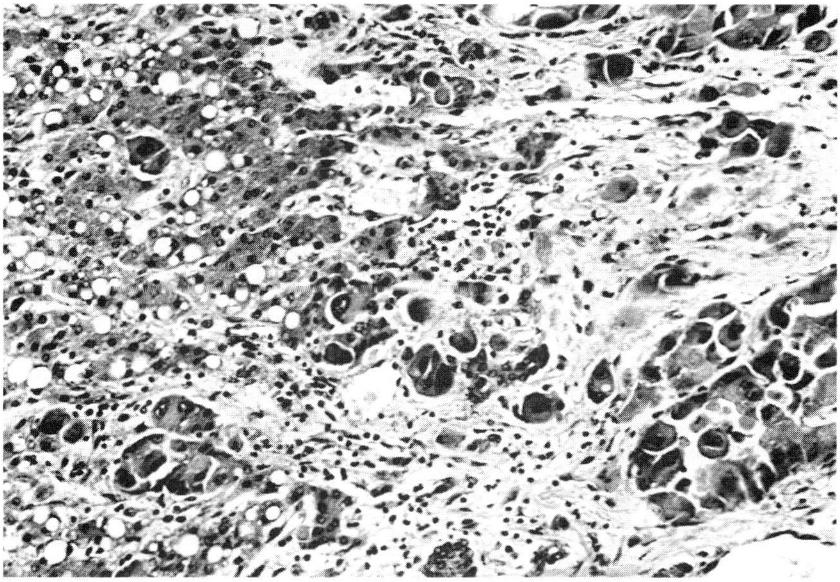

Fig. 13.6 Hepatocellular carcinoma (bottom and right) invading liver (HE, × 180).

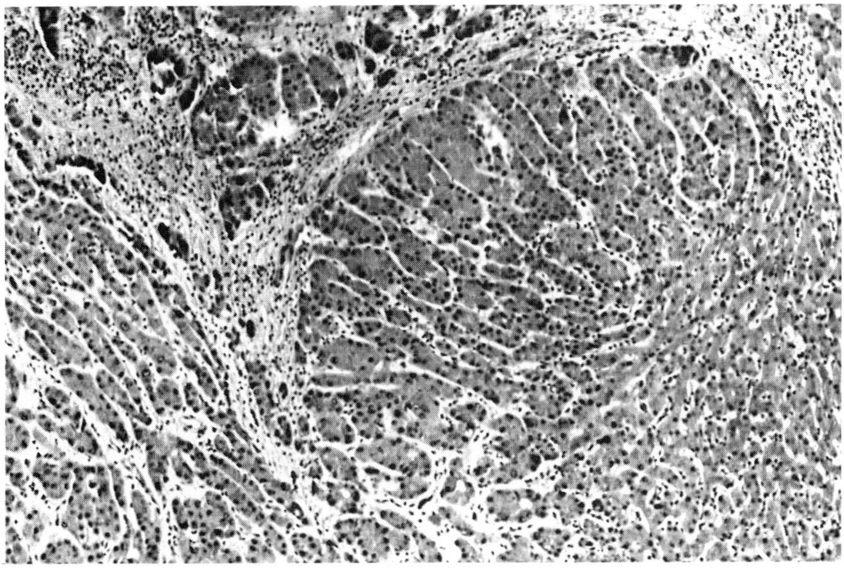

Fig. 13.7 Hepatocellular carcinoma showing a fairly well defined transition with normal liver (lower right) (HE, × 90).

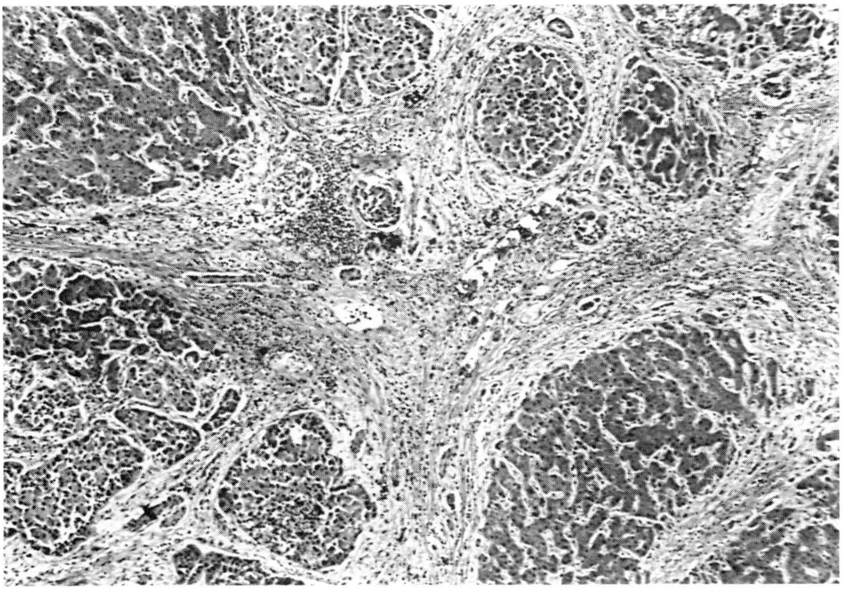

Fig. 13.8 Hepatocellular carcinoma in a cirrhotic liver. Discrete tumour nodules are seen (upper centre and lower left) (HE, × 72).

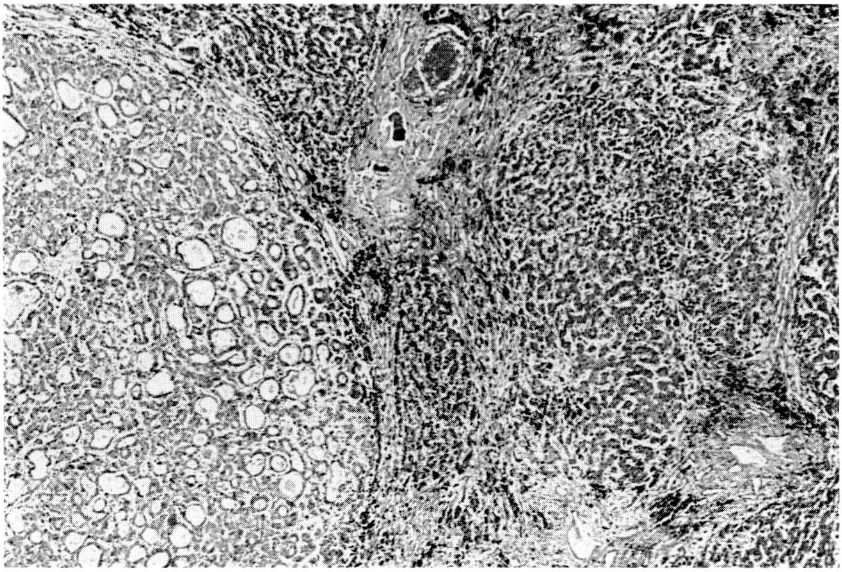

Fig. 13.9 Hepatocellular carcinoma complicating haemochromatosis. The tumour nodule on the left consists of many small acini and has no haemosiderin pigment (Perls' method for haemosiderin, × 45).

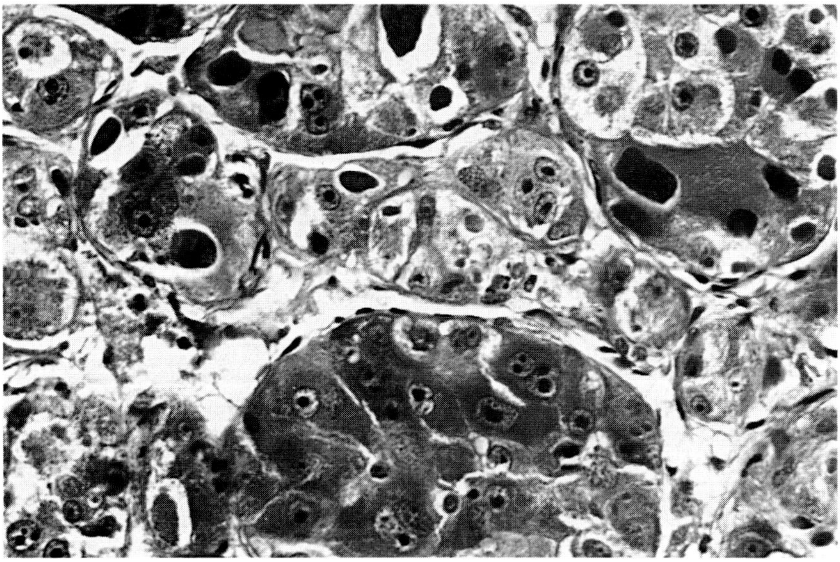

Fig. 13.10 Hepatocellular carcinoma. Many tumour cells contain large clearly defined inclusion bodies (HE, × 450).

cytoplasmic hyaline inclusion bodies (Fig. 13.10). Frequently, these represent alpha-1-antitrypsin and there is a clear association between alpha-1-antitrypsin deficiency and both hepatocellular carcinoma and cholangiocarcinoma. Presumably the inclusion bodies are the product of the tumour cells, but it is possible that their components are produced elsewhere and ingested by the tumour. Immunohistochemical staining for alpha-1-antitrypsin can reveal a more diffuse distribution of the protein in tumour cell cytoplasm. Mallory bodies are also described but are seen less frequently and usually in cases complicating alcoholic cirrhosis (Fig. 13.11). Glycogen may be demonstrated in tumour cell cytoplasm but this is not peculiar to hepatoma, being found, for example, in renal and occasionally in adrenal cortical tumours. Some tumour cells have profuse mitochondria, and this oncocytic variety is referred to below (p. 313). Parts of the tumour are frequently degenerate or necrotic. Fatty change when present (Fig. 13.12) is often irregular in distribution.

In view of the association between hepatocellular carcinoma and hepatitis B virus infection, it is of interest to search for this in biopsy as well as autopsy specimens. Both surface and core antigens may be found by histochemical methods, especially in otherwise normal or dysplastic

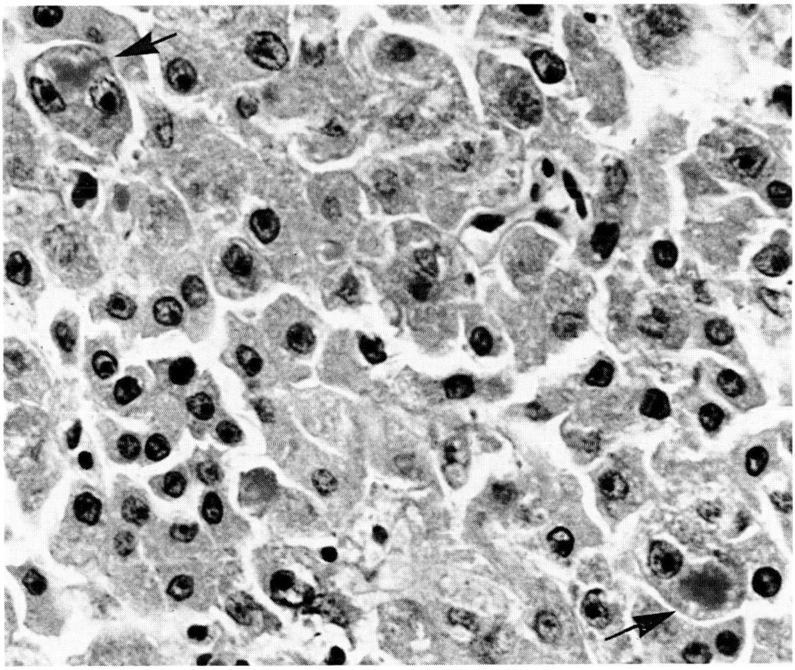

Fig. 13.11 Hepatocellular carcinoma from an alcoholic patient. Two tumour cells (arrows) show cytoplasmic hyaline degeneration (Mallory bodies) (HE, × 720).

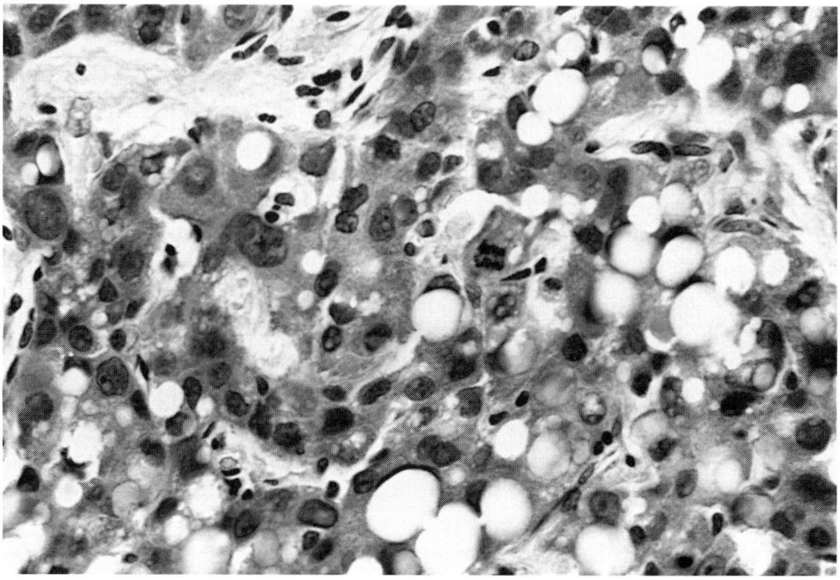

Fig. 13.12 Hepatocellular carcinoma showing fatty change in tumour cells (HE, × 490).

hepatocytes close to tumour deposits. Surface antigen is seen less frequently in tumour cell cytoplasm but this should not be mistaken for positive staining of non-neoplastic cells surrounded by tumour. Hepatitis B infection varies in incidence in different parts of the world; our experience in the West of Scotland gives positive results in about 10% of hepatocellular cancer patients (Turbitt *et al.*, 1977).

The assessment of very poorly differentiated cases can be difficult. In these the characteristic cohesiveness between adjacent tumour cells is lost and there may be numbers of multinucleated cells (Fig. 13.13), sometimes arranged in cords radiating from blood vessels. Clear cell varieties are described (Figs 13.14 and 13.15), some consisting of relatively small rounded cells, which usually retain the characteristic nuclei and nucleoli of hepatocellular carcinoma. Their true nature may be suspected also by a trabecular pattern and by the presence in other parts of the specimen of tumour cells more characteristic of liver cell cancer.

In most hepatocellular carcinomas the trabeculae consist of multiple plates of cells some of which may be arranged around bile canalicular structures. A more distinct acinar pattern is seen not infrequently and these often contain bile or bile-stained secretion (Fig. 13.16). Occasionally such acinar structures are lined by columnar cells with basal nuclei and a mucoid secretion but sufficient cytological detail may be retained to

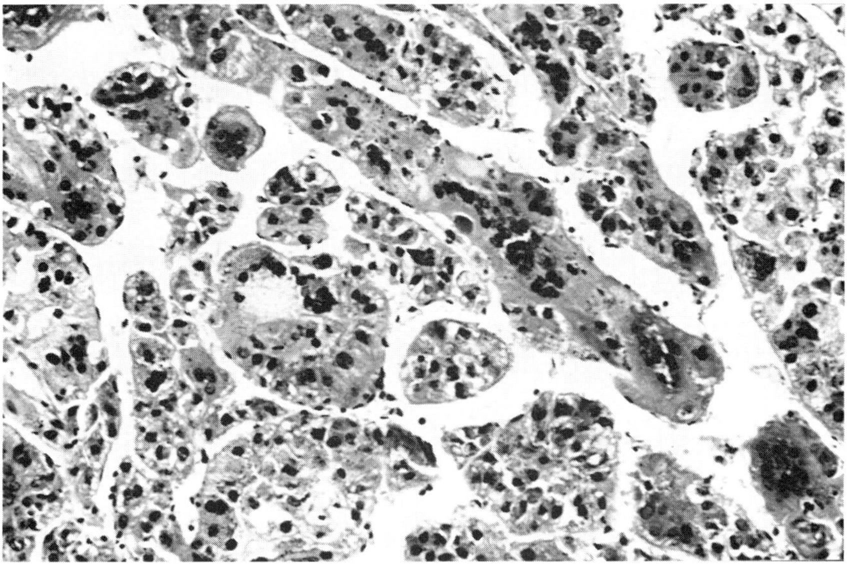

Fig. 13.13 Hepatocellular carcinoma. There are many giant tumour cells but the trabecular pattern is maintained (HE, × 180).

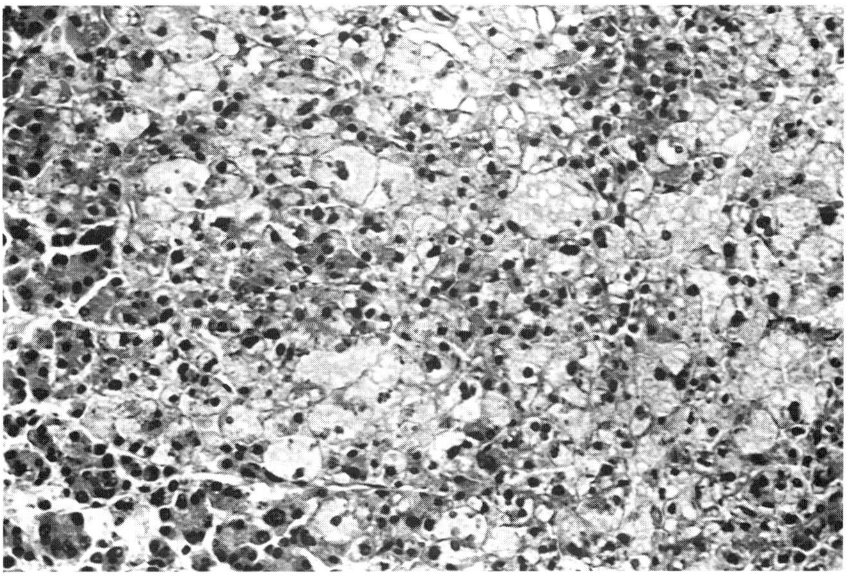

Fig. 13.14 Hepatocellular carcinoma. Large clear-cell variety (HE, × 225).

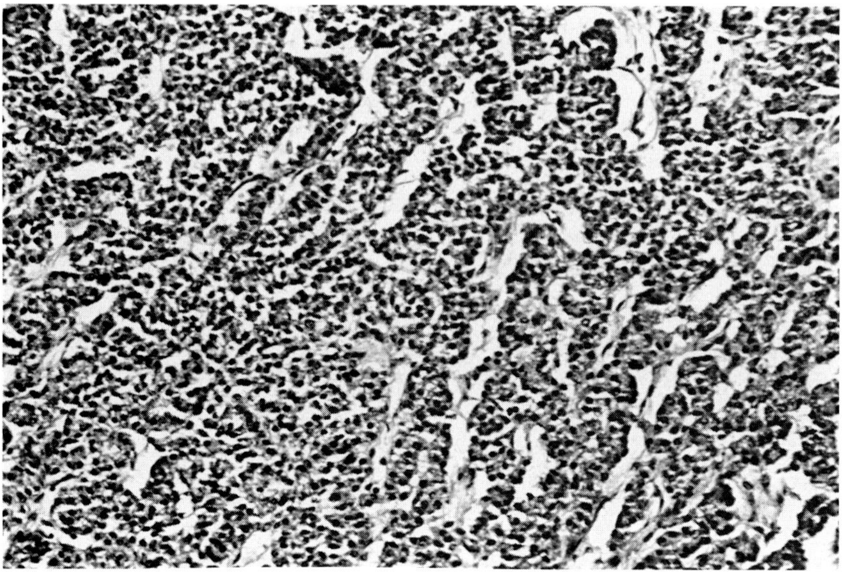

Fig. 13.15 Hepatocellular carcinoma. Small cell variety (HE, × 180).

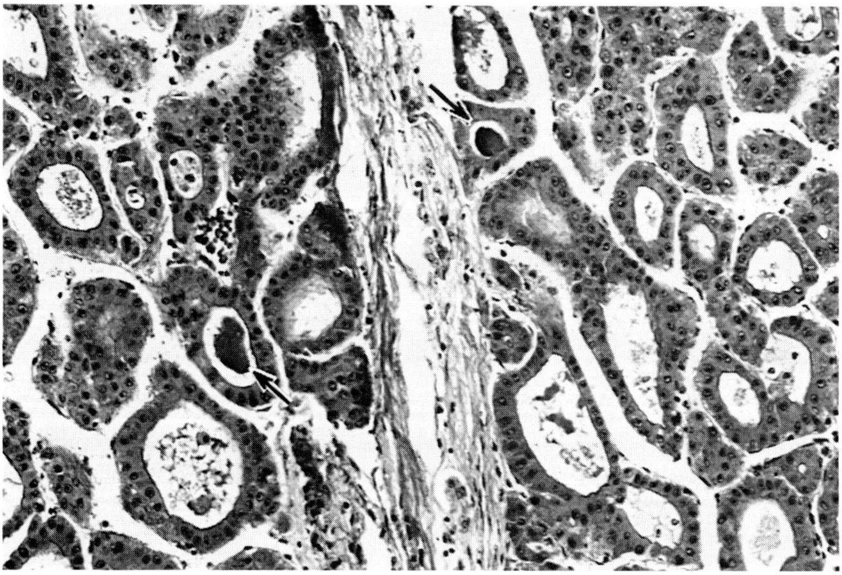

Fig. 13.16 Hepatocellular carcinoma. Tumour cells form acinar structures and some of these contain bile secreted by the tumour. This appears as a dense inspissated pigment (arrows) or as loose flocculent material (HE, × 180).

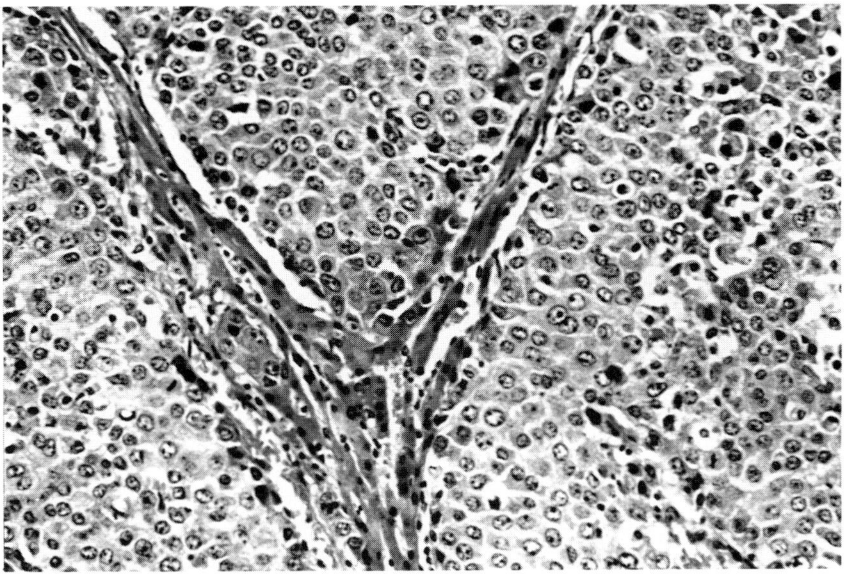

Fig. 13.17 Hepatocellular carcinoma. Solid 'cobblestone' variety (HE, × 225).

distinguish them from bile duct cancers. Other tumours consist of solid sheets of cells, the so-called 'cobblestone' variety (Fig. 13.17).

A minority of hepatocellular cancers, like some benign liver cell adenomas, are noted for their unusual vascularity and should not be mistaken for angiosarcomas. There is irregular gross dilatation of sinusoids situated between cords of tumour cells (Fig. 13.18). Less frequently liver adjacent to the tumour has similar vascular features (see Fig. 10.26) with some resemblance to peliosis hepatis but in comparison to that condition the vascular lesions are less well defined and less spherical. There is an association with long-term androgenic-anabolic steroid therapy, as in cases of Fanconi and aplastic anaemia (Farrell *et al.*, 1975), and rarely with oral contraceptive intake. Severe intra-abdominal bleeding may arise.

Primary liver cancers have a distinct tendency to invade portal and hepatic veins (Fig. 13.19) as well as lymphatics, and the venous involvement can cause fatal portal hypertension before the development of metastases. Rare cases present as obstructive jaundice because of invasion by tumour of the bile channels with extension into the common bile duct, and in one such case in our experience there was involvement also of nerves in the region of the porta hepatis (Fig. 13.20).

Connective tissue stroma in the majority of cases is not abundant and in large areas of tumour it may be quite inconspicuous. As in normal liver

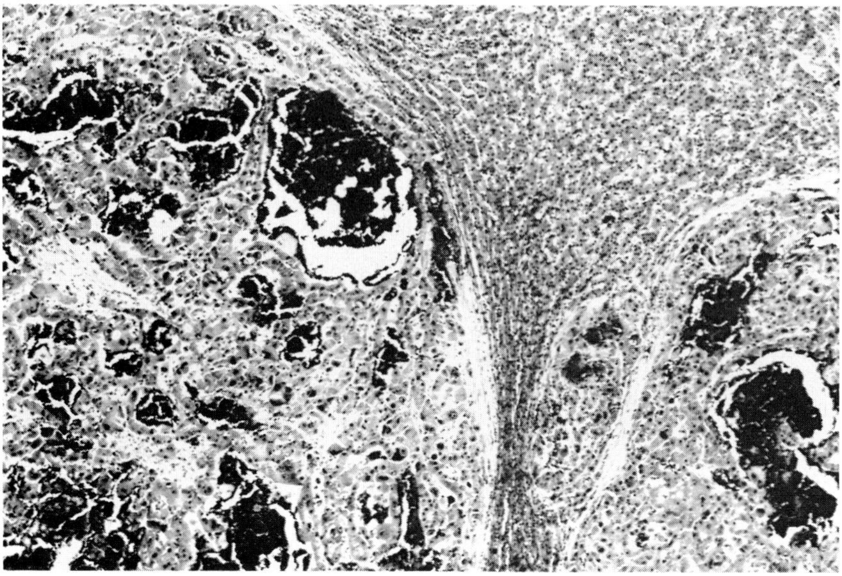

Fig. 13.18 Hepatocellular carcinoma. Parts of two nodules showing prominent dilated vascular channels filled with blood (HE, × 72).

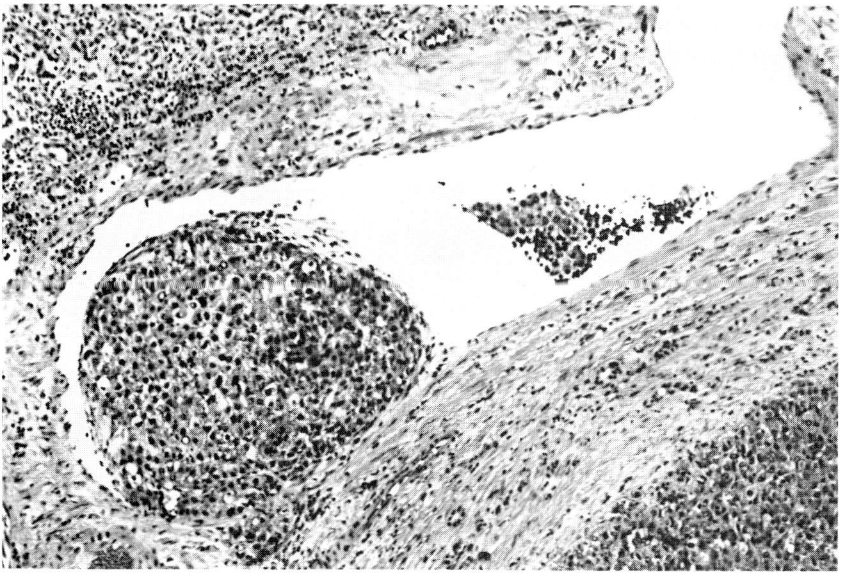

Fig. 13.19 The large portal vein branch contains deposits of hepatocellular carcinoma. Extravascular tumour is also present (bottom right) (HE, × 90).

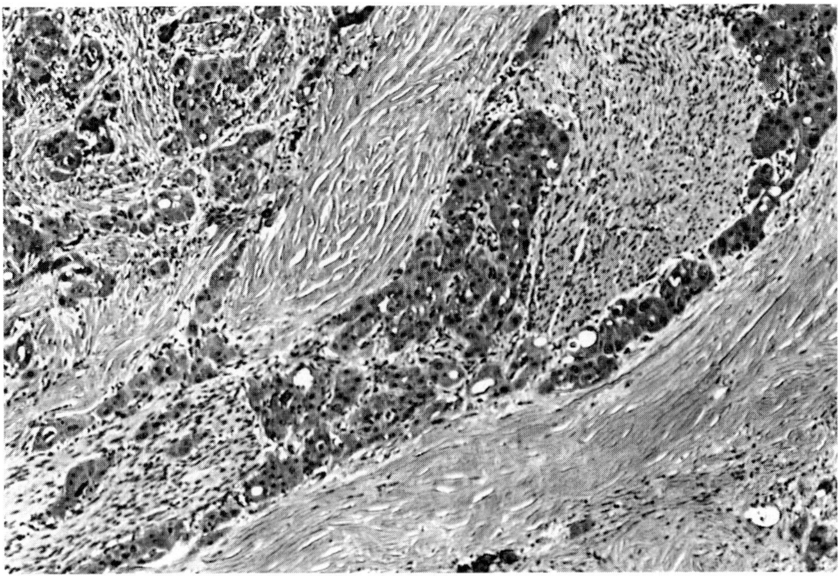

Fig. 13.20 Hepatocellular carcinoma invading nerve (HE, × 90).

there may be no basement membrane between cords of epithelial cells and sinusoidal lining cells, and this may help to differentiate a glandular arrangement of liver cancer cells from bile duct tumours. Some hepatocellular cancers, however, have a substantial stroma and there may be solid scirrhous areas with narrow cords of rather degenerate tumour cells. Calcification may occur and be detected radiologically. An association has been noted between sclerosing tumours and pseudohyperparathyroidism, and the same is true of some unusually scirrhous forms of bile duct cancer.

Fibrous tissue is abundant in the *fibrolamellar type* of primary liver cancer (Berman *et al.*, 1980; Craig *et al.* 1980), where it is seen mainly as irregular parallel lamellae of mature collagen separating trabeculas of large polygonal cells with abundant granular eosinophilic cytoplasm and prominent nuclei and nucleoli (Fig. 13.21). These tumour cells are rich in mitochondria and can be regarded as components of malignant oncocytoma. Fibrolamellar carcinoma tends to occur in children and young adults of both sexes. The affected liver is non-cirrhotic and the prognosis is better compared with other forms of primary liver cancer.

Hepatic tissue adjacent to the tumour should of course be assessed for the presence of cirrhosis and other disease. Tiny deposits of tumour or even single tumour cells may be seen in veins or sinusoids and this is clearly of importance in the examination of resected specimens. Dark

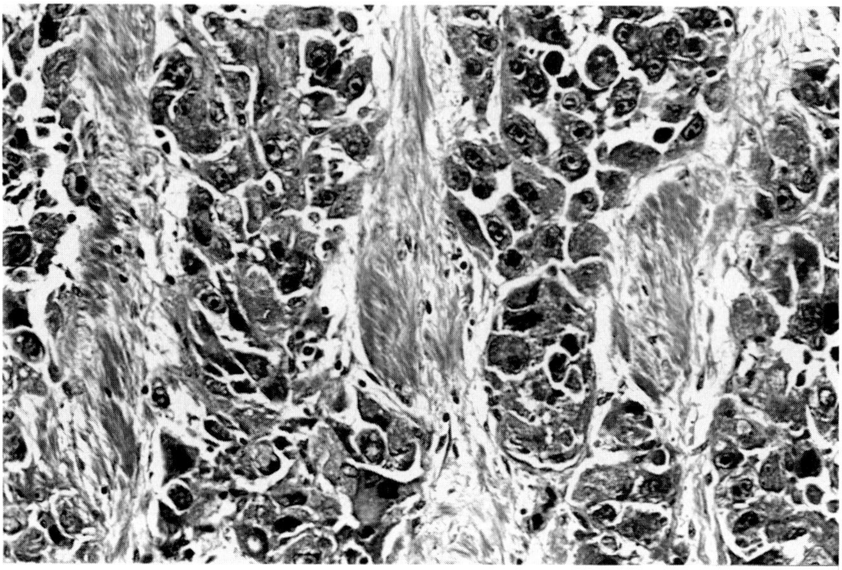

Fig. 13.21 Hepatocellular carcinoma of fibrolamellar type. Cords of tumour cells with prominent eosinophilic cytoplasm are separated by thick bands of fibrous connective tissue (HE, × 247).

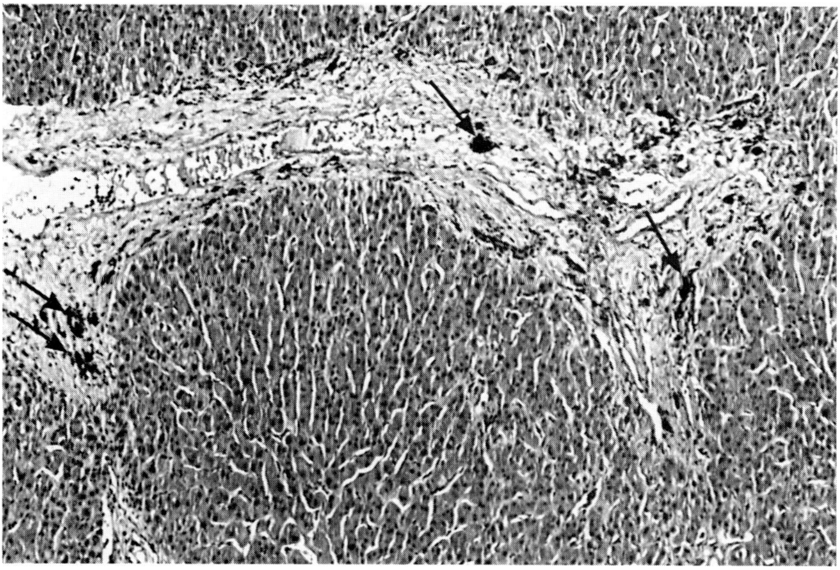

Fig. 13.22 Portal fibrosis and septum formation induced by thorotrast which forms black granules (arrows) (HE, × 90).

refractile granules in fibrous scars should raise the suspicion of thorotrast liver damage (Fig. 13.22), because this radioactive material is associated with hepatocellular carcinoma and cholangiocarcinoma as well as angiosarcoma (Fig. 13.23). Its presence can be confirmed by electron-probe analysis or by autoradiography which will demonstrate alpha tracks emanating from deposits of radioactive material (Fig. 13.24).

Foci of dysplastic hepatocytes may be recognized as indistinct nodules within hepatoma-bearing livers, especially those with cirrhosis, and this association with cancer suggests that it may be a premalignant lesion. This has been the experience of pathologists in East Africa where it is seen frequently and where there is also an association with hepatitis B infection (Anthony *et al.*, 1973). We have failed to find the presence or subsequent development of cancer in many cases of dysplasia seen in this country. Dysplastic nodules maintain a normal arrangement of cell plates but differ from the surrounding hepatic tissue with which they blend by possessing more cell cytoplasm and slightly larger, hyperchromatic, folded or multiple nuclei (Fig. 13.25). The cytoplasm may be less basophilic than that of neighbouring cancer cells. In the cirrhotic liver they may comprise whole nodules or parts of nodules, and in haemachromatosis they resemble true hepatoma by lacking haemosiderin. The distinction from well-differentiated hepatocellular carcinoma can be

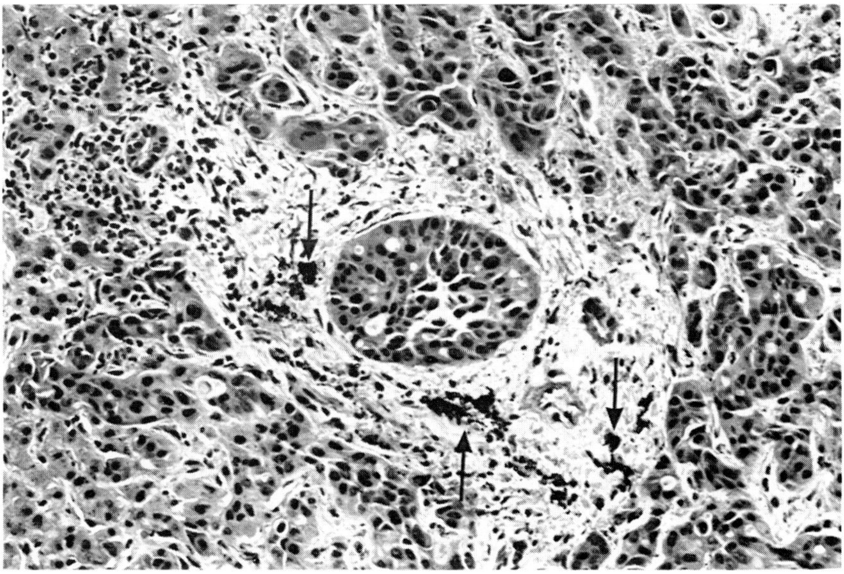

Fig. 13.23 Granules of thorotrast (arrows) lie within the fibrous stroma of a hepatocellular carcinoma (HE, × 225).

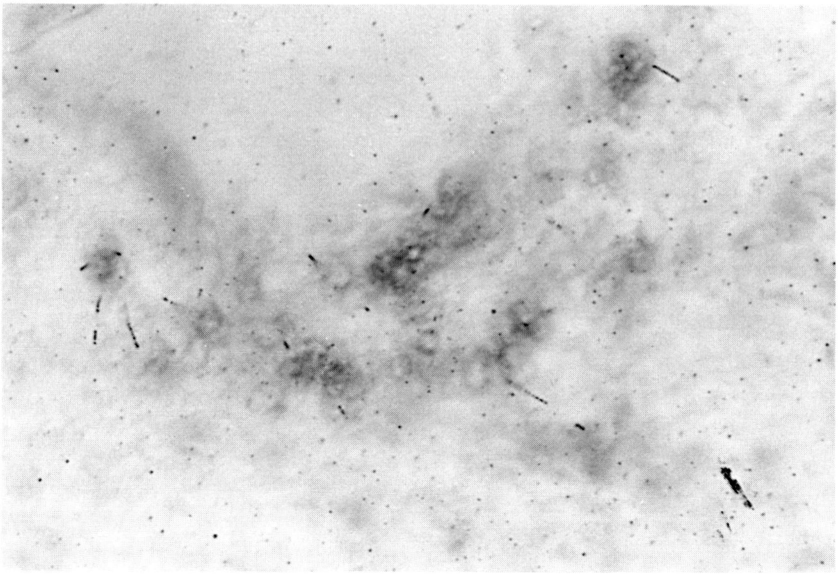

Fig. 13.24 Autoradiograph of liver section containing thorotrast. Short alpha-tracks are seen radiating from deposits of the radioactive material (HE, × 565).

difficult in small specimens. Occasionally we have noted non-cirrhotic nodular regenerative hyperplasia in tumour-bearing livers (Fig. 13.26), but cannot state whether this more diffuse lesion, which is usually associated with chronic venous congestion and rheumatoid disease, is also a premalignant condition.

(b) Differential diagnosis of hepatocellular carcinoma

The distinction between highly differentiated cancers, benign adenomas, and focal hyperplastic and dysplastic lesions has already been discussed. Poorly differentiated tumours may be confused with bile duct cancers or with secondary deposits from other sites. Points in favour of hepatocellular carcinoma are the tendency for the cells to be arranged in a trabecular fashion along sinusoidal channels, the prominent cytoplasm, and nuclei with distinct nuclear membranes and nucleoli. Other characteristics such as the presence of cytoplasmic glycogen and bile secretion are less important, as these occur most often in more highly differentiated tumours which present no diagnostic problem. Some portions of the lesion may be better differentiated than others and where the opportunity exists repeat biopsy or biopsy from multiple sites may be helpful. The presence of established cirrhosis can also be a point in favour of the

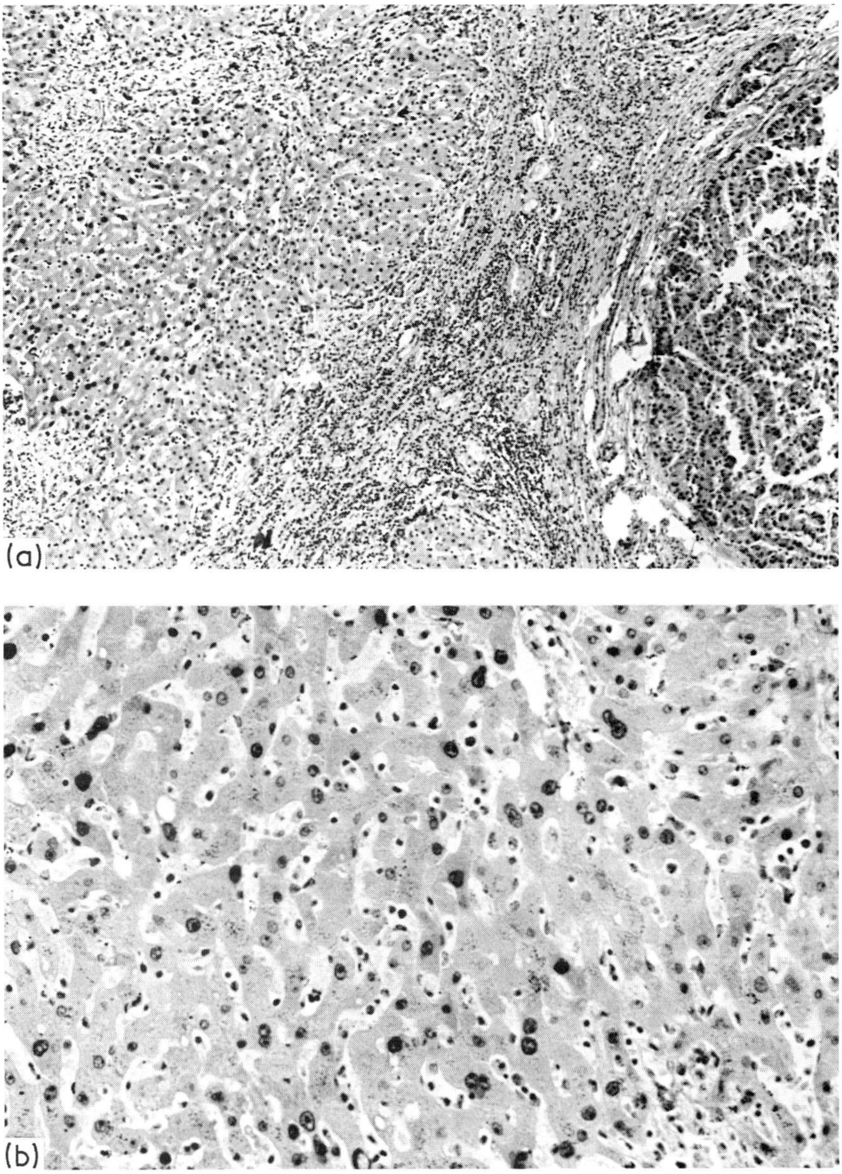

Fig. 13.25 Dysplasia and neoplasia in a cirrhotic liver: (a) There is a nodule of hepatocellular carcinoma (right), and many dysplastic cells with prominent pleomorphic nuclei occupy part of a large cirrhotic nodule (left). (b) Dysplastic area at higher magnification. ((a) HE, × 72; (b) HE, × 225.)

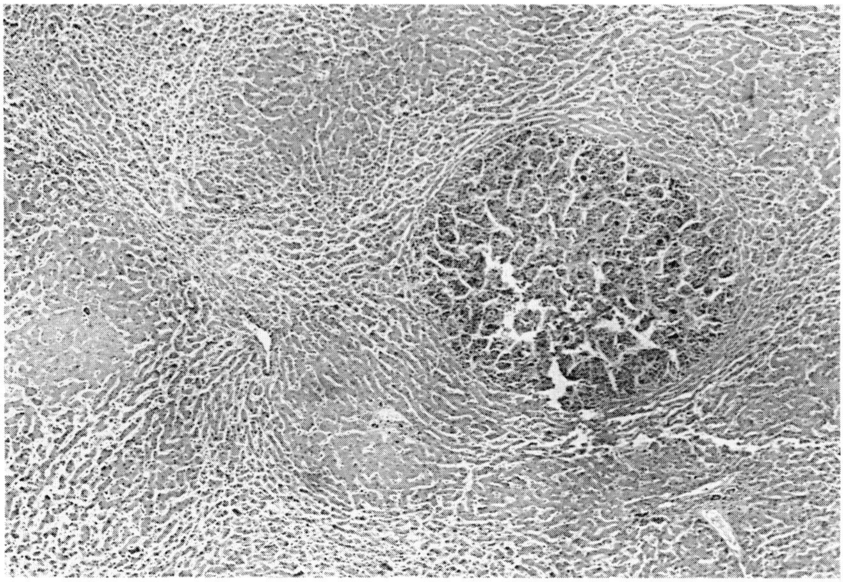

Fig. 13.26 Small deposit of hepatocellular carcinoma in liver showing nodular regenerative hyperplasia (H and van Gieson's stain, × 45).

diagnosis of hepatocellular carcinoma as, in our experience, secondary cancer is uncommon in the cirrhotic liver. The presence of α-fetoprotein in tumour cells and blood, and excessive amounts of prolyl hydroxylase in the serum, are also helpful in establishing the diagnosis.

(c) Prognosis

In general the prognosis in hepatocellular carcinoma is very bad, only a few cases being confined to one or other lobe and suitable for resection. These are usually well-differentiated cases arising in a non-cirrhotic liver but, with the possible exception of the fibrolamellar type, it is very difficult to assess prognosis from any single histological feature. Death frequently ensues within months of establishing the diagnosis. It may be due to extensive destruction of the liver by tumour, to the effects of portal hypertension caused by associated cirrhosis or by occlusion of the portal vein by tumour and thrombus, to extensive intra-abdominal bleeding, or to metastatic spread which most commonly involves the abdominal lymph nodes, lung, vertebrae and right adrenal. Rarely the inappropriate production of hormones, such as insulin-like products or erythropoietin, by the tumour cells may have serious or even fatal effects.

Prognosis can be better in some rare cases associated with prolonged steroid therapy. It is reported that these cancers may grow slowly and may even regress, as do some benign liver tumours, when hormone administration is discontinued (Farrell *et al.*, 1975).

13.2 Tumours arising from bile duct epithelium

13.2.1 Bile ductular adenoma (benign cholangioma; tubular adenoma)

These small subcapsular tumours are uncommon and are usually an incidental finding at laparotomy. They consist of small ductular structures and a variable amount of fibrous stroma. Unlike von Meyenburg complexes they never contain inspissated bile and need not be related to portal triads, although the specimen illustrated does show this feature (Fig. 13.27).

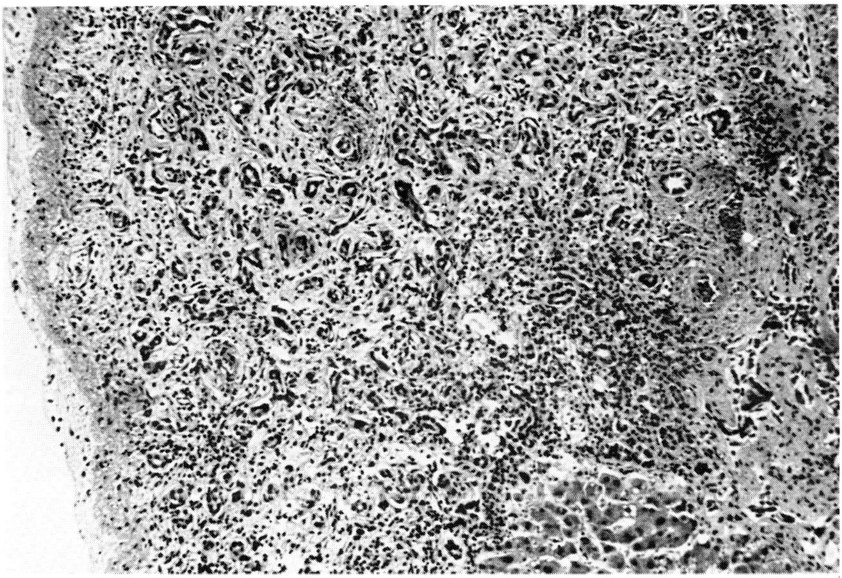

Fig. 13.27 Bile ductular adenoma situated under Glisson's capsule (HE, × 100).

13.2.2 Bile duct papilloma

Papilloma arising in bile ducts within the liver is also extremely uncommon. We have seen one example diagnosed initially by the presence of small fragments within bile recovered from the common duct (Fig. 13.28). This tumour was resected but caused death one year later from local recurrence with suppurative cholangitis. At autopsy there was no evidence of invasive spread or metastatic deposits. Local recurrence has

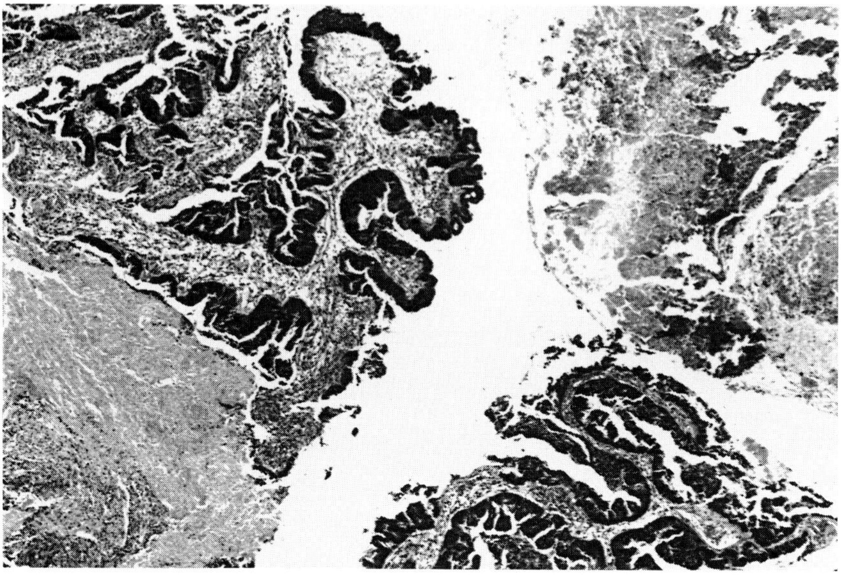

Fig. 13.28 Fragments of bile duct papilloma (possible low-grade papillary carcinoma) recovered from bile in common bile duct (HE, × 45).

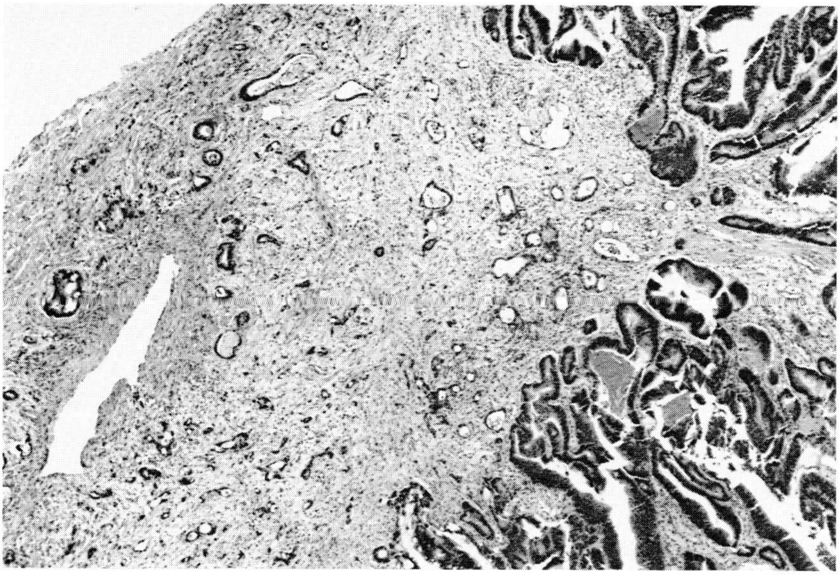

Fig. 13.29 Papillary cystadenocarcinoma of intrahepatic bile duct. The base is scirrhous and contains numerous small tumour acini, some with a mucus secretion (HE, × 45).

been recorded by others, and it seems possible that invasive papillary bile duct cancer (Fig. 13.29) could arise from such tumours.

13.2.3 *Cystadenoma of liver* (Ishak *et al.*, 1977)

This is another uncommon benign lesion. It is a multiloculated cystic tumour which can be quite large, weighing several kilograms. Multiple lesions may be present and the condition should not be mistaken for polycystic disease which can coexist with cholangiocarcinoma. The cysts have a lining of folded tall columnar epithelium and there is abundant connective tissue which resembles ovarian stroma sometimes infiltrated with lymphocytes and macrophages containing prominent amounts of haemosiderin or lipofuscin pigment. The cysts contain serous or mucinous fluid which may be blood-stained. Some tumours may be ductal but in others the site of origin cannot be clearly established. Malignant change may occur.

13.2.4 *Cholangiocarcinoma (cholangioma; malignant cholangioma; adenocarcinoma; alveolar carcinoma) of the main intrahepatic ducts* (Weinbren and Mutum, 1983)

These are usually hilar, arising at the junction of the right and left hepatic ducts in the porta hepatis. Accordingly, most cases present with obstructive jaundice when the tumour is quite small. The remainder of the liver usually shows cholangiohepatitis and multiple abscess may be mistaken for secondary tumour on naked-eye examination. The tumour cells do not of course secrete bile and the pallor of the lesion offers a sharp contrast to the surrounding bile-stained parenchyma.

Histologically they are adenocarcinomas sometimes papillary and with acini lined by cuboidal or columnar epithelium (Figs 13.29 and 13.30). There may be a mucoid secretion, but this is often scanty, although pools of extracellular mucinous material may be demonstrated by Alcian blue staining. Compared with hepatocellular carcinoma the stroma is usually more abundant and some tumours are very scirrhous with acinar structures lying in the midst of dense connective tissue. Some tumour cells appear to line lymphatic channels and form small cysts into which project short blunt papillae (Fig. 13.31). Occasionally there is a more readily defined papillary cystadenomatous pattern (Fig. 13.32). Compared to hepatocellular cancers, cholangiocarcinomas have less tendency to invade the large veins and when portal hypertension is present it is usually due to pressure on the main vein by the tumour or to pyelophlebitis complicating cholangitis. However, the tumour has a tendency to spread peripherally and may be seen in portal tract lymphatics at some distance from the main mass (Fig. 13.33). Periportal zones

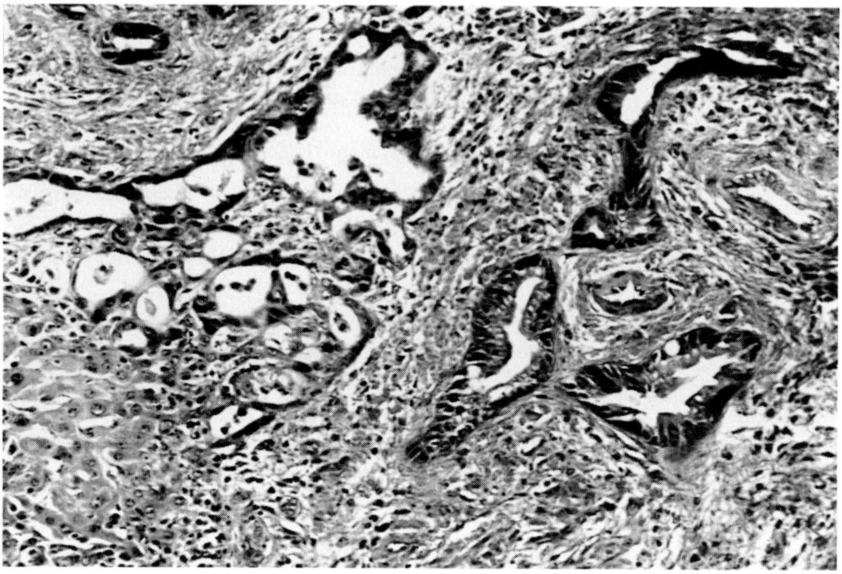

Fig. 13.30 Cholangiocarcinoma arising within the liver. The tumour acini are lined by columnar, cuboidal or flattened cells and lie in a profuse fibrous stroma (HE, × 180).

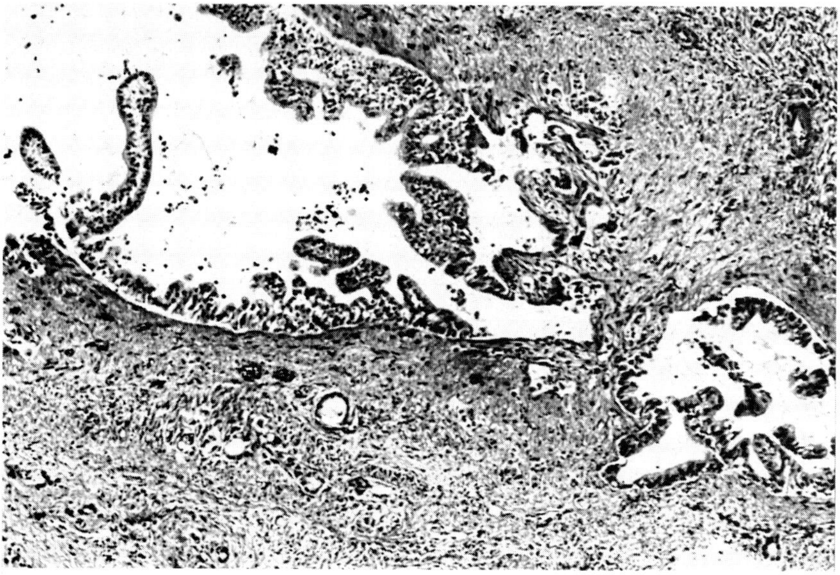

Fig. 13.31 Cholangiocarcinoma arising within the liver at the junction of right and left hepatic ducts. It is very scirrhous and tumour cells with blunt papillae line cystic spaces (HE, × 90).

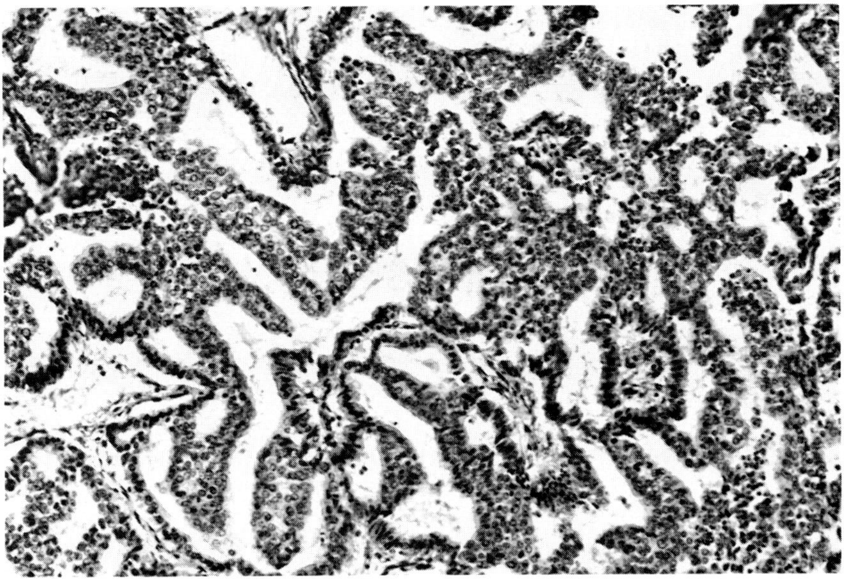

Fig. 13.32 Cholangiocarcinoma with a papillary cystadenomatous pattern (HE, × 180).

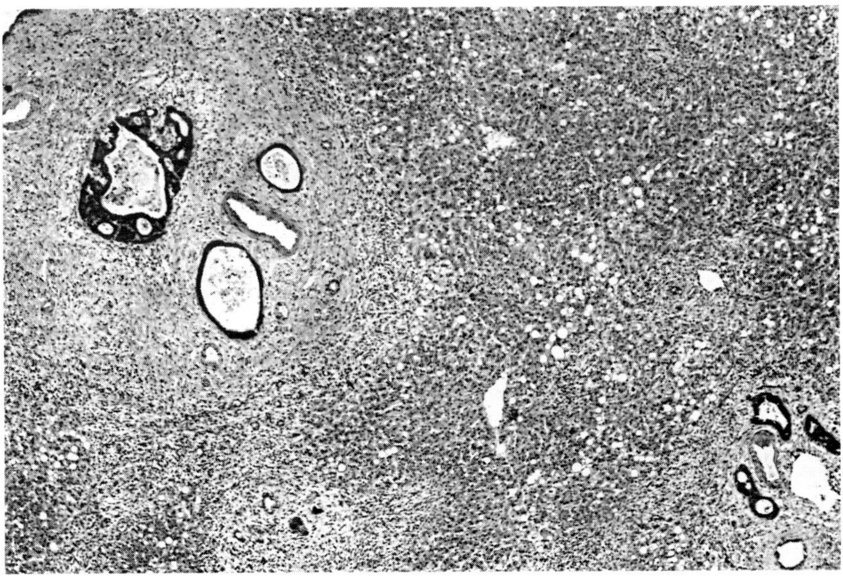

Fig. 13.33 Cholangiocarcinoma showing invasion of portal lymphatic vessels at some distance from the primary tumour (HE, × 45).

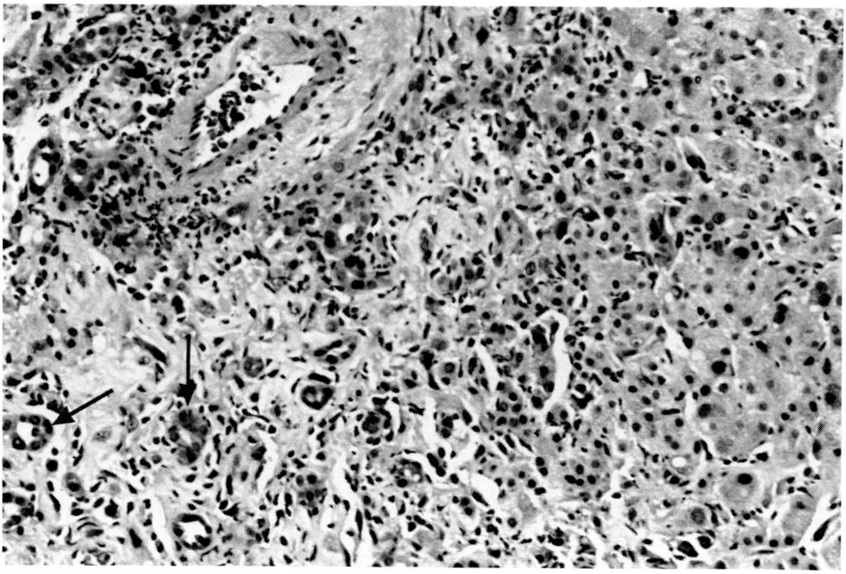

Fig. 13.34 Cholangiocarcinoma. The expanded portal area contains small acinar structures, some of these are neoplastic (arrows). Small tumour deposits are present also in adjacent sinusoids (HE, × 225).

may have tumour spreading into sinusoids (Fig. 13.34) as well as proliferating non-neoplastic bile ductules. Invasion of nerves and perineural tissue in the porta hepatis is common.

The differential diagnosis from other adenocarcinomas either invading the liver directly from extrahepatic bile ducts, gall bladder or pancreas, or appearing as blood-borne metastases can be impossible. In small biopsies the lesion may be mistaken for sclerosing cholangitis, although both sclerosing cholangitis and cholangiocarcinoma can coexist as complications of ulcerative colitis. These tumours are often located early by a combination of ascending and transhepatic cholangiography, but prognosis is not good because of difficulties in resection.

13.2.5 Cholangiocarcinoma of peripheral intrahepatic ducts

The distinction is made from larger duct tumours principally on clinical grounds. Whereas large duct tumours often cause obstructive jaundice when the lesion is small, the peripheral duct tumour is much more like hepatocellular carcinoma in its mode of presentation and also in the gross appearance of the liver. There is, however, no association with cirrhosis and the sex distribution is not predominantly male as in hepatocellular cancer cases. Where chronic liver disease coexists it is frequently of biliary type, but could be a consequence of obstruction by tumour already present in the liver.

Microscopically these are again adenocarcinomas with tubular or acinar structures lined by cuboidal or columnar cells. Compared with hepatoma the cells have less cytoplasm and more deeply staining rounded nuclei, but nucleoli can be prominent in both. There may be mucous secretion but never any bile. The fibrous stroma is usually substantial and there may be dense scirrhous areas, especially in the centres of nodules. Ductular structures have a distinct basement membrane. The tendency is to lymphatic permeation rather than invasion of veins.

The differential diagnosis from hepatocellular carcinoma has already been referred to. Distinction from secondary carcinoma arising outwith the liver may be impossible. The prognosis is similar to hepatocellular cancer.

13.2.6 *Cholangiolocellular carcinoma of liver* (Steiner and Higginson, 1959)

These rare tumours allegedly arise from the smallest bile ductular radicles in the liver, i.e. the canals of Hering. They consist of double-layered cords of small cells morphologically resembling ductular epithelium and are set in a fibrous stroma (Fig. 13.35). They may form tiny acinar structures. They behave in a similar fashion to tumours of the larger ducts.

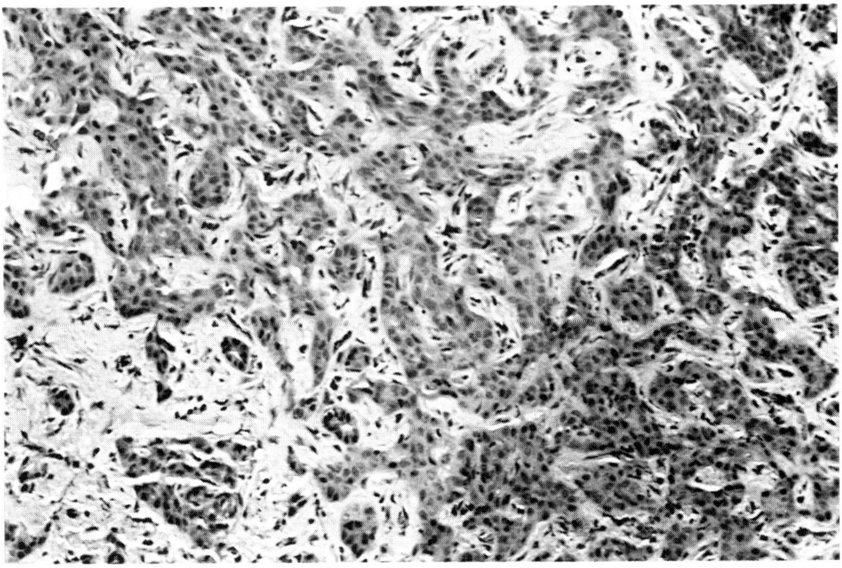

Fig. 13.35 Cholangiolocellular carcinoma of liver (HE, × 180).

13.2.7 *Primary carcinoid tumour of liver*

Rarely, such a tumour may be found in the liver in the absence of a similar lesion elsewhere. Presumably it originates from argentaffin cells in the walls of bile ducts and so may be classified as a variety of bile duct cancer. Carcinoid syndrome can occur.

13.2.8 *Squamous cell carcinoma of liver*

Squamous and adenosquamous cancers are described which appear to arise from bile duct epithelium or from foci of squamous metaplasia in cyst lining. A mucoepidermoid variant is also recorded.

13.3 Mixed hepatocellular and cholangiocellular carcinoma

Hepatocytes and bile duct epithelium have a common embryonic origin and it is possible for tumours to develop which possess components of both types of cell.

Reference has already been made to a glandular pattern of hepatocellular carcinoma which might be mistaken for bile duct cancer, and careful examination of most hepatomas reveals similar areas in which a clear distinction is not possible. However, most pathologists would reserve the diagnosis of mixed type to either poorly differentiated tumours, in which both classes of cell can just be detected (Fig. 13.36), or to more highly differentiated tumours in which the hepatoma and cholangioma component both make a substantial contribution (Fig. 13.37); it is uncertain whether these have a common cellular origin or whether the hepatoma arises independently of the cholangioma.

In some cases of hepatocellular carcinoma complicating cirrhosis there is considerable proliferation of bile ductules or apparent metaplasia of hepatocytes to a bile duct type of epithelium at the margins of cirrhotic nodules (Fig. 13.38). This should not be confused with the cholangioma component of a mixed cell tumour.

13.4 Vascular tumours of liver

13.4.1 *Haemangioma*

Cavernous haemangioma is common and small subcapsular examples are found incidentally at laparotomy. Unusually large tumours may cause abdominal pain, swelling and intraperitoneal bleeding. These patients may suffer also from coagulation defects with thrombocytopenia or hypofibrinogenaemia.

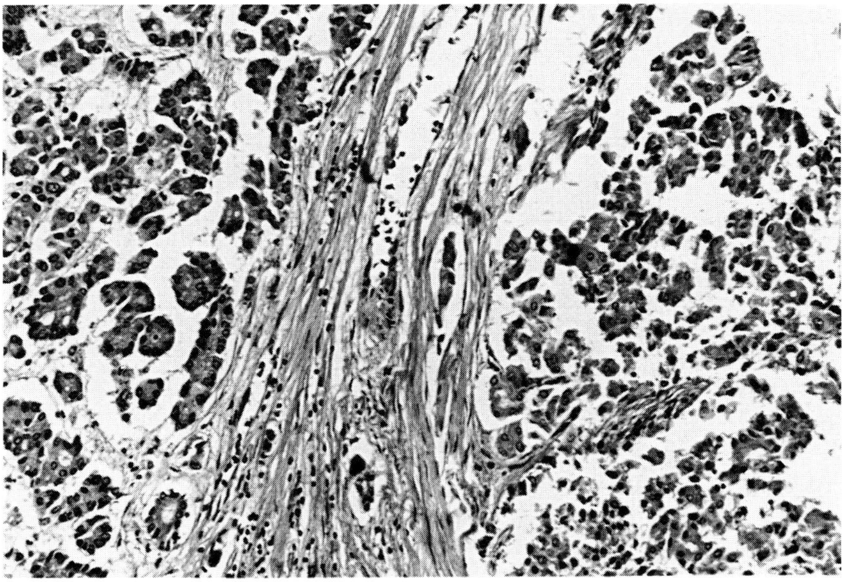

Fig. 13.36 Primary liver cancer of mixed type. A poorly differentiated example in which some nodules (left) resemble bile ductules and others (right) resemble hepatocytes with a little bile secretion (not shown) (HE, × 180).

The tumours consist of blood-filled spaces lined by flattened endothelial cells and with a variable amount of stroma (Fig. 13.39). Some are multiple and may coexist with benign cysts, von Meyenburg complexes or focal nodular hyperplasia. Rarely they are pedunculated. Some may undergo sclerosis which spreads outwards from the centre of the lesion, and calcification may be present.

It is important to consider other forms of vascular tumour and peliosis hepatis in particular cases. Where there are multiple tumours, and especially if other organs are involved, it may not be possible to make a clear distinction from hereditary telangiectasia. Angiomatoid lesions of hairy cell leukaemia should be recognized from the lining of vascular spaces by tumour cells.

Capillary haemangiomas (Fig. 13.40) are very uncommon unless in young children in whom they may really be components of haemangioendothelioma.

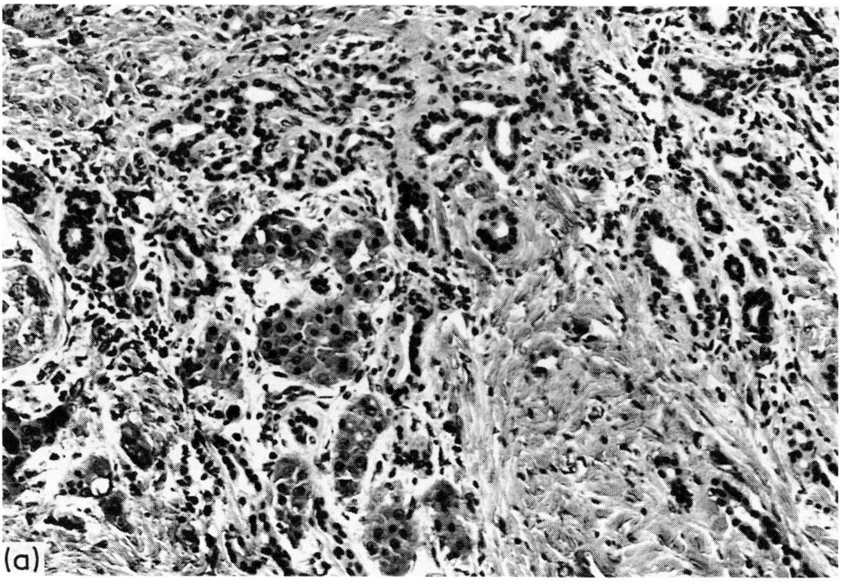

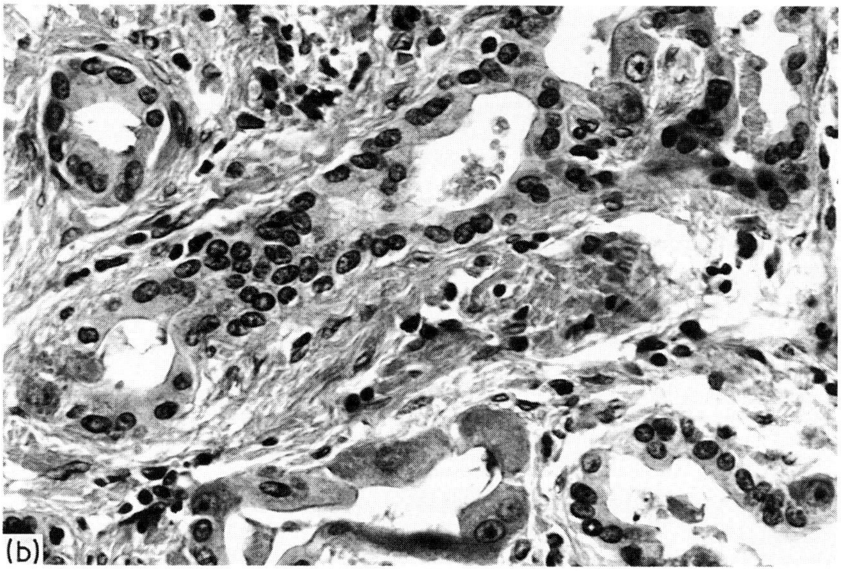

Fig. 13.37 (a) and (b) primary liver cell cancer of mixed type. A more highly differentiated example with cholangiocellular and hepatocellular elements, both of which may contribute to the formation of single acini. (a) HE, × 180; (b) HE, × 450.)

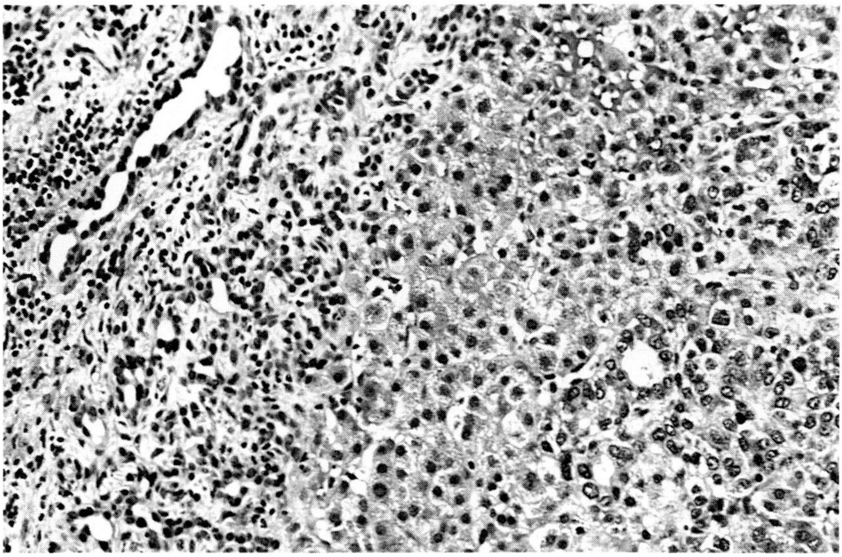

Fig. 13.38 Prominent non-neoplastic bile ductular proliferation (left) in a cirrhotic liver containing hepatocellular carcinoma (right). The neoplasm is forming a few acini. Non-neoplastic parenchyma is present (centre) (HE, × 180).

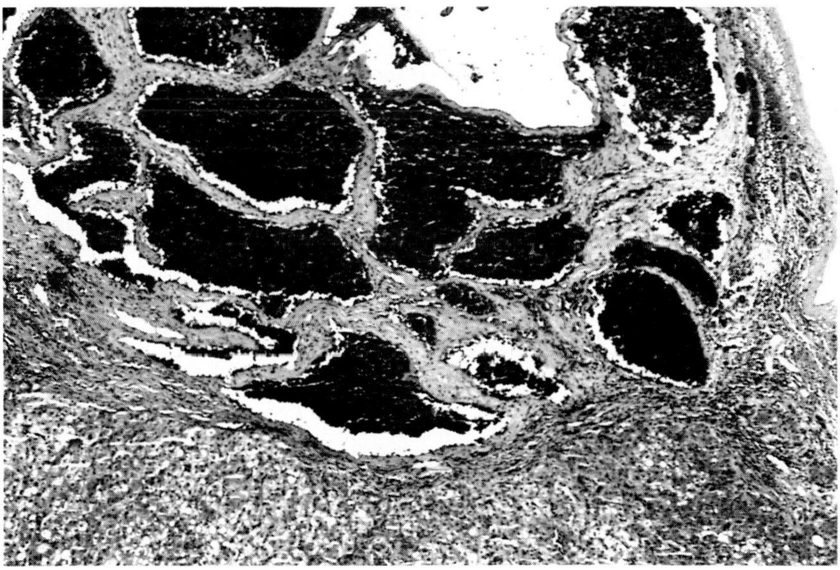

Fig. 13.39 Subcapsular cavernous angioma of liver (HE, × 180).

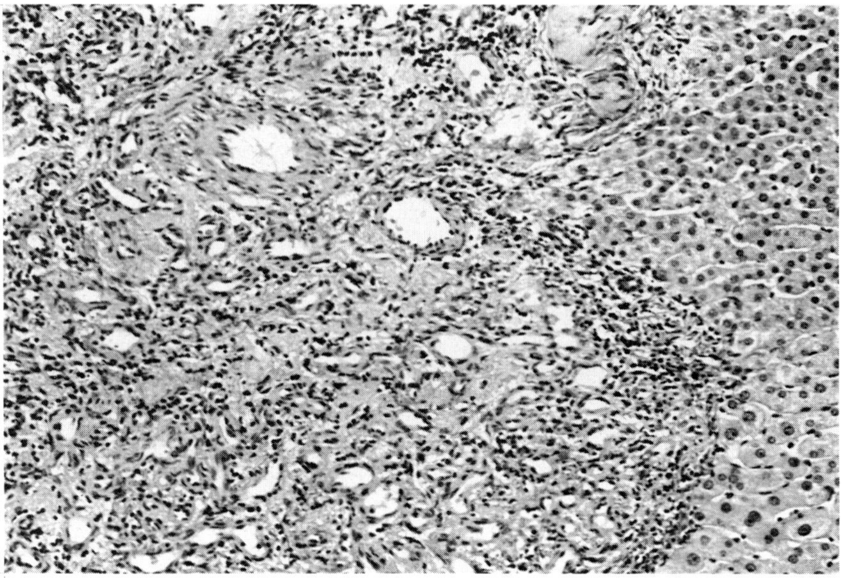

Fig. 13.40 Capillary angioma of liver (HE, × 180).

13.4.2 *Infantile haemangioendothelioma (haemangioma; malignant haemangioma; haemangioendothelial sarcoma; multiple alveolar angiosarcoma)*

These tumours form pale circumscribed nodules with congested centres, which are often multiple and grow to a large size. Although cytologically benign, they can cause fatal liver destruction or high output cardiac failure that is frequently mistaken for congenital heart disease. Other organs such as the skin may be involved also.

Microscopically the tumour consists of vascular channels lined by one or more layers of plump endothelial cells (Fig. 13.41). These channels are usually of capillary size and empty of blood but there may be cavernous areas. Mitotic activity is uncommon. Foci of haematopoiesis are sometimes seen (Fig. 13.42) and the fibrous stroma which is variable in amount often contains rather prominent fibroblasts. A few cases appear to undergo malignant change with infiltration of cells alongside sinusoids and atrophy of hepatocytes. In the children who survive, tumour growth will stop and the lesion gradually undergo sclerosis. Benign multiple tumours with similar histological features have been noted in an adult.

True cavernous haemangioma and focal areas of sinusoidal dilatation can also occur in infant liver and are probably distinct from haemangioendothelioma. Vascular components of mixed and mesenchymal harmartomas should also be differentiated.

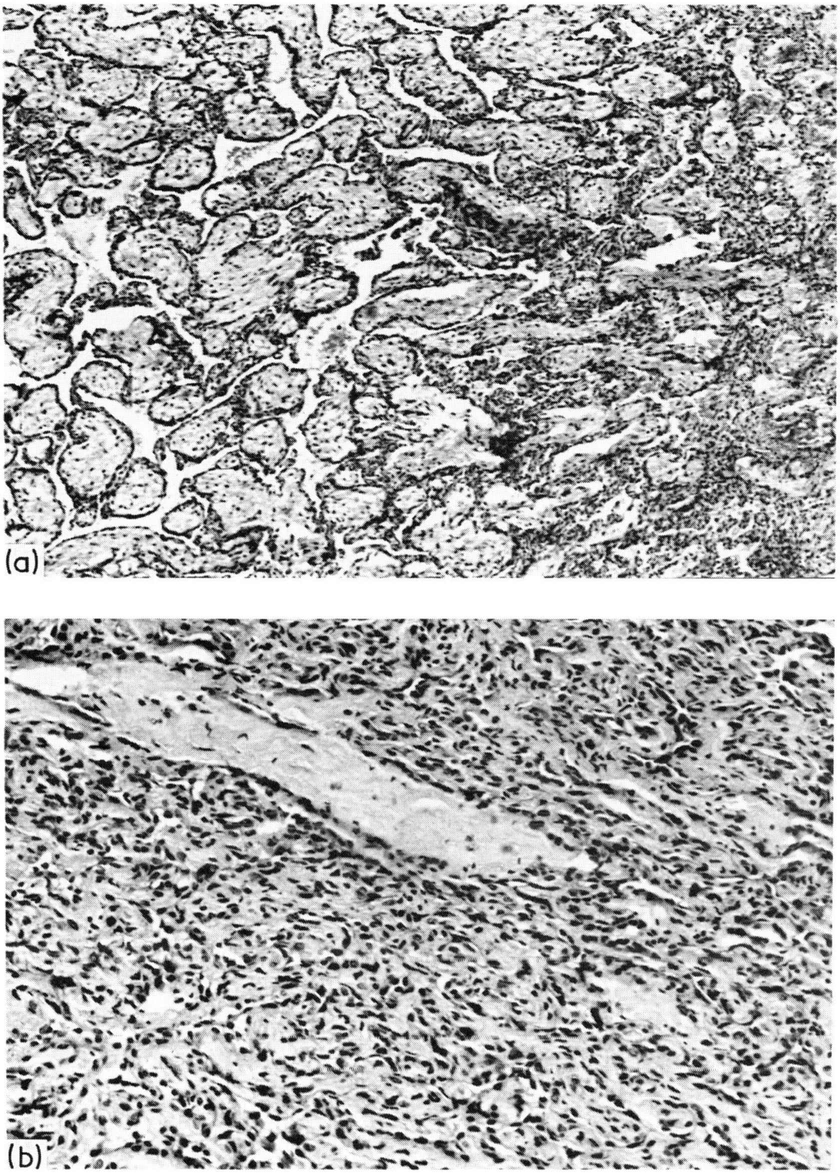

Fig. 13.41 (a) and (b) haemangioendothelioma of infant liver. Two fields show the variable morphology of this tumour which replaced most of the liver. ((a) and (b) HE, × 225.)

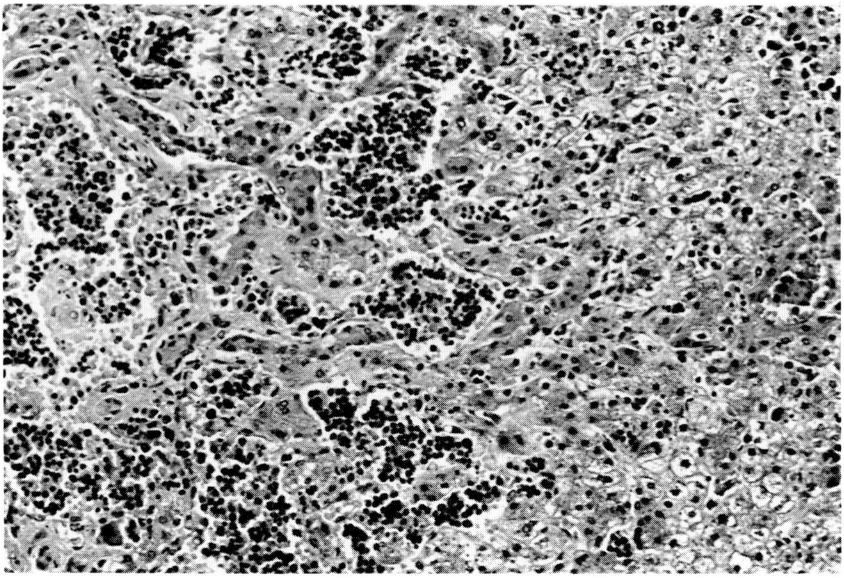

Fig. 13.42 Haemangioendothelioma of infant liver. Vascular spaces contain deeply staining foci of haemopoiesis (HE, × 180).

13.4.3 *Angiosarcoma (haemangioblastoma; haemangioendothelial sarcoma; Kupffer cell sarcoma; endothelioblastoma; reticuloendothelioma)*

Unlike the last tumour these are seen only in adults and are very uncommon. The clinical features are similar to those of hepatocellular carcinoma with a distinct tendency to serious and fatal internal haemorrhage. As in large benign angiomas there may be a blood coagulation defect and microangiopathic haemolytic anaemia.

Some cases follow the administration of thorotrast or arsenical drugs over many years. More recently an association has become apparent with industrial exposure to vinyl chloride. Here again many years elapse between the initial exposure and the development of tumour. Persons living in close proximity to vinyl chloride polymerization and fabrication plants can be at risk. When angiosarcoma is suspected from a history of such exposure it is wise to avoid aspiration needle biopsy because of risk of haemorrhage, and to seek confirmation of the diagnosis from angiography and wedge biopsy obtained at laparotomy.

The liver is usually extensively involved with multiple haemorrhagic nodules and blood-filled cystic spaces of variable size. It is often impossible to assess coexisting liver disease; cirrhosis is uncommon but we have seen one case in an alcoholic patient. Microscopically the tumour cells are spindle shaped with hyperchromatic and pleomorphic nuclei (Fig. 13.43).

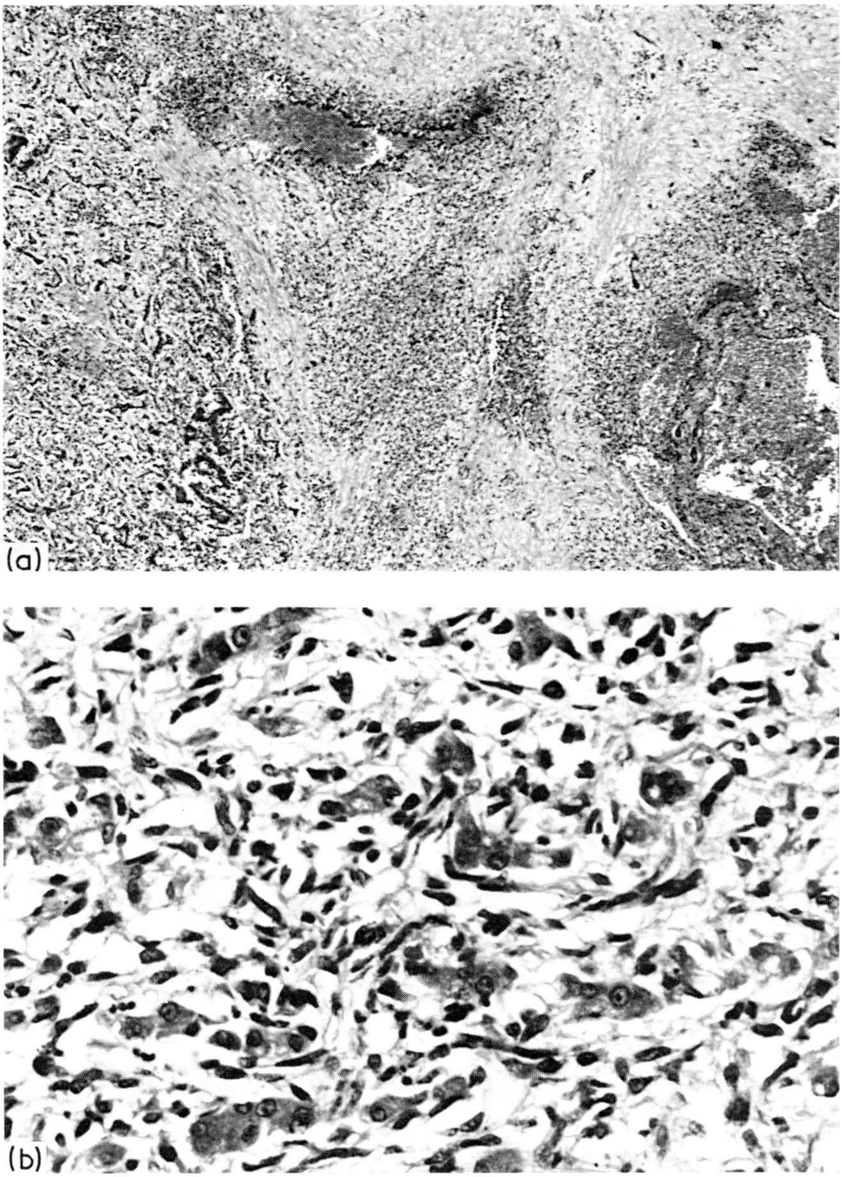

Fig. 13.43 Angiosarcoma of liver: (a) cavernous angiomatous areas are present (top and right) and there is a wide band of loose stroma; (b) a higher magnification of part of the tumour seen in (a) (left), showing single cords of atrophic hepatocytes separated by spindle-shaped tumour cells. ((a) HE, × 45; (b) HE, × 450.)

They are seen in small clusters within sinusoids and form a lining for these channels, at first without any destruction of hepatic plates. Later the liver cells disintegrate and are replaced by collagen bundles while the tumour-lined sinusoids dilate and fill with blood. Small groups of liver cells surrounded by tumour may persist and appear to be suspended within cavernous spaces (Fig. 13.44), but we have seen this phenomenon occasionally in hepatocellular carcinoma (Fig. 13.45). As well as a sinusoidal and cavernous arrangement of tumour cells there may also be solid masses which could be difficult to interpret in small biopsies (Fig. 13.46); vascular channels are indistinct and reticulin stains are not very helpful in outlining these. Occasionally there are foci of haematopoiesis.

Like hepatocellular carcinoma these tumours have a tendency to invade veins. Similar lesions found in spleen, bone marrow and other sites could be metastases or multiple foci of primary tumour formation. We have observed in three patients with primary angiosarcoma of breast, tumour deposits in liver which could not have been differentiated from a primary lesion on histological appearances alone.

Screening of workers exposed to vinyl chloride for many years occasionally uncovers cases of angiosarcoma which are still clinically silent. In addition, liver biopsy may show, even in the absence of established tumour, focal dilatation of sinusoids with perisinusoidal

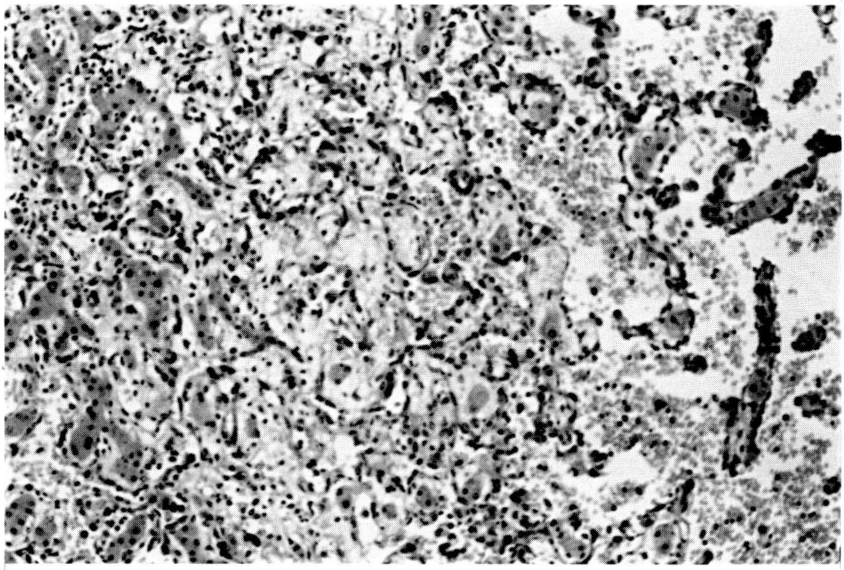

Fig. 13.44 Angiosarcoma of liver showing tumour cells lining sinusoids and surrounding small groups of surviving hepatocytes in the cavernous space (right) (HE, × 135).

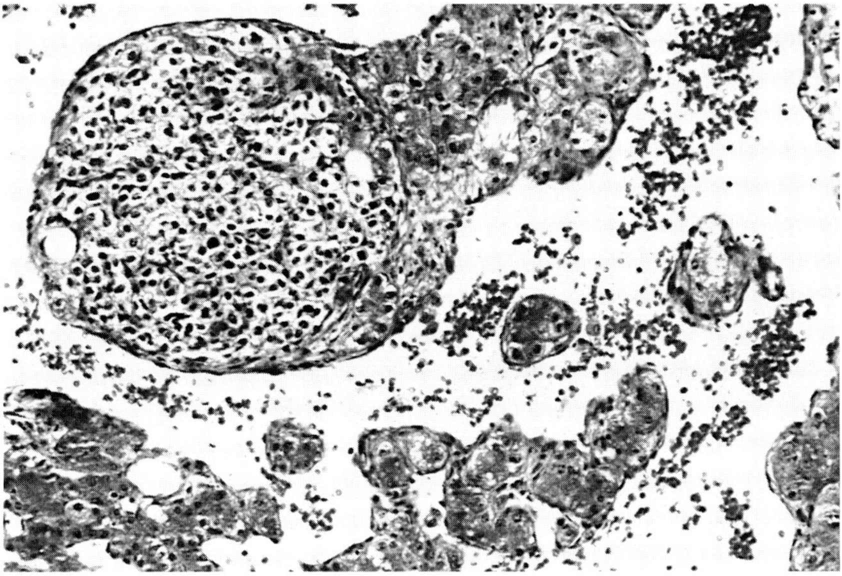

Fig. 13.45 Hepatocellular carcinoma. An area of gross sinusoidal dilatation with small groups of cancer cells surrounded by endothelium and lying in cavernous vascular spaces (HE, × 180).

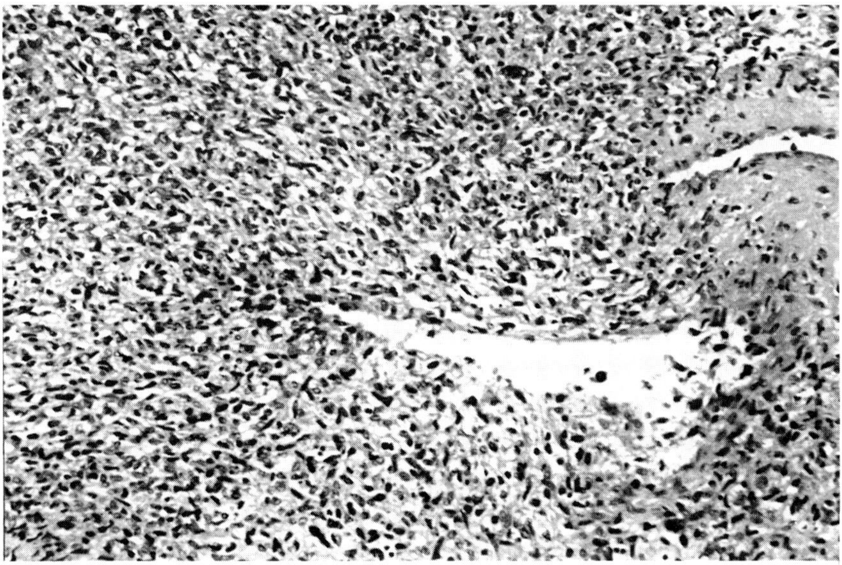

Fig. 13.46 Angiosarcoma of the liver. A solid area consisting of malignant endothelial cells (HE, × 180).

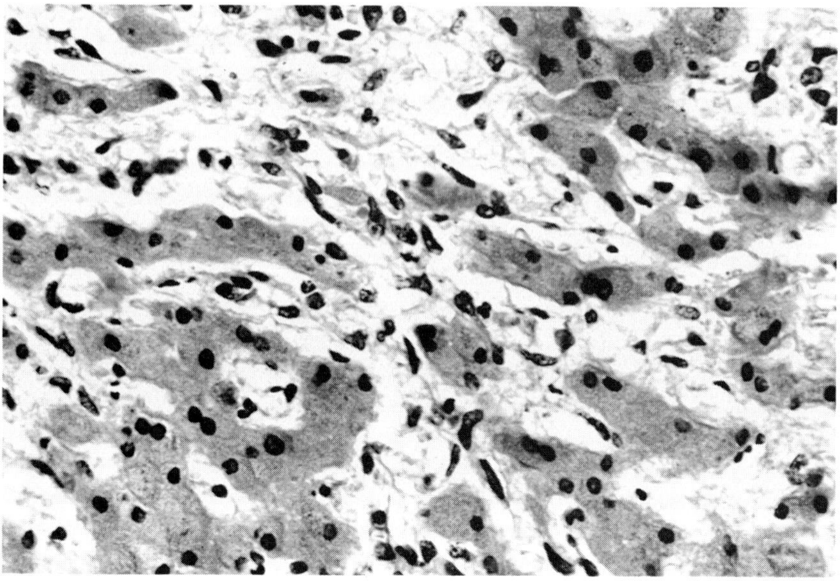

Fig. 13.47 Several dysplastic sinusoidal lining cells which are enlarged and spindle shaped (HE, × 450).

fibrosis and prominence of lining cells (Fig. 13.47) which may be a premalignant change (Popper *et al.*, 1978). This fibrosis is probably a manifestation of hyperactivity of perisinusoidal cells. Adjacent hepatocytes are also hyperplastic. In addition, fibrosis is prominent under the capsule and in portal and periportal regions from which may extend fibrous septa into the liver parenchyma bridging adjacent portal areas and central veins. There may be associated splenomegaly and portal hypertension. Identical fibrotic lesions are seen after prolonged exposure to thorotrast and arsenical drugs. There is also some resemblance to idiopathic non-cirrhotic portal fibrosis (p. 229), but the portal venous thickening and obliteration of that condition are not seen. Prognosis in angiosarcoma is bad with survival rarely more than a few months from the time of establishing the diagnosis.

13.4.4 *Epithelioid haemangioendothelioma* (Ishak *et al.*, 1984)

The liver is a possible site of origin of this unusual malignant vascular tumour which forms multiple pale nodules of firm consistency. Unlike angiosarcoma there are no vascular spaces evident on naked-eye examination, and the prognosis is better because, although there is a tendency to invade sinusoids and portal and hepatic veins, most cases grow slowly and undergo fibrosis sometimes with calcification in the centre of nodules. Microscopically two types of tumour cells are described:

(a) Dendritic cells (Fig. 13.48(a))

These are elongated or irregular in shape with a vacuolated cytoplasm which may contain erythrocytes but no mucin. Fusion leads to the formation of small vascular channels lined by flattened dendritic cells on a basement membrane. There is extracellular accumulation of a mucinous gel which later becomes fibrotic and sometimes calcified; veins may be invaded and obstructed by tumour, and the production here of small neoplastic vascular channels could be mistaken for recanalization of a venous thrombus.

(b) Epithelioid cells (Fig. 13.48(b))

These are rounded and have a non-vacuolated eosinophilic cytoplasm and hyperchromatin nucleus. They may be seen singly or in groups in liver sinusoids or veins.

Small vascular channels formed by the tumour, especially if they contain few or no erythrocytes, may be mistaken for proliferating bile ductules or even a sclerotic cholangiocarcinoma or secondary adenocarcinoma. The diagnosis of epithelioid haemangioendothelioma should be considered when there is any suggestion of vasoformative features and when there is venous invasion and extensive fibrosis. In some cases the tumour cells produce factor VIII-related antigen and immunohistochemical demonstration of this material is diagnostic of its vascular nature (Fig. 13.48(c)).

13.4.5 Lymphangioma

This tumour is rarely seen in liver (Fig. 13.49).

13.5 Other mesenchymal tumours of liver

13.5.1 Fibroma, lipoma, leiomyoma and mesothelioma

These are all reported but are exceedingly rare. As in the adrenals, some lipomas may contain myeloid tissue. True lipoma (Fig. 13.50) differs from small encapsulated masses of mature and degenerate adipose tissue lying on the surface of the liver. It has been suggested that these 'pseudolipomas' represent appendices epiploicae displaced during a

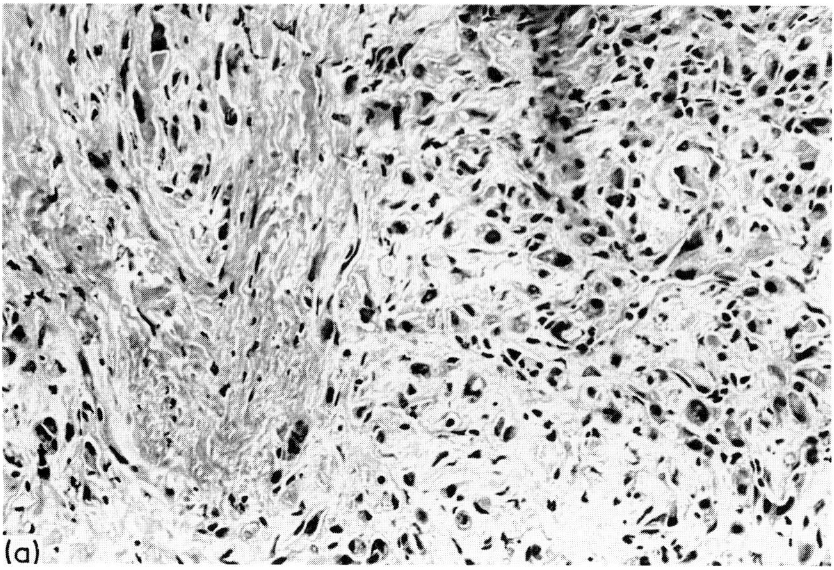

Fig (a)

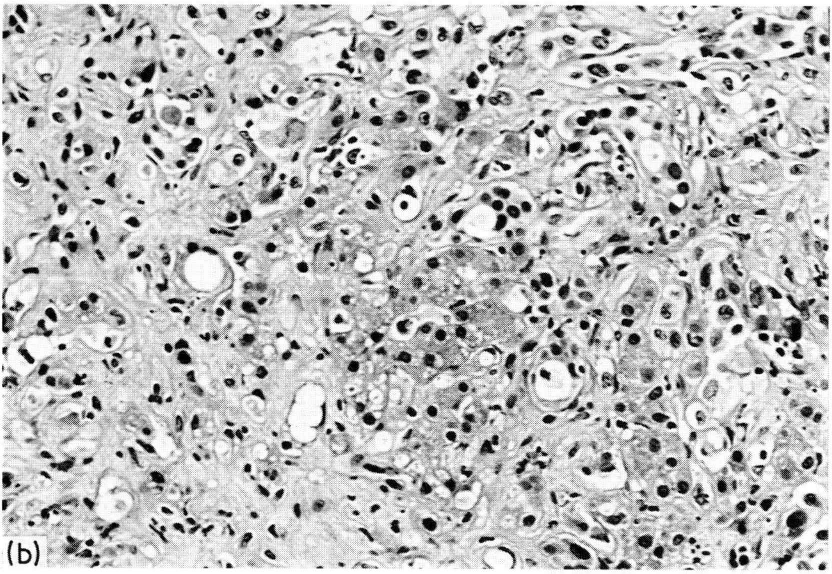

Fig. 13.48(b)

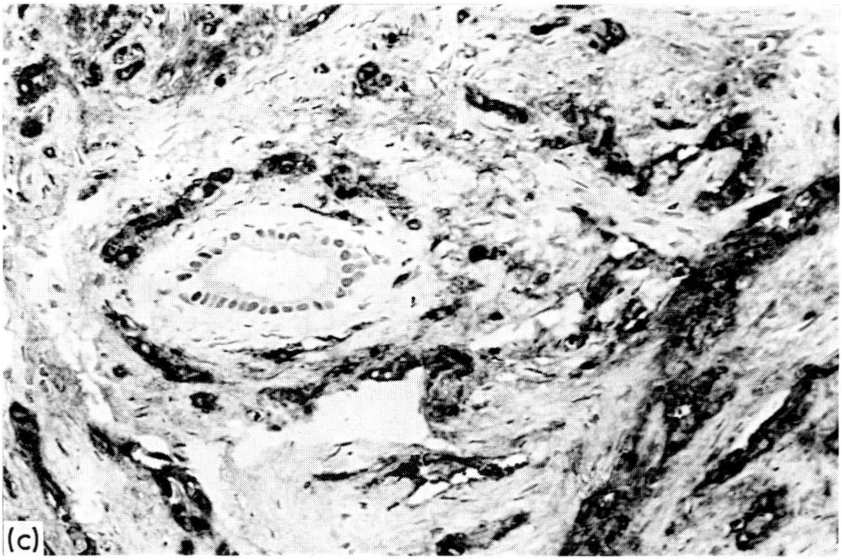

Fig. 13.48 Epithelioid haemangioendothelioma. (a) The hepatic venule (left) is occluded by tumour including several prominent dendritic cells and there is elastosis of this vessel wall. Tumour also invades adjacent liver sinusoids. (b) There are several vacuolated tumour cells (centre and left) and solid cords of epithelioid cells in sinusoids (right). There is some fibrosis (bottom left). (c) Factor VIII-related antigen is present in tumour cells in this expanded portal tract but not in bile duct epithelium. ((a) and (b) HE, × 247; (c) immunoperoxidase, × 247).

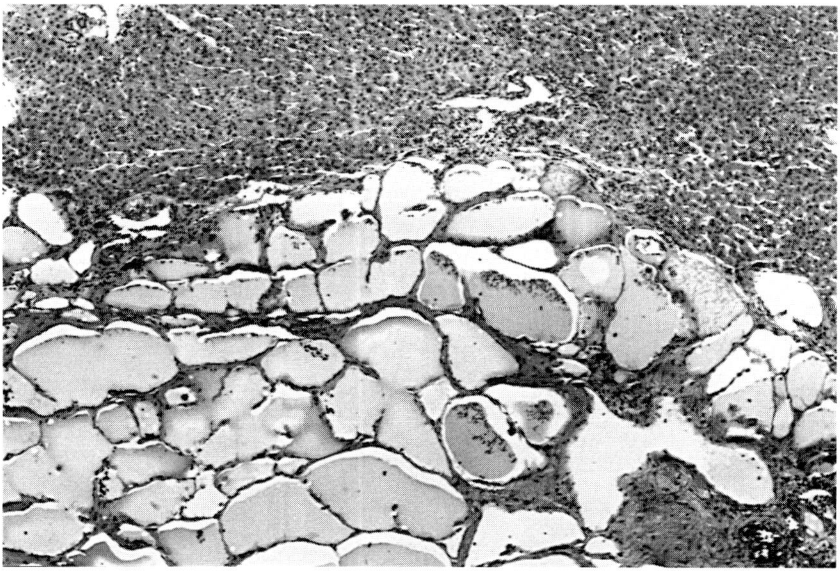

Fig. 13.49 Lymphangioma of liver (HE, × 66).

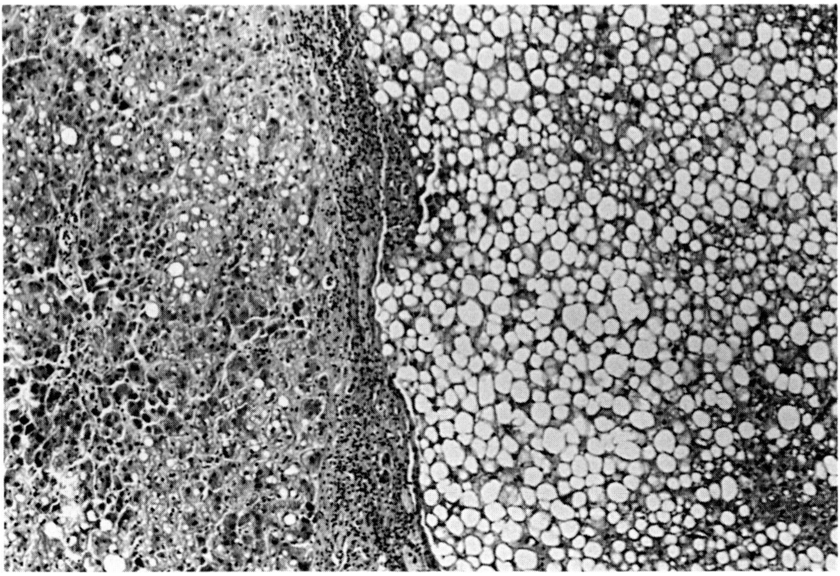

Fig. 13.50 Lipoma of liver (HE, × 90).

previous surgical operation. Severe fatty change may be confined to a few nodules in a cirrhotic liver or hepatocellular carcinoma and these may resemble lipoma. Some Zahn infarcts may also apper as focal areas of fatty change. Occasionally mesenchymal tumours are large enough to cause abdominal pain and swelling or be a source of serious haemorrhage.

13.5.2 Fibrosarcoma, leiomyosarcoma and malignant Schwannoma

It is difficult to exclude the possibility that many of these are secondary tumours in the liver. Some leiomyosarcomas appear to rise from the ligamentum teres or the wall of a large hepatic vein. Hypoglycaemia caused by the inappropriate production of insulin-like material is a possible manifestation of these tumours. The histological appearances are identical to those of similar tumours arising in other tissues and require no special description.

13.6 Adrenal rest tumour

Portions of adrenal tissue and even the whole gland may appear as a congenital anomaly within the liver, usually close to the capsule and adjacent to the normal site of the right gland. They can give rise to adenomas and carcinomas (Fig. 13.51), which are difficult to distinguish

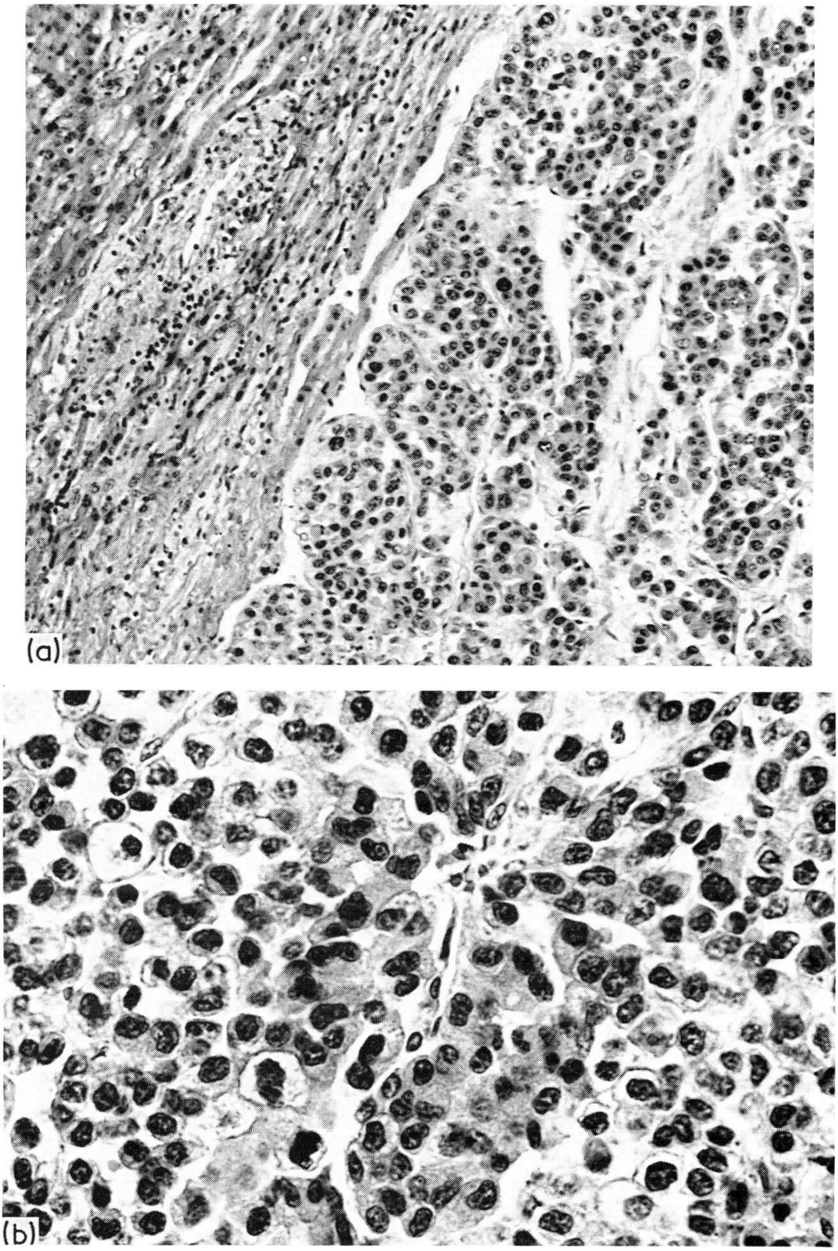

Fig. 13.51 Carcinoma arising in an intrahepatic adrenal rest. (a) Many tumour cells are arranged in a characteristic alveolar pattern and thin-walled vessels are numerous. (b) A more solid area of the same tumour. Unlike typical hepatocellular carcinoma there are no linear sinusoids, nucleoli are less conspicuous and some cells have faintly staining vacuolated cytoplasm. ((a) HE, × 180; (b) HE, × 450.)

from similar tumours arising in adjacent normally situated adrenal or in kidney. Unlike hepatomas they lack the characteristic trabecular arrangement of liver cell plates and never reproduce canalicular structures or secrete bile. On the other hand, their functional activity when present may indicate the correct diagnosis.

13.7 Other hepatic mixed tumours

13.7.1 *Hepatoblastoma (hepatic mixed tumour; hepatic teratoid tumour; hepatic embryoma)*

This is an embryonic tumour of liver found usually in infants and young children, although a few slow-growing examples have been noted in adults. Conversely, hepatocellular carcinoma may be found occasionally in children. Many hepatoblastomas grow quite rapidly, with malignant invasion of surrounding tissue sufficient to cause clinical evidence of hepatic failure in addition to abdominal swelling. Some are less aggressive and the successful resection of these is possible sometimes. A few are associated with hyperglycaemia, osteoporosis, lipid histiocytosis and sexual precocity which is, presumably, the result of gonadotrophin production by tumour cells. Various congenital anomalies in other organs may coexist also.

These are typically large single tumours which may be cystic and which may have foci of bone formation which can be detected macroscopically. Two main histological variants are described:

(a) Epithelial type

This consists mainly of fetal and embryonic liver cells and both may be seen in the same specimen (Fig. 13.52). The foetal component consists of irregular plates, two cells thick, separated by sinusoids but there is no normal lobular pattern. The tumour cells are smaller than normal hepatocytes but have a higher nuclear : cytoplasmic ratio. The nuclei may be somewhat pleomorphic but mitotic activity is unusual. The cytoplasm is often clear because of its content of fat and glycogen but it may stain deeply with eosin. Granules of bile pigment may be present. Embryonic-type cells are less well differentiated, more pleomorphic or elongated, and show more mitotic activity. They form solid sheets, ductular structures or rosettes. Sinusoids are indistinct or there may be lakes of blood lined by tumour cells. Mature fibrous tissue surrounds these tumours and forms septa containing prominent vessels and immature bile ducts. This probably represents tumour capsule and stroma rather than a mesenchymal neoplastic element. However, in addition prominent foci

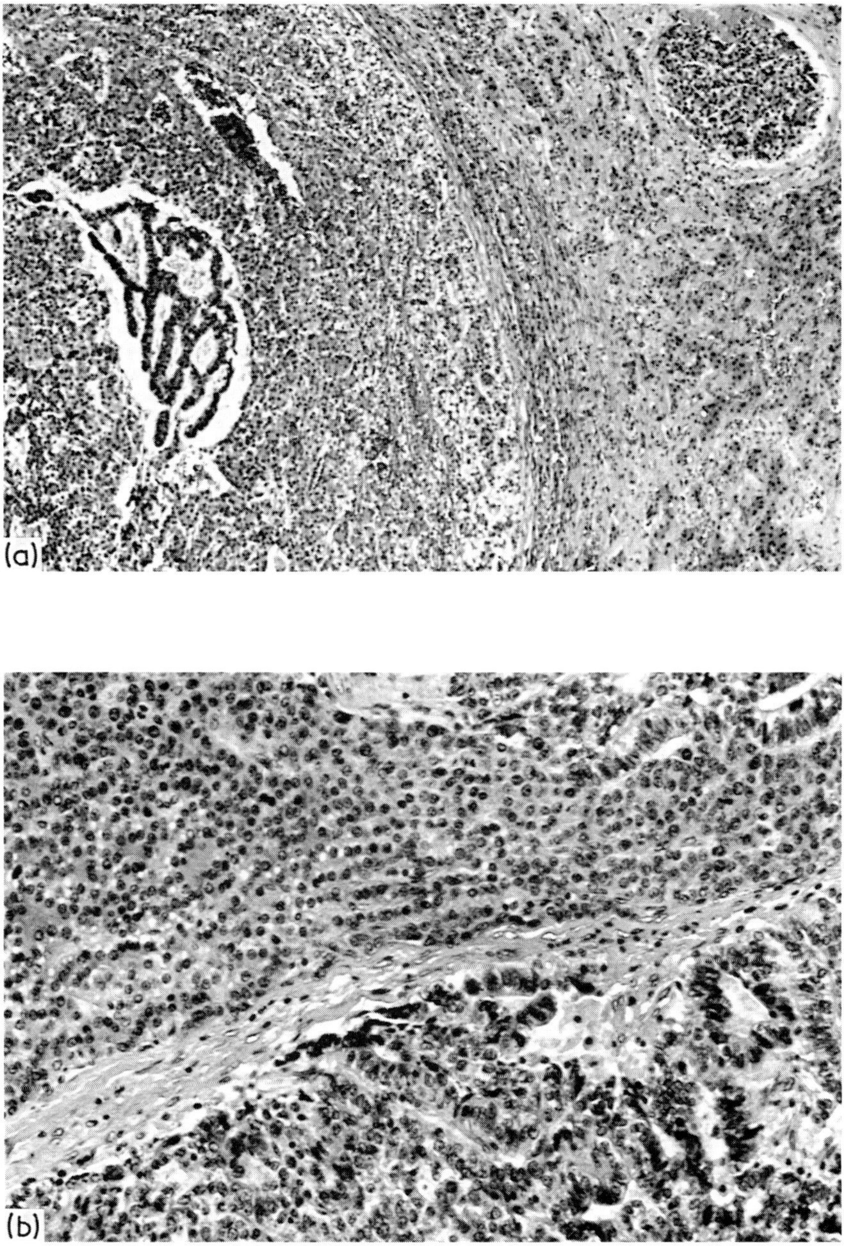

Fig. 13.52 Hepatoblastoma of infant liver, epithelial type. (a) Tumour nodules consist mainly of fetal hepatocytes. The larger nodule has cystic structures lined by embryonic hepatocytes. (b) Part of the same tumour showing embryonic cells (lower) and foetal cells (upper). ((a) HE, × 90; (b) HE, × 225.)

of erythropoiesis are present in parts of fetal cell tumour in many cases.

(b) Combined epithelial and mesenchymal type

The epithelial component is as described above. In addition, there may be foci of squamous cell carcinoma with keratin formation. The mesenchymal component often consists of closely packed spindle-shaped cells with little intercellular material. In some areas these cells may be sufficiently differentiated to be recognizable as active fibroblasts surrounded by reticulin or collagen fibres. Some parts may show myxomatous change (Fig. 13.53). Islands of osteoid tissue may be present and are sometimes calcified (Fig. 13.54). Cartilage is very uncommon unless in the rare examples found in older patients. Unlike nephroblastoma, skeletal muscle is an unusual component.

The degree of differentiation of hepatoblastoma varies considerably from case to case and in different parts of the same tumour. Sometimes it is only the epithelial element which appears to behave in a malignant fashion. A few purely benign forms have been described, some containing epithelial-lined cysts.

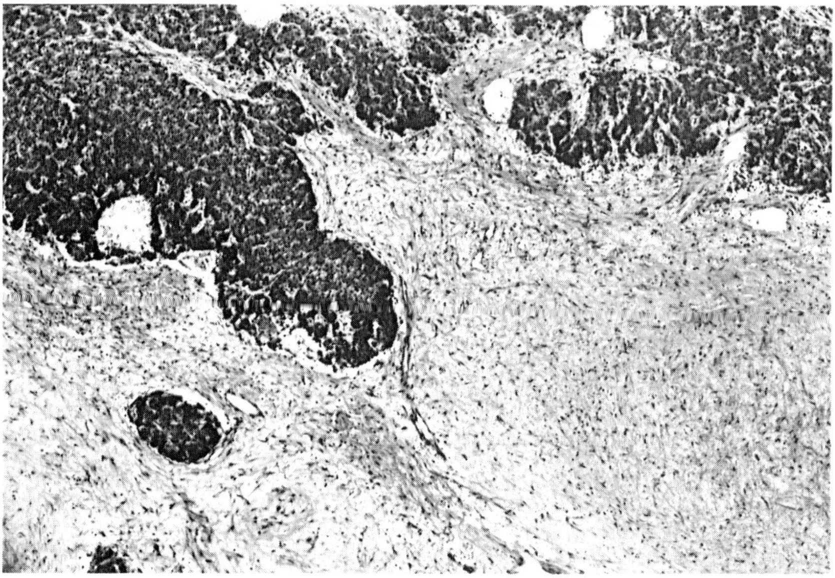

Fig. 13.53 Hepatoblastoma of infant liver, combined type. There is a large mesenchymal component consisting of pale myxomatous tissue. The epithelial component forms deeply staining nodules (HE, × 56).

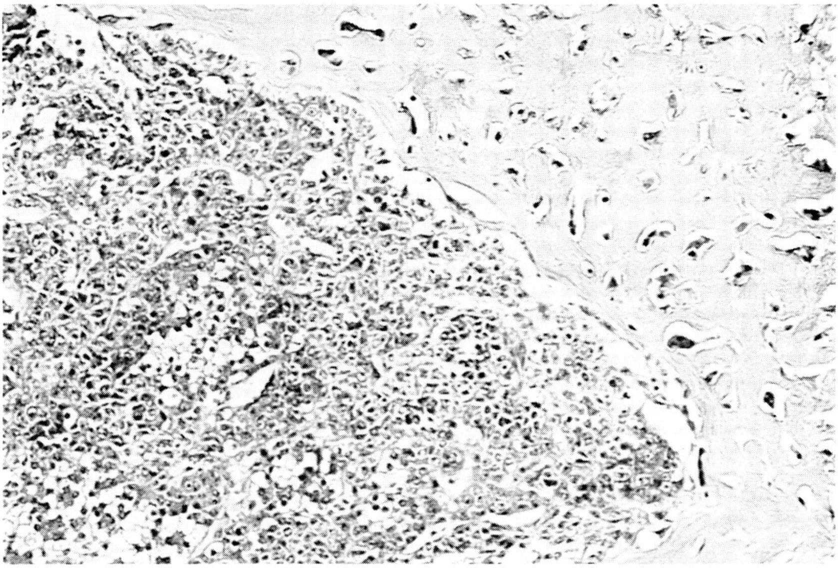

Fig. 13.54 Osteoid tissue in hepatoblastoma (HE, × 300).

13.7.2 Mesenchymoma (undifferentiated embryonal sarcoma) (Stocker and Ishak, 1978)

This rare primary liver tumour can also occur in children. Compared with hepatoblastoma it tends to affect an older age group (usually 5–10 years) and there is no convincing epithelial neoplastic component. Unlike mesenchymal hamartoma it is malignant and pulmonary metastases may appear. Histologically it consists of pleomorphic malignant cells with areas of myxomatous and cystic degeneration, and haemorrhagic necrosis. Some examples have features of malignant fibrous histiocytoma (Keating and Taylor, 1985). There may be small clusters of acinar structures, possibly remnants of bile ducts, and an epithelial lining to some cysts. Many tumours are quite undifferentiated, but it is possible that embryonic rhabdomyosarcoma, which can arise in the liver in children (Fig. 13.55), is a form of mesenchymoma. Alpha-1-antitrypsin inclusion bodies have been found in some cases, both extracellularly and within tumour cells.

Mesenchymoma can occur also in adult liver when it is usually better differentiated, e.g. parts may consist of osteogenic sarcoma or chondrosarcoma. Angiolipoma and angiomyolipoma can be regarded as benign mesenchymoma also arising in adult liver: it is suggested that some of these tumours could originate in perisinusoidal cells or myofibroblasts.

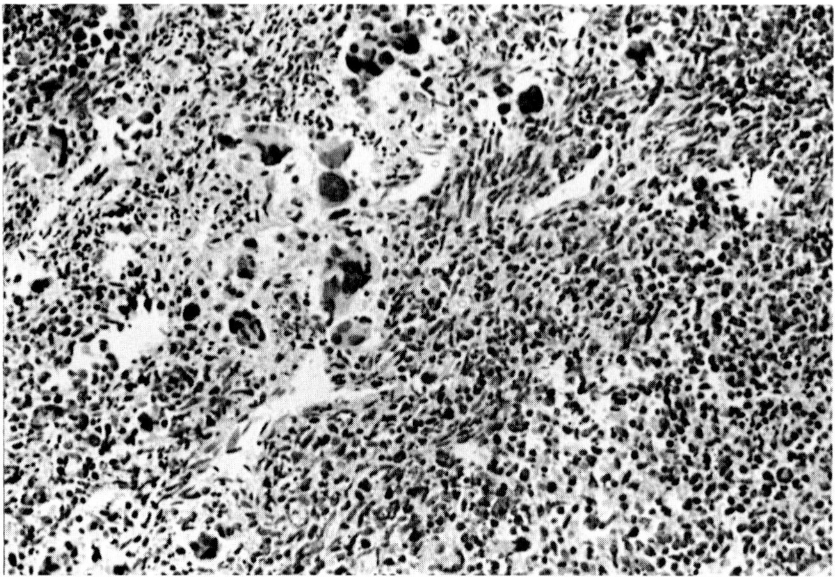

Fig. 13.55 Rhabdomyosarcoma. Part of a large tumour of fairly uniform appearance in a child's liver. No metastases were found at autopsy (HE, × 45).

13.7.3 Teratoma

A few cases are on record, usually in adults, of large cystic tumours of liver which have the structural characteristics of benign teratoma of ovary. A malignant form has been recognized in a young child.

13.7.4 Mixed hamartoma

This term is generally used to designate large well-defined nodular tumours occurring in infant liver; they consist of small foci of mature hepatocytes separated by wide strands of cellular connective tissue containing very large numbers of small bile ductules and thick-walled arteries and veins. Focal nodular hyperplasia is a similar condition occurring in older patients (Figs 13.2–4).

13.7.5 Mesenchymal hamartoma

These also occur in infant liver and consist of masses of greyish oedematous tissue, often with areas of necrosis or cystic degeneration which are not clearly demarcated from the surrounding tissue. They may grow quickly to form lesions of large size but they are not invasive. Microscopically they consist of loose areolar tissue in which can be recognized small islands of normal hepatocytes or portal triads. Angiomatous areas may be present.

13.8 Secondary tumours of liver

These are very common and need no special description. Interest in this subject has increased recently because partial hepatectomy offers a means of treatment for solitary nodules, the latter usually derived from primary carcinoma of the colon. In most cases, however, metastases are multiple and although blood borne, in contrast to primary liver cancer, their presence in portal veins is not usually recognized. Features other than cytological detail are sometimes helpful in assessing the source of these tumours: thus there may be calcification in the stroma of cancers derived from colon and haemorrhage around metastasis derived from pancreas (Fig. 13.56). Some continue to spread by infiltrating lymphatic channels in the portal triads and some may invade the whole of the liver like the diffuse form of hepatocellular carcinoma. In our experience secondary cancer in a cirrhotic liver is uncommon (Fig. 13.57), although this is not generally accepted. Other difficulties in differential diagnosis from primary liver tumours have already been referred to. Needle biopsies from suspected cases which at first sight appear negative should be scanned

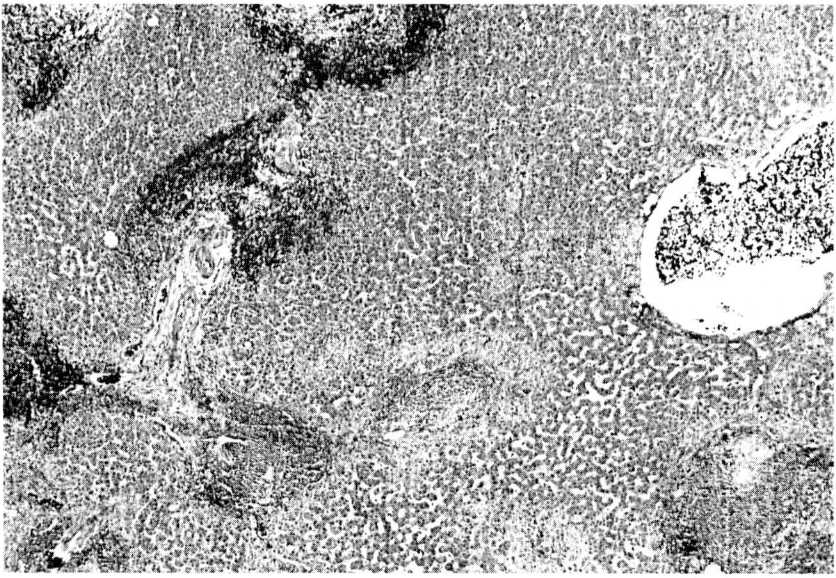

Fig. 13.56 Secondary carcinoma from primary pancreatic tumour in liver. There is haemorrhage around each tumour deposit (HE, × 45).

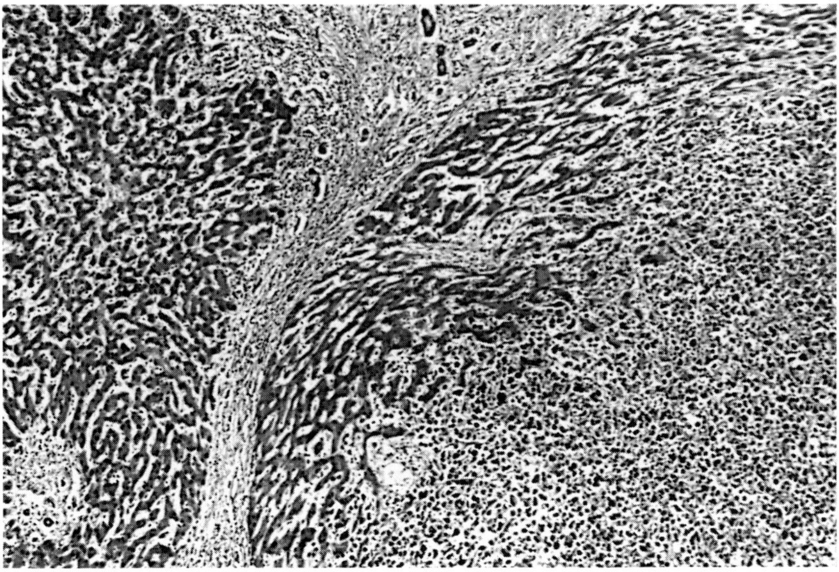

Fig. 13.57 Secondary tumour (bottom right) from bronchus in cirrhotic liver (HE, × 225).

with some care for tiny metastases lying within sinusoids or portal venous radicles (Fig. 13.58).

The liver can be involved also by secondary sarcoma and various lymphoreticular diseases such as leukaemia, lymphoblastic tumours, Hodgkin's disease, Letterer–Siwe disease, etc. Some of these are illustrated (Figs. 13.59–63). (Very rarely malignant lymphoma has been described as a primary liver tumour because of the absence at autopsy of involvement of other organs.) Infiltration with malignant tissue is usually most marked in the portal triads which are sometimes obscured by these cells, and there may be obstruction of the small biliary vessels. This is probably a commoner cause of obstructive jaundice than compression of ducts in the porta hepatis by enlarged lymph nodes. Nevertheless, there are many cases, particularly of Hodgkin's disease, which could be explained by some humoral effect on bile metabolism initiated by tumour cells. Hodgkin's disease in liver is often of the nodular sclerosing type, but Reed–Sternberg cells can be scanty. There may be an Indian file arrangement of neoplastic cells in sinusoids with little or no hepatocyte atrophy, especially in chronic myeloid leukaemia and myeloma. A similar distribution of mature lymphocytes is seen in tropical and non-tropical splenomegaly. Normal sinusoidal lining may be replaced by 'hairy-cell' leukaemia cells followed by focal dilatation of the affected vascular

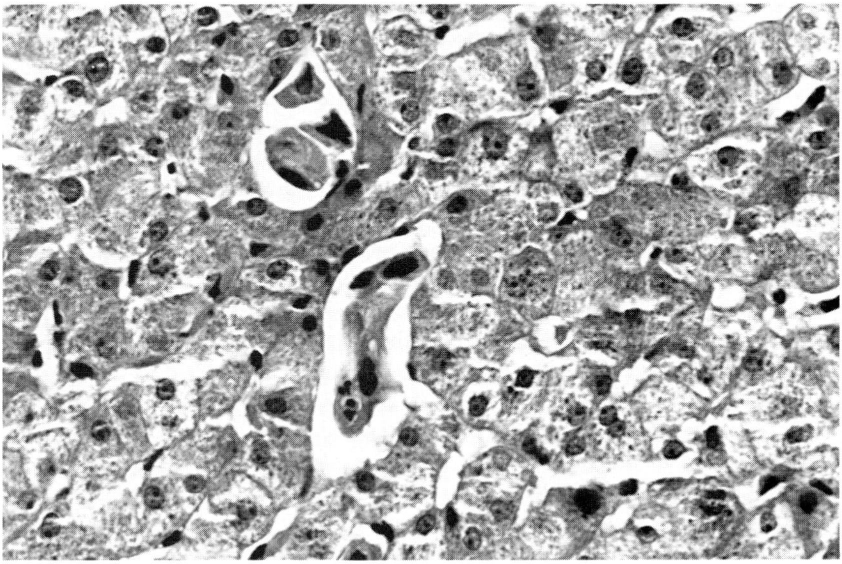

Fig. 13.58 Small deposits of secondary cancer in liver sinusoids (HE, × 450).

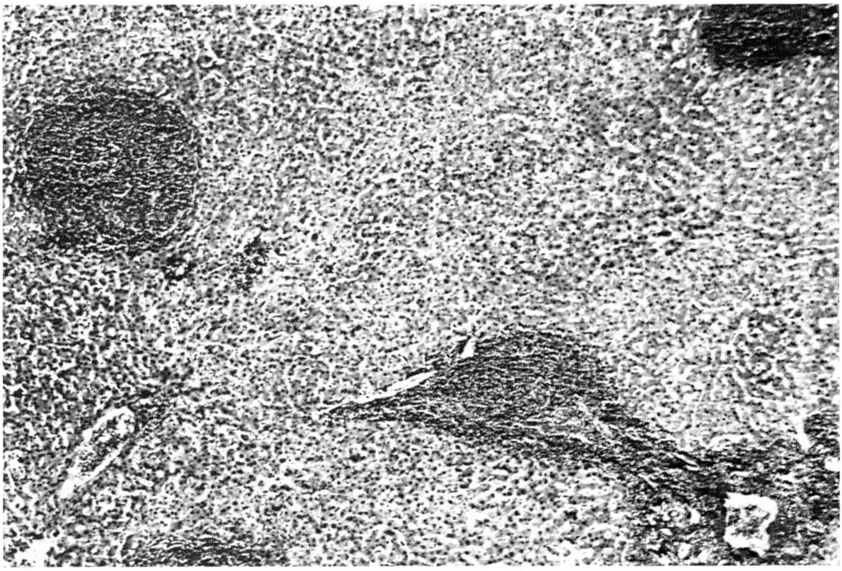

Fig. 13.59 Lymphocytic lymphoma involving liver (HE, × 56).

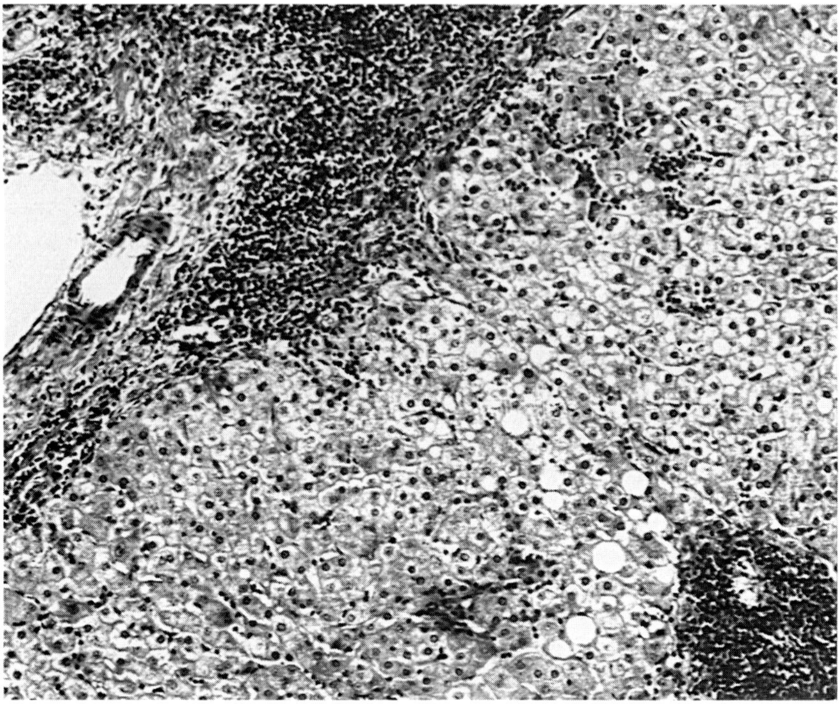

Fig. 13.60 Lymphatic leukaemia involving liver (HE, × 180).

channels, especially in periportal zones (Nanba *et al.*, 1977). While Kupffer cell hyperplasia is seen in many systemic infections, it may be sufficiently prominent to raise suspicion of histiocytic reticulosis (Strayer *et al.*, 1980). Aggregates of non-malignant lymphocytes are found occasionally in the liver in cases of Hodgkin's disease and non-Hodgkin's lymphoma occurring at other sites. The pathogenesis of these lesions is not known but they should not be overinterpreted as neoplastic (Leslie and Colby, 1984).

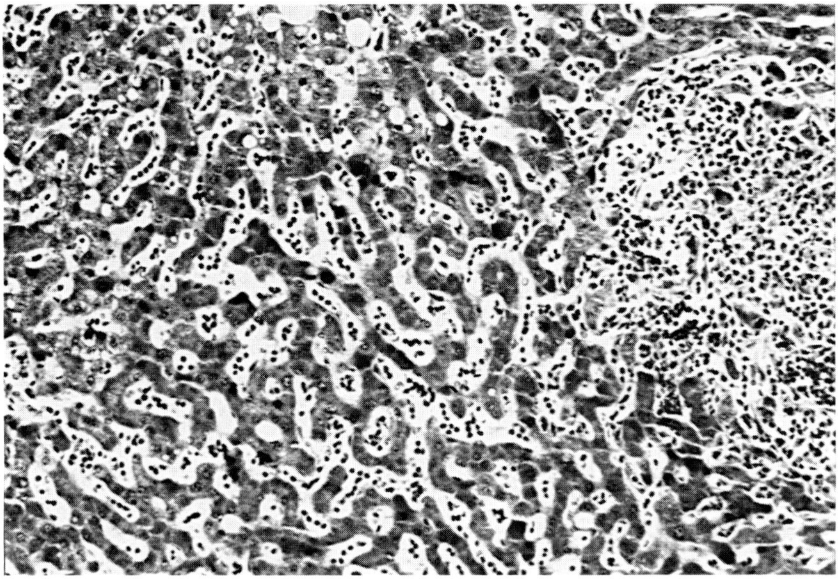

Fig. 13.61 Myeloid leukaemia involving liver. Note the 'Indian file' arrangement of leukaemic cells in sinusoids (HE, × 180).

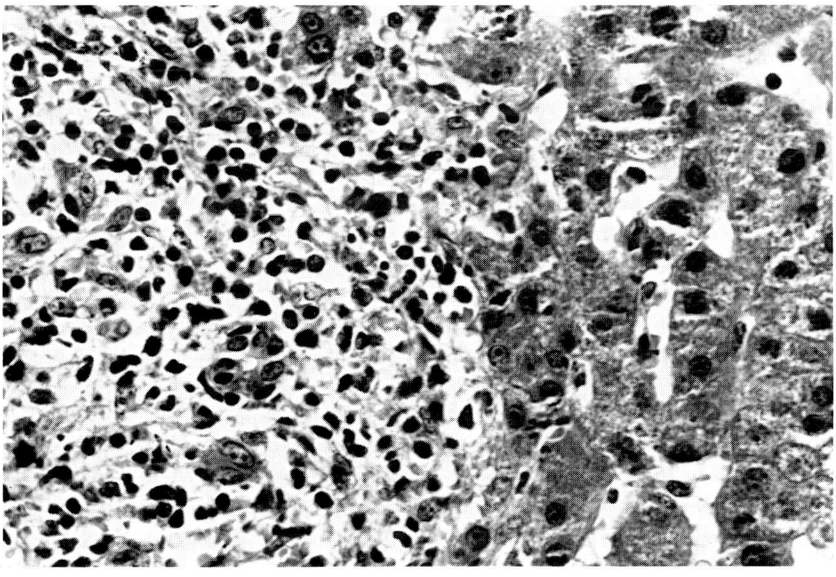

Fig. 13.62 Hodgkin's disease involving liver (HE, × 495).

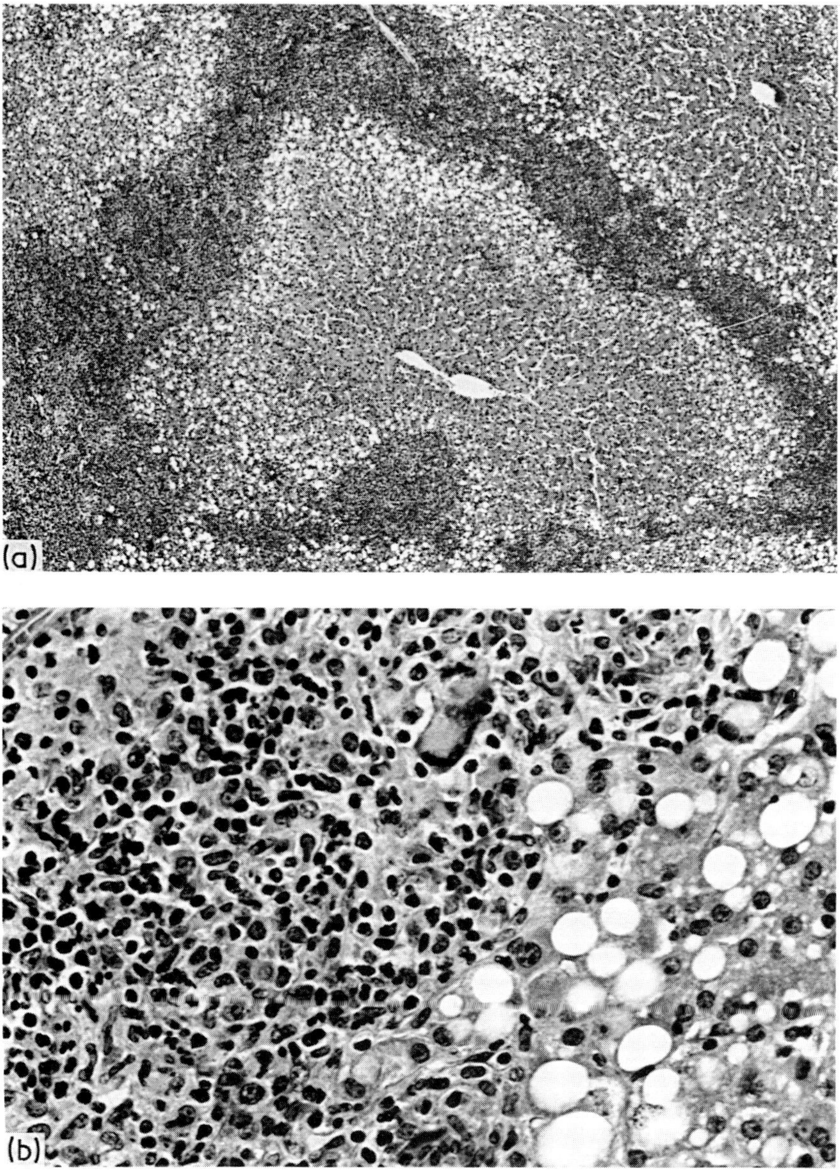

Fig. 13.63 Liver in Letterer–Siwe disease. ((a) HE, × 45; (b) HE, × 450.)

References

Anthony, P. P., Vogel, C. L. and Barker, L. F. (1973), Liver cell dysplasia: a premalignant condition. *J. Clin. Pathol.*, **26,** 217–23.

Berman, M. M., Libbey, N. P. and Foster, J. H. (1980), Hepatocellular carcinoma: polygonal cell type with fibrous stroma. *Cancer*, **46,** 1448–55.

Craig, J. R., Peters, R. L., Edmondson, H. A. *et al.* (1980), Fibrolamellar carcinoma of the liver. *Cancer*, **46,** 372–9.

Edmondson, H. A., Henderson, B. and Benton, B. (1976), Liver-cell adenomas associated with the use of oral contraceptives. *N. Engl. J. Med.*, **294,** 470–2.

Edmondson, H. A., Reynolds, T. B., Henderson, B. *et al.* (1977), Regression of liver cell adenomas associated with oral contraceptives. *Ann. Intern. Med.*, **86,** 180–5.

Farrell, G. C., Joshua, D. E., Uren, R. F. *et al.* (1975), Androgen-induced hepatoma. *Lancet*, **i,** 430–2.

Flejou, J-F., Barge, J., Menu, Y. *et al.* (1985), Liver adenomatosis: an entity distinct from liver adenoma. *Gastroenterology*, **89,** 1132–8.

Gold, J. H., Guzman, I. J. and Rosai, J. (1978), Benign tumours of the liver: pathologic examination of 45 cases. *Am. J. Clin. Pathol.*, **70,** 6–17.

Ishak, K. G., Sesterhenn, I. A., Goodman, Z. D. *et al.* (1984), Epithelioid hemangioendothelioma of the liver. *Hum. Pathol.*, **15,** 839–52.

Ishak, K. G., Willis, G. W., Cummins, S. D. *et al.* (1977), Biliary cystadenoma and cystadenocarcinoma: report of 14 cases and review of the literature. *Cancer*, **39,** 322–38.

Keating, S. and Taylor, G. P. (1985), Undifferentiated (embryonal) sarcoma of the liver. *Hum. Pathol.*, **16,** 693–9.

Klatskin, G. (1977), Hepatic tumours: possible relationship to use of oral contraceptives. *Gastroenterology*, **73,** 386–94.

Knowles, D. M., Casarella, W. J., Johnson, P. M. *et al.* (1978), The clinical, radiologic and pathologic characterization of benign hepatic neoplasms. *Medicine*, **57,** 223–7.

Leslie, K. O. and Colby, T. V. (1984), Hepatic parenchymal lymphoid aggregates in Hodgkin's disease. *Hum. Pathol.*, **15,** 808–9.

London, W. T. (1981), Primary hepatocellular carcinoma: etiology, pathogenesis and prevention. *Hum. Pathol.*, **12,** 1085–97.

Nanba, K., Soban, E. J., Bowling, M. C. *et al.* (1977), Splenic pseudosinuses and hepatic angiomatous lesions: distinctive features of hairy cell leukaemia. *Am. J. Clin. Pathol.*, **67,** 415–26.

Okuda, K., Musha, H., Nakajima, Y. *et al.* (1977), Clinicopathologic features of encapsulated hepatocellular carcinoma: a study of 26 cases. *Cancer*, **40,** 1240–5.

Popper, H., Thomas, L. B., Telles, N. C. *et al.* (1978), Development of hepatic angiosarcoma in man induced by vinyl chloride, thorotrast and arsenic. *Am. J. Pathol.*, **92,** 349–76.

Steiner, P. E. and Higginson, J. (1959), Cholangiolocellular carcinoma of the liver. *Cancer*, **12,** 753–9.

Stocker, J. T. and Ishak, K. G. (1978), Undifferentiated (embryonal) sarcoma of the liver. *Cancer*, **42,** 336–48.

Strayer, D. S., Reppun, T. S., Levin, M. *et al.* (1980), Primary lymphoma of the liver. *Gastroenterology*, **78,** 1571–6.

Sumithran, E. and MacSween, R. N. M. (1979), An appraisal of the relationship between primary hepatocellular carcinoma and hepatitis B virus. *Histopathology*, **3**, 447–58.

Turbitt, M. L., Patrick, R. S., Goudie, R. B. *et al.* (1977), Incidence in South-West Scotland of hepatitis B surface antigen in the liver of patients with hepatocellular carcinoma. *J. Clin. Pathol.*, **30**, 1124–8.

Wanless, I. R., Mawdsley, C. and Adams, R. (1985), On the pathogenesis of focal nodular hyperplasia of liver. *Hepatology*, **5**, 1194–200.

Weinbren, K. and Mutum, S. S. (1983), Pathological aspects of cholangiocarcinoma. *J. Pathol.*, **139**, 217–38.

Zaman, S. N., Melia, W. M., Johnson, R. D. *et al.* (1985), Risk factors in development of hepatocellular carcinoma in cirrhosis: prospective study of 613 patients. *Lancet*, **i**, 1357–60.

Further reading

Okuda, K. and MacKay, I. (eds) (1982) *Hepatocellular Carcinoma* (*UICC Technical Report Series*, Vol. 74), International Union Against Cancer, Geneva.

Weinberg, A. G. and Finegold, M. J. (1983), Primary hepatic tumors of childhood. *Hum. Pathol.*, **14**, 512–37.

14 Clinicopathological correlations of liver biopsy

Liver biopsy is usually undertaken to further the investigation of one of the following clinical observations and assessments occurring singly or in combination: (1) liver parenchymal dysfunction; (2) cholestatic jaundice; (3) portal hypertension; (4) hepatomegaly. In addition, it may form part of the examination of certain pyrexial and metabolic disorders, not necessarily with clinical evidence of liver involvement, and of structural liver abnormalities discovered by chance during laparotomy. Many microscopic changes can be found within each one of these categories, but their number is smaller than the number of possible causes because the ways in which the organ may respond to injury are limited. In many cases, therefore, the diagnosis cannot be made without consideration of the results of clinical examination and of biochemical and biophysical investigations in addition to the liver biopsy appearances.

14.1 Liver parenchymal dysfunction

Almost any form of liver disease may be associated with clinical evidence of hepatic functional disorder ranging in severity from mild alterations detected only by standard biochemical tests up to hepatic coma. A correlation between clinical and pathological severity would be expected, but sometimes this is rather poor. The histological features of hepatocyte degeneration or necrosis and of hepatitis are usually seen singly or together. Most cases will be due to viral hepatitis, alcoholic liver damage, drug-induced liver injury, CAH or cirrhosis. It is not possible to present a concise differential diagnosis because of the wide range of causes and histological appearances, but a few of the more important features are summarized below.

14.1.1 Hepatitis

The distribution and characteristics of the inflammatory cell infiltration should be noted. In viral hepatitis it is often rather diffuse, imparting a distinct impression of increased cellularity on low-power examination. The vast majority of cells are lymphocytes, in contrast to alcoholic hepatitis in which some neutrophil polymorphs are always seen and may be prominent. Conspicuous numbers of eosinophils suggest drug-induced liver injury but are seen in only a minority of these cases; many others present a non-specific reactive hepatitis pattern, and a distinct resemblance to viral hepatitis is sometimes found. The portal and periportal infiltration of chronic inflammatory cells including plasma cells is characteristic of CAH. Granulomatous hepatitis may arise from a number of infections and liver toxins.

14.1.2 *Hepatocyte degeneration*

Of the various types described, liver dysfunction is most often associated with hydropic and hyaline degeneration, especially when these changes are extensive. Acidophilic degeneration is prenecrotic but seems to be less important, probably because relatively few hepatocytes show this change in any one specimen. Hyaline degeneration is characteristic of alcoholic hepatitis, while the others are not related to any particular form of liver damage but are usually prominent in viral infections. Fine droplet fatty change is a possible manifestation of severe toxic liver injury.

14.1.3 *Necrosis of liver*

All varieties as described in Chapter 2 are associated with liver functional abnormality. Distinct zonal necrosis without hepatitis is indicative of acute toxic liver injury; with coexistent and prominent hepatitis a viral or alcoholic aetiology is likely, and in these cases it may spread to form bridging or submassive necrotic lesions. Centrilobular necrosis may be seen also in severe hepatic congestion.

Focal loss of single hepatocytes or small groups of these cells occurs from many causes, while piecemeal necrosis is the characteristic feature of CAH. A similar loss of periportal cells may be seen in association with biliary infection and would help to account for the biochemical evidence of parenchymal liver damage in some cases of obstructive jaundice. Irregular areas of submassive necrosis are generally manifestations of vascular occlusion or viral hepatitis.

14.1.4 *Cirrhosis*

Liver dysfunction is frequently present and may be marked where there is continuing CAH or where there is ischaemic damage within regenerating nodules. The severity of functional disturbance is not related to the histological or aetiological type of cirrhosis.

14.2 Cholestatic jaundice

In most cases there will be intrahepatic cholestasis, usually centrilobular but sometimes periportal and often accompanied by some degree of hepatocyte degeneration and low-grade inflammation. More severe portal inflammation with oedema, dilatation and proliferation of ducts, bile infarcts and bile lakes indicates obstruction to an extrahepatic duct which will require surgical intervention. The distinction between these extrahepatic cases of obstructive jaundice and intrahepatic cases associated with primary parenchymal disorders is fully discussed in the Chapter on biliary disease.

14.3 Portal hypertension

Liver biopsy is essential for the investigation of these cases and the following classification is based on the possible histological findings.

14.3.1 Cirrhosis

In the majority of cases liver biopsy will reveal or confirm established hepatic cirrhosis. The severity of portal hypertension is not related to the histological pattern or aetiology of the cirrhotic condition.

14.3.2 Non-cirrhotic liver disease

Developmental anomalies such as congenital hepatic fibrosis, extensive cystic disease and partial nodular transformation (Chapter 12) can cause severe portal hypertension. The last of these may go undetected in liver biopsy as it is usually confined to the porta hepatis; it should not be mistaken for focal nodular hyperplasia (focal cirrhosis) which is not a cause of portal hypertension.

Portal fibrosis with maintenance of normal liver architecture is seen from time to time. Various terms have been used to describe this condition, such as non-cirrhotic portal fibrosis hepatoportal fibrosis, chronic septal hepatitis, Banti's syndrome, etc. The portal veins in these lesions may have thick fibrous walls or be obliterated by scar tissue. The condition, which is sometimes related to the intake of certain toxic substances, is described in Chapter 10. Also described in that Chapter are other vascular conditions which cause portal hypertension – Budd–Chiari syndrome and veno-occlusive disease where there is blockage of hepatic venous outflow.

Central hyaline sclerosis is an important cause of severe portal hypertension in certain cases of alcoholic hepatitis (p. 139); in this condition the centrilobular venous radicles of the liver become obliterated by scar tissue.

Portal hypertension is freqent in hepatocellular carcinoma and angiosarcoma, even in the absence of cirrhosis, because of the tendency to thrombosis and to invasion of the portal and hepatic venous system by tumour.

14.3.3 *Portal hypertension in the presence of normal liver structure*

Angiography may reveal some extrahepatic obstructive lesion such as tumour, intrinsic thrombosis or replacement of normal portal veins by a system of small venular or capillary channels. This 'cavernomatous transformation' may be a developmental anomaly, but many are probably the result of umbilical vein sepsis in infancy with thrombosis and collateral channel formation. Persistence of the umbilical vein in the falciform ligament (Cruveilhier–Baumgarten syndrome) is another unusual extrahepatic cause of portal hypertension.

Certain chronic blood diseases and other reasons for gross splenomegaly may be associated with portal hypertension because of augmented venous outflow from the enlarged organ. Liver biopsy may be normal or show some evidence of the underlying condition such as extramedullary haemopoiesis or leukaemic infiltration.

Reference has already been made to partial nodular transformation when biopsies from the superficial part of the liver often show no abnormality.

14.4 Hepatomegaly

Like liver parenchymal dysfunction, hepatomegaly is found in so many different types of liver disease, including many of those conditions already mentioned, that its use as a basis for classification would have little value. Not infrequently it is the principal clinical feature and sometimes the only sign of liver disorder and the following paragraphs refer mainly to such cases.

14.4.1 *Irregular enlargement of the liver*

This is usually due to secondary or primary malignant tumour readily diagnosed on biopsy; occasionally, however, both may involve the liver diffusely. Some benign tumours can grow to a large size and cause abdominal distension, especially in infants. Other causes include developmental and acquired cystic and nodular hyperplastic lesions.

14.4.2 *Diffuse liver enlargement*

Fatty change is a frequent cause and is readily detected in biopsy specimens. The causes and morphological types have been reviewed and are discussed in Chapter 9.

Diffuse hepatomegaly is a feature of many other liver diseases, but in most cases there will be clinical and histopathological manifestations of the underlying disorder. The conditions referred to include venous congestion, viral and alcoholic hepatitis, leukaemic infiltration, amyloid degeneration, glycogen storage disease, tropical splenomegaly, etc.

It is not uncommon to find no structural abnormality in liver biopsy despite a clinical assessment of large liver. This may have been mistaken for downward displacement of the organ but it seems that liver size may fluctuate over short periods of time depending on its content of lipid, glycogen or sinusoidal blood. Drugs such as phenobarbitone can cause hypertrophy which may not be readily evident in biopsy specimens (p. 171).

14.5 Other indications for liver biopsy

Biopsy is occasionally undertaken in the absence of good clinical evidence of liver disease for the investigation of chronic illness and pyrexias of unknown origin when conditions such as sarcoidosis and unsuspected malignancy may be revealed. It is frequently part of the staging procedure to measure the extent of malignant lymphoreticular disease. Laparotomy carried out for some other purpose may reveal a focal lesion or abnormal appearance of the liver which leads to the submission of a wedge biopsy. It may be done in patients with persistent itching and xanthelasma to investigate the possibility of primary biliary cirrhosis.

Index

Figures in *italics* refer to pages with illustrations: **bold** figures indicate major entries.